THEPOLITICS OFBONES

J. TIMOTHY HUNT

100

McCLELLAND & STEWART

Hardcover edition published 2005
Trade paperback edition published 2006

Library and Archives Canada Cataloguing in Publication

Hunt, Timothy
The politics of bones : Dr. Owens Wiwa and the struggle for
Nigeria's oil / Timothy Hunt.

ISBN 13: 978-0-7710-4158-7
ISBN 10: 0-7710-4154-3 (bound).–ISBN 10: 0-7710-4158-6 (pbk.)

1. Wiwa, Owens. 2. Ogoni (African people) – Government relations.
3. Shell Oil Company. 4. Nigeria – Politics and government – 1993–.
5. Political activists – Nigeria – Biography. I. Title.

DT515.83.W58H8 2005   966.905'3'092   C2005-902303-1

We acknowledge the financial support of the Government of Canada through the Book Publishing Industry Development Program and that of the Government of Ontario through the Ontario Media Development Corporation's Ontario Book Initiative. We further acknowledge the support of the Canada Council for the Arts and the Ontario Arts Council for our publishing program.

Typeset in Minion by M&S, Toronto
Printed and bound in Canada

This book is printed on acid-free paper that is 100% recycled, ancient-forest friendly (100% post-consumer recycled).

McClelland & Stewart Ltd.
75 Sherbourne Street
Toronto, Ontario
M5A 2P9
www.mcclelland.com

1   2   3   4   5   10   09   08   07   06

THEPOLITICSOFBONES

To Ogoni ("The land and the people are one")

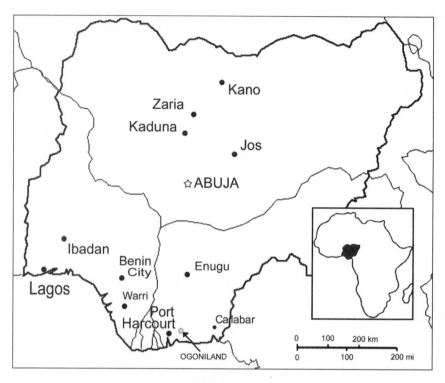

Nigeria

Ogoniland

ACKNOWLEDGEMENTS

Over the span of five years and hundreds of hours of interviews, I have come to know Dr. Monday Owens Wiwa on a level of familiarity one rarely attains with another human being – and yet, each time I speak with him, Owens finds ways to surprise me, impress me, and deepen my respect for him. It has been a great honour to have the opportunity to chronicle his astonishing life. Owens Wiwa's story is far from over; I am certain this book will need a sequel.

I would like to extend my thanks to Ken Wiwa, Jr., for allowing me to reprint two of his father's poems, and to the entire Wiwa and Saro-Wiwa family for their support and co-operation in this project. I am proud and grateful to have been able to spend time with Owens's late parents, Chief Jim and Jessica Wiwa, whom I feel rank among the most important people of the twentieth century. I send thanks to Owens's siblings, Barine, Comfort, and Letam, for their welcome, their kindness, and their memories; and many thanks to the citizens of Bane and to all the Ogoni people who showed me such warmth during my visits to Nigeria.

As for Diana Wiwa, I have to confess I harbour a great deal of affection for this remarkable woman. Courageous, funny, intelligent, beautiful – she is the perfect literary heroine. Thank goodness she's real; no novelist could have dreamed her up. Her mother, Dora, and her father, Professor Cosy

Nuaba Barikor, are amazing as well and I send my thanks to them for their help and hospitality.

I am deeply indebted to Paul Wilson, who edited the original magazine feature "The Politics of Bones" in *Saturday Night*; to Gary Ross at Macfarlane Walter & Ross, who guided me through the first draft of this book and shaped its development; and to Chris Bucci at McClelland & Stewart, who inherited the project and took it to the finish line with skill and infectious enthusiasm. I have been blessed with the best agent in the world, Perry Zimel of Oscars Abrams Zimel & Assoc. My thanks to you all.

I would also like to thank (in alphabetical order) Kaye Bishop, Shelley Braithwaite, Lynn Chukura, Judith DeWitt, Roger Diskey, Glenn Ellis, Malcolm Hamilton, Alfred Ilenre, Doris Lessing, Lekue Laa Loolo, Hauwa Madugu, Andrew Male, Sister Majella McCarron, Flora MacDonald, Steven Mills, Jeanne Moffat, Dr. Ben Naanen, Gerald Ohlsen, Sam Olukoya, Dr. Osaro-Edee, and the staff at UNPO for their time and assistance.

Thanks to Greenpeace Canada for graciously permitting us to reprint photos from its archives.

Thank you to the Canada Council for the Arts. This project was made possible with its generous support.

Lastly, and most importantly, this book would not exist at all were it not for Dr. Morton Beiser. It was Morley who introduced me to the Wiwas, who made all my trips to Nigeria possible, who held my hand and supported me from the first draft of the original magazine story to the final proofreading of the book galleys. His presence is felt on every page of this work. I can't imagine my life without him.

                                                    J. Timothy Hunt

THEPOLITICSOFBONES

PROLOGUE

Dr. Owens Wiwa walked behind his brother's coffin worrying like an old woman. On this sweltering Monday morning in April 2000, it seemed as if all of Ogoniland had come to witness the funeral of Ken Saro-Wiwa – tens of thousands of people, an excitable, militant throng, jockeying for a glimpse of the casket. Ken had wanted a small, private funeral; this was definitely not what he'd had in mind.

The procession slowly parted the crowds in Bane as Ken's daughter Zina, holding a large crucifix of hibiscus flowers, led the cortege. Behind the pallbearers, her twin sister, Noo, held aloft a large colour photograph of their father in a golden frame. Ken Junior followed, in his role of chief mourner, leading his uncle Owens and the rest of the immediate family toward the church.

The little cinder-block chapel in Bane had been packed for more than an hour with members of the media and groggy villagers who had stayed up all night dancing, singing, and drumming. Tribesmen crowded around the open windows and doors and spilled into the adjacent fields, fanning themselves with copies of the hastily printed church bulletin that stated the ceremony would commence with the "Reception of the Corpse at the entrance of the West Door."

But everyone knew there was no corpse. The coffin that was carried through the village and into the chapel was empty except for two books (*On a Darkling Plain* and *Pita Dumbrok's Prison*) and the curved-stem pipe Ken smoked before his hanging.

Since November 10, 1995, the bones of Ken Saro-Wiwa, the celebrated author and Nobel Peace Prize nominee, had lain in a secret, unmarked grave, reportedly mingled with the remains of the eight other men who were hanged with him. Before being charged with murder, Ken had spent seventeen years leading non-violent protests against the destructive exploitation of Ogoniland by multinational oil companies like Royal Dutch/Shell.

The paradox of multinational petroleum production is that it generates great wealth while generating even greater poverty. No place in the world illustrates this paradox more clearly than Nigeria. Despite being the wealthiest country on the African continent and despite wielding significant political, economic, and military influence over its neighbours, Nigeria's economy has contracted rather than expanded over the past three decades. The country's absolute poverty rate (the percentage of the population living on less than a dollar a day[1]) soared from 9 per cent in 1970 to 46 per cent in 1998. Yet while most Nigerians have been sliding into destitution, the political and economic elites of the country have grown ever richer – in most cases, obscenely so.[2]

Nearly every penny of Nigeria's ill-dispersed wealth comes from petroleum, and, at present, the oil industry is the only part of the Nigerian economy that hasn't largely collapsed.[3] Oil accounts for 90 per cent of Nigeria's foreign exchange and 80 per cent of government revenue. Fully half of this oil revenue is generated by the Shell Petroleum Development Company (SPDC), a joint venture of Royal Dutch/Shell and the government-owned Nigerian National Petroleum Corporation (NNPC).

Not surprisingly, the Ogoni people objected to wallowing in poverty while multinational corporations and government elites grew rich on resources drained from Ogoni lands. The Ogonis also publicly and vociferously accused Shell of permanently devastating their environment.[4] Although Shell acknowledges environmental problems in the Niger delta and the oil industry's contribution to them, the company dismisses as "distorted and inaccurate" all allegations of environmental devastation.[5] Shell says that it is deeply concerned about the situation in Ogoniland. According to the company, the Ogonis' real problems stem from ethnic conflicts between neighbouring communities and the infamous Nigerian federal government.

"It is totally unjustified to suggest that Shell, by virtue of endeavouring to carry out its legitimate business of oil exploration, is in some way responsible for such conflicts or the level of the Nigerian government's response to them. . . . Private companies have neither the right nor the competence to become involved," said David Williams, a Shell International spokesman, in a media interview.[6]

Yet the fact remains that for half a century, Shell has played a significant role in Nigeria's economy and politics, and the repressive political climate of Nigeria has directly benefited the economic performance of SPDC.

The execution of Ken Saro-Wiwa and the rest of the "Ogoni Nine" sparked international outrage and led to Nigeria's suspension from the Commonwealth. Since his brother's death, Owens Wiwa had lobbied hard with Nigeria's new democratically elected regime to have Ken's remains returned to the family for a proper burial. But his quest was as frustrating and futile as his efforts to save his brother's life. "There still are some individuals who have very strong feelings against my brother," said Owens in an interview for this book. "These are very powerful individuals. Some are in government, some are businessmen who deal with oil companies, and some are the traditional elites and rulers who don't want to change. It's been very taxing emotionally, the way people have behaved. And for what? For bones."

"The politics of bones" is how Bill Haglund, a forensic anthropologist from Physicians for Human Rights, had wryly summed up the strange combination of silence, misdirection, myth, greed, and fear surrounding the death and secret burial of Ken Saro-Wiwa. So far, all Owens's fight had come to was a symbolic funeral with an empty coffin.

Owens took his seat near the front of the church and watched aghast as independent film crews making documentaries for the CBC and PBS clambered around the altar with boom mikes and video cameras. Vigilant members of MOSOP, the Movement for the Survival of the Ogoni People, patrolled the chapel, poking the many exhausted mourners who had nodded off. Reporters from the Nigerian press hurried up and down the aisles taking photographs and asking for interviews during prayers. With all the commotion going on, the long, loopy eulogy delivered by Archdeacon Ven. Dr. S.O. Amadi was frequently drowned out.

But Owens wasn't listening to the sermon, or to the faint voices of Ogoni women in the distant fields singing folk songs about his brother's heroic deeds. He was staring at his hands and thinking about the intensely private part of the ceremony to come. Could it even be private, in the circumstances? Honouring Ken's wishes, the Wiwa family had prepared a concrete tomb in a sandy, one-acre lot behind the village. At one end of the field was a lush mangrove swamp and, beyond that, the sea.

Just as Owens feared, the family found it impossible to keep the crowds from following the coffin to the tomb. Everyone wanted to see the last symbol of Ken as it was lowered into the ground. It seemed as if every Ogoni man, woman, and child wanted to go to the grave with him. The archdeacon uttered, "Ashes to ashes, dust to dust," as Ken Junior shovelled three spadefuls of dirt onto his father's coffin.

After the graveside ceremony, all the Wiwa family except Owens departed through the crowd. When the attendants started to seal the crypt with concrete slabs, he asked if they would wait a moment. Before the coffin was hidden forever, he wanted a final moment with Ken, if only in spirit. They had shared so much: the same mother, birthday, love of the Ogoni culture, land, and people, the same vision for Ogoniland. Often, each brother had known what the other was thinking. Alone at the tomb, Owens realized he lacked the sense of closure. Although Ken's memorial service had befitted a man of greatness, the exercise had been hollow. It also confirmed Ken's prediction that the Nigerian government and Shell would deny his remains their six feet of earth.

No, Owens vowed, they would not. Staring at the empty wooden box, he promised his brother that he would not rest until the body of Ken Saro-Wiwa was located, exhumed, and laid properly at peace.

CHAPTER**ONE**

Ma's lantern signalled the start of day.

"Wake up, boys! Go get water."

Each morning at five, Jessica Wiwa sent her children to the stream half a kilometre from town. Fetching water was a job her two youngest loved. The stream was clear and deep with placid pools for swimming. There was a little bridge for diving. In the early morning light, the water was prismatic, each grain of riverbed sand magnified in sharp focus. Schools of tiny brown fish undulating lazily against the current scattered at the first child's plunge.

The stream was where the children took their morning baths and filled large jugs with fresh water. On this particular morning, six-year-old Owens Wiwa and his little brother, Letam, were joined as usual by the neighbourhood children from the Igbo tribe. "*Ututu òma!*" the Igbos greeted. *Good morning.*

"*Kèdú,*"[1] Owens replied.

He had picked up a bit of the language while playing with the Igbos after school, kicking balls with them on the sandy road or pestering anthills in the forest. Sometimes Owens would join the older Igbo boys on their slingshot hunts.

"Letam, come!" Owens called as he struggled to hoist his container. "Ma is waiting."

5

With the jug in one hand and his brother's paw in the other, Owens struggled up the riverbank, through the trees, and back to the house. Monday through Friday, his family lived in Bori, a thriving town at the centre of Ogoniland in Nigeria. The divisional headquarters of the six Ogoni kingdoms, Bori was a bustling enclave of dirt streets and ramshackle dwellings surrounding an open-air market, a handful of churches, a police station, a clinic, and an elementary school. Pa Wiwa was the market master for the region and lived with his large family in a comfortable cinder-block house just opposite the Methodist church.

Returning home from the stream, Owens delivered his little brother and jug of water to his mother before going to get dressed for school. Just before he got to his room, however, he stopped. A radio was playing in an unused bedroom down the hall. Electricity had not come to Bori. What was a radio doing in his house? Curious, he crept toward the source of the sound.

A young man holding a cup of water and a sort of small stick with bristles came to the doorway. "You, come here," he said. "Come."

Owens went closer, mainly to get a peek at the radio. The stranger was an adult, short and slight, perhaps twenty years old. There was something familiar and yet unfamiliar about him. He told Owens to open his mouth and clench his teeth. Owens complied.

"Your teeth are dirty," the stranger said. "Why don't you clean your teeth?"

"I do," said Owens. "I use a chewing stick."

"Oh, really? Let me see again." As Owens bared his teeth for another inspection, the young man squeezed white paste from a tube onto the bristles and began to brush the little boy's teeth.

"Ow!" cried Owens. It was the first time he'd had a toothbrush in his mouth and his gums bled.

"There," said the young man, "that's better," and laughed as the child hightailed it back down the hallway.

The taste of blood and mint lingered in Owens's mouth. Before leaving for school, he asked his mother about the stranger with the radio and the toothbrush. Jessica looked at him oddly. "Don't you know your own brother? That is Ken."

|||||||

The Wiwas' lives revolved around the market. As market master, Chief Jim Beeson Wiwa ("Pa" to the family) was the government administrator in charge of the scores of open-air community markets throughout Ogoniland. People from all over the Niger delta came to Ogoni markets to buy the yams, corn, leaves, and other foodstuffs the fertile Ogoni fields produced in abundance. Before Pa Wiwa took over, Ogoni markets had been haphazard gatherings of buyers and sellers conducting business on makeshift tables, upturned barrels, or dusty mats. Jim taught the villagers how to build vending huts and airy stalls, organized the vendors, and created a coherent and sanitary marketplace.

Pa Wiwa also owned an old flatbed lorry that he used to ferry Ogoni traders to Aba, a hundred kilometres to the north. Sometimes Jessica ("Ma") Wiwa made the trip to Aba, taking tender pumpkin leaves she had purchased from Ogoni women to barter for spices, salt, and other bulk goods. A successful entrepreneur in her own right, Ma had her own yam, cassava, and palm oil farms as well as a lucrative business selling produce in the local markets. From October to January, she also bought and sold agriculturally improved yam seedlings and corn for Ogoni farmers to plant for the coming season.

Pa had four wives back in the family village, but because of her thriving business, Ma was the only one who came with him to town during the workweek. Ma was not Jim's first wife, but she was the most powerful wife of the clan. Ma had been the one to give Pa his first child – a son. She also had the most money and was the most literate of Pa's wives, able to speak some English and to read both Khana and Igbo.

After their morning baths, the older Wiwa children would typically help Jessica lug her goods to the market square, carrying items on their heads or in a small wheelbarrow. Once Jessica's stall was set up, the children would then head off to school.

Owens went to the Khana County Council School in Bori. His full name was Monday Owens Wiwa; at school, he went by Owens, but to his family he was always "Mon." Owens spoke Khana at home, and Igbo with the neighbourhood children. School was taught in English.

Until the morning of the toothbrush, Owens had been only dimly aware that somewhere in the world he had a much older brother. Ken had turned sixteen the very day Owens was born and had graduated secondary school when Owens was barely a toddler. For most of Owens's life, Ken had been away at university. Late in the afternoon of the day of the toothbrushing, Owens was playing during school recess when he saw Ken come bicycling up the road. Stopping at the edge of the schoolyard, Ken called, "Mon! Come here." When Owens ran over, Ken gave him a silver coin – sixpence, the most money he'd ever seen. Ken told him to spend it on *akara*, a type of sweet bean cake. Later that day, handing out bean cakes, Owens was king of the school.

By the time he got home, eager to see this mysterious big brother, Ken had left to return to the University of Ibadan, five hundred kilometres west of Ogoniland.

|||||||

It has been claimed that the Ogonis were the last people in Nigeria to come into contact with Europeans, although this is somewhat disputable.[2] During the last half of the nineteenth century, the powerful King Jaja of Opobo controlled most of the trade and politics in a large part of the Niger delta and prevented most European penetration into the region.[3] This suited Jaja's neighbours, the Ogonis, a fiercely independent people who wanted little to do with the outside world. For generations, the swampy terrain of Ogoniland had shielded them from outsiders and had proven the perfect refuge from the region's flourishing slave trade. According to Ogoni lore, no Ogoni man or woman was ever taken into slavery.[4] Although the influence of King Jaja of Opobo did not extend into Ogoniland, after Nigeria became a British colony in 1914, the new administration placed the Ogonis within the newly created Opobo Division. The Ogoni people protested immediately, demanding their own self-administered body, which only served to put Ogoniland in direct conflict with the British.

Its boundaries amorphous and disputed, Ogoniland is a small section of the Niger River delta bounded by the big bend of the Imo River in the north, the Andoni flats in the south, the mouth of the Imo River in the east, and the Aba-Port Harcourt highway to the west. It occupies about four hundred square miles of the delta's gently sloping coastal plain terrace.

Home to three population groups, Ogoniland is divided into six kingdoms. The Khana, largest of the three groups, live in the eastern and northern parts of the area. The Gokana, who occupy the south central part of the land, speak a language as closely related to Khana as, say, Yiddish is to German. The third Ogoni subgroup, the Eleme, inhabit a kingdom to the west and speak a language that cannot be understood by the other two groups.[5] Although Christian denominations abound in Ogoniland, many people remain animists, worshipping various deities and honouring the fruits of the land, particularly yams, in festivals. The Ogonis consider their rivers to be sacred for they provide fish for food and water for life.[6]

The Wiwa family traces its roots to the early 1800s and to a Khana ancestor named Gbenedorbi. It is difficult, if not impossible, to determine the birthdates of many of the Ogonis who lived prior to the twentieth century. Until fairly recently, the Ogonis were an oral society with no written language.[7] Before British colonization, calendars (as Westerners have constructed them) were unheard of.[8]

Jim Wiwa was born sometime during 1904 in the tiny Ken-Khana village of Bane (pronounced *BAH-neh*). When he was five, his widowed mother sent him out of Ogoniland to live in the nearby town of Opobo, just across the Imo River. It was common for Ogoni parents to billet their children with more prosperous relations or to apprentice them with area merchants.

Although turn-of-the-century Opobo was largely a mud and thatch city, it held an irresistible allure as one of the first "civilized" communities in the Niger delta. After the British deported King Jaja to the West Indies in 1883, Opobo became a base for British Christian missionaries, and the town fell under their modernizing influence. In 1909, five-year-old Jim was apprenticed to Harry Jaja, a relative of the fabled king and a prosperous Opobo merchant. Although still a small child, Jim was put to work in Jaja's storehouse and was occasionally asked to keep an eye on the customers in the shop. Jim enjoyed his work and served Jaja with a diligence and honesty that endeared him to his master.

Business seemed to be booming but Harry Jaja's profit margins kept falling. Jaja was mystified, but to young Jim the reason was obvious. Jaja had several boys working for him as indentured servants; they were stealing their master blind. Jim had caught them taking goods and money and the boys had encouraged the young Ogoni to do the same. Jim would have no part of it. "I am Ogoni man, honest man," he told them. "I don't thief."

This proud, intelligent, and rigidly moral child was exceedingly unpopular with his fellow indentured servants. Nevertheless, by the time he had reached adolescence, Jim had effectively put an end to employee theft at Jaja's store and business was prospering. Touched by Jim's devotion, Jaja asked the twelve-year-old what he would like as a reward for his seven years of faithful service. Jim said, "I want to go to school."

In 1916, there were only two schools in this part of the Niger delta – the Primitive Methodist Mission Elementary School at Kono and St. Paul's, the Anglican Church school in Opobo.[9] Harry Jaja chose the Anglican school and paid the missionaries to give the boy a proper British education. Jim graduated seven years later, fluent in English, fully literate, and deeply Christian. At nineteen, he returned to his family compound in Bane.

Because he was adept at speaking English, Khana, Igbo, and Ibibio, young Jim was hired in 1925 as a shop interpreter by the McIver Company, a successful European trading firm. The same year, Jim married a local teenager named Wale. Her name in Khana meant "fine," a name Jim liked to use in English when he spoke about his young bride. Sadly, Fine Wiwa died during their first year of marriage, leaving Jim with a broken heart and no male heir. Shortly after her death, he married again. His second wife, Wayii, had difficulty becoming pregnant and then miscarried every child they managed to conceive.

After Opobo, Jim found life in his native village primitive, but Bane was not without opportunity. The jungles of Ogoniland had an abundance of *Elaeis guineensis*, the oil palm tree, a remarkable resource. The bright orange, vitamin-rich oil came from the palm's plum-sized fruit. Each bunch, around 2,000 individual fruits, weighed from 10 to 20 kilograms. The trees reached maturity in only three years and continued producing for another 35, yielding between 4 and 4.5 tonnes of oil per hectare. Resigning from his interpreter's job in 1927, Jim went into business for himself as a trader, bartering palm oil and palm kernel for tobacco, salt, and thread, items in great demand in his home kingdom.

Jim soon became one of Bane's most prosperous residents. He was also one of the earliest Christians in the village – but not for long. Shortly after his homecoming from Opobo, Jim Wiwa had a small Anglican church built behind his family compound and the congregation steadily grew.

Despite modern religious proscriptions, all Nigerian ethnic groups, whether followers of Christianity or Islam, practise polygyny. Jim Wiwa, for one, had no difficulty reconciling his people's traditional polygynous lifestyle with his adopted Christian faith. Frustrated at Wayii's inability to produce children, Jim added another wife to his household, Jessica Widu Yaaka, a young woman of keen business sense.

Traditionally, an Ogoni patriarch places his house in the centre of his family compound and his wives build their homes around his. In Bane, Jim constructed the first "storey house" in the history of Ogoni, a cement-block structure with an upper floor. As for Jessica, she was soon wealthy enough through her trade to build herself a comfortable single-storey house under a roof of zinc.

During the Depression, Jim lost his palm oil trading business and, in 1933, took a British government job as a forest ranger guarding the Ogoni bush. Jim was just as loyal to the British as he had been to Harry Jaja in Opobo. As a reward, the British government promoted him to the newly created post of market master.

On October 10, 1941, Jessica gave Jim Wiwa his first child, Kenule Beeson Saro-Wiwa.[10] (In Ogoni culture, "Saro" is a title of honour bestowed on the eldest boy of a family; the eldest girl is the "Sira.") As the first surviving child – and a boy – Ken was doted on by his parents, sent to the best schools, and treated as a prodigy. For the next seven years, the Wiwas' adoration of their son only increased as he remained an only child.

Over the next sixteen years, Jim added three more wives to his clan and Jessica produced three more children: two daughters, Comfort and Barine, and another son, Goteh James ("Jim"). On Ken's sixteenth birthday, October 10, 1957, Jessica gave birth to her fifth child, Monday Owens Wiwa.

The two brothers may have shared a birthday, but they seemed to share little else. Ken had been an active, curious child, full of opinions, delighting in the attention he received as the Saro. He was gregarious, loquacious, and charming. Owens was the opposite – unobtrusive, contemplative, unquestioningly obedient, and thoroughly unremarkable.

At the time of Owens's birth, Ken was away on a scholarship at one of the best secondary schools in Nigeria, Government College at Umuahia, where British expat instructors followed a curriculum modelled on Eton, which included Latin, the sciences, and fine arts. In this transplanted English public school, Ken became an anglophile, playing cricket, drinking tea, and developing a passion for British literature, especially Dickens and Shakespeare. Later, at the University of Ibadan, Ken majored in English, edited two campus newspapers, and served as president of the dramatic society. His play *The Transistor Radio*, a satirical piece that went on to win a BBC World Service radio drama competition,[11] received its first performance at Ibadan. Ken also began writing political columns in the school newspapers and letters to the editors of local tabloids. He never lacked subject matter for his commentaries for this was a crucial period of Nigerian history. Nigeria was coming into a new age – the age of politics and oil.

IIIIIII

In 1937, a Dutch company called Shell D'Archy was given a concession by colonial authorities to begin oil exploration throughout the entire Nigerian territory.[12] After nearly twenty years of searching, oil in commercial quantities was finally located in a Niger delta town called Oloibiri. Shell-BP (a joint venture of Shell D'Archy with British Petroleum) produced its first barrel of Nigerian crude in 1956 and by February 1958 was exporting six thousand barrels per day.[13] Until 1959, Shell-BP was the only company producing oil in Nigeria. By 1960, however, Shell-BP's success began attracting the attention of French, Italian, and American companies eager to exploit the imminent Nigerian oil boom.

Concurrent with the development of Nigeria's oil industry, the movement to free Nigeria from colonialization had been launched toward the end of World War II. After more than a decade of intense political wrangling among a wide assortment of Nigerian interests, Britain finally granted the country independence in 1960. Because this momentous act was achieved non-violently, it was famously stated that Nigeria won its independence "on a platter of gold." Some pointed out bitterly, however, that this platter was extended only to the Yoruba, the Hausa-Fulani, and the Igbo people, while hundreds of other minority groups in Nigeria were left empty-handed.

In truth, the founders of this new nation all but ignored the fact that their country had originally been cobbled together by the British out of more than three hundred distinct ethnic groups with differing cultures, languages, and religions. Independent Nigeria was structured by its constitution into three regions with boundaries drawn according to the concentrations of Nigeria's three dominant tribes – the Yoruba in the west, the Hausa-Fulani in the north, and the Igbo in the east. In theory, the three groups were to pool their traditions and resources to create a unified

Nigeria. Far from promoting national unity, however, Nigeria's regional arrangement merely emboldened the Yoruba, Hausa-Fulani, and Igbo to assert their dominance and to exploit the hundreds of minorities the constitution had placed under their respective jurisdictions.

The regional arrangement also whetted the appetites of the three dominant tribes to establish their own independent nations. At one point, the Hausa-Fulani demanded that the Nigerian Federation be dissolved and the three regions kept linked together by a central agency of a "non-political character." The Yoruba proclaimed that they would not be relegated to the background in Nigerian politics and declared that their objective was to "create and foster the idea of a single nationalism through Yorubaland." More ominously, the Igbo announced that "the God of Africa has created the Igbo nation to lead the children of Africa from the bondage of the ages." The spectre of Igbo secession was worrisome for many, for the Niger delta was the location of all the new nation's oil reserves.

At independence, Nigeria's former governor general, Dr. Nnamdi Azikiwe, became the ceremonial head of state. A six-foot-tall, stately man, Azikiwe (known as "Zik") was the father of modern Nigerian nationalism and the chief architect of the country's independence. Three years later, in October 1963, when Nigeria officially became a republic, Azikiwe was unanimously appointed president by the federal parliament. At the same time, the Yoruba's Western Region was subdivided to create the Mid West Region.

All semblance of democracy in Nigeria ended abruptly on January 15, 1966, in an Igbo-led military *coup d'état*. Nigeria's federal prime minister, Sir Abubakar Tafawa Balewa, was killed along with two of the three regional heads, namely the Yoruba premier from the Western Region and the Hausa-Fulani premier from the North.[14] Azikiwe escaped harm and went into virtual isolation.[15] Since the coup leaders were Igbo and the succeeding military dictator, Major-General John Aguiyi-Ironsi, was Igbo as well, the Hausa-Fulani people turned their revenge on the Igbo people, leading to the widespread slaughter of Igbos living in the Northern Region. A few months later, in July 1966, the dictator, Ironsi, was kidnapped, executed, and replaced by Lieutenant Colonel Yakubu Gowon. Surprisingly, the new head of state was neither a Hausa-Fulani nor a Muslim, but a Christian from the Middle Belt.[16] Gowon was chosen because he was the most senior-ranking military officer from the North left alive.

Only four days after Gowon's counter-coup, Colonel Odumegwu Ojukwu, an Igbo, became military governor for Eastern Nigeria. Ojukwu – an intransigent, egotistical, political neophyte – immediately transferred all political power in the Eastern Region from the elected politicians to his coterie of Igbo military cronies and to the predominantly Igbo police.

The Hausa, Igbo, and Yoruba were close to breaking the federation into three separate kingdoms. Alarmed at this prospect, the three hundred-plus minority groups argued that the regional system needed to be scrapped and the country partitioned into a number of states following the U.S. model. Surprisingly, the minorities won the day in the legislature and the four regions became twelve states.[17] For the Igbos, led by Colonel Ojukwu, this meant a dilution of their overwhelming superiority in the east. Refusing to accept the new order, the Igbos seceded from the federation. On May 30, 1967, they declared themselves the new Republic of Biafra.[18]

IIIIII

A year before Biafra's creation, Ken Saro-Wiwa came home to his father's house to stay. And he did not come alone. With him was his seventeen-year-old[19] bride, Maria Bayor, a relative of the influential Orage clan of Gokana. The beautiful Maria was delighted to find herself married into one of Ogoniland's pre-eminent families. Pa Wiwa considered it a desirable marriage as well. The Orages had long-established ties to the Wiwas, and the union transformed the friendship between the Orages and the Wiwas into a family relationship.[20]

Ken had graduated from the University of Ibadan with a B.A. (Honours) in English,[21] subsequently working as an instructor, first at Port Harcourt's Stella Maris College, then at Government College at Umuahia. Now that he was back at Pa's house in Bori, his transistor radio was on all the time. Early every morning, Owens would hear the distant drone of the BBC News followed by the news in French and then the bombastic local Nigerian news. From the newlyweds' room also came the sound of Ken bellowing at the radio, delighted by good news, indignant at bad.

Ken saw to it that his little brother brushed his teeth every morning before school. But Owens was no longer going to school every day. Khana County Council School was often closed; furthermore, his brother Jim and teenage sisters Barine and Comfort had inexplicably come home from boarding school. Something was happening in the Niger delta and Owens guessed that his senior brother was involved.

He was right. Ken Saro-Wiwa was one of a group of young men responsible for publishing a September 14, 1966, memorandum calling for the creation of Rivers State.[22] Ken felt that if his tribe's corner of the country were to be divided into a smaller region, the Ogonis would have a stronger voice at the federal level. Ken's memorandum took the nearby oil metropolis of Port Harcourt by storm; the Igbos reacted to it with ridicule, resentment, dismay, and alarm.[23]

A few months later, Ken again left home to serve as a graduate assistant at the University of Nigeria at Nsukka, near the southeastern city of Enugu. He was not to remain long. Civil war mania had gripped the campus, with most of the staff and students rooting for Ojukwu to carry through with his threat to secede and create an Igbo nation. Remaining impartial was nearly impossible. Anyone who dared disagree with Ojukwu's position was hauled away to prison, never to be heard from again. Opposed as he was to the prospect of Ogoniland's becoming part of a new Igbo-dominated country, Ken had no choice but to resign his position at Nsukka and return home.[24]

When Ken got back to the delta, he found some solace in manual labour, accompanying his mother to her farm to harvest her cassava crop and then to turn the cassava into *gari*. *Gari* is a staple of the Ogoni diet. The tuber-like cassava root is grated into a coarse meal and packed in burlap bags that are left in the sun. Leaching the cassava of its juices is essential because the liquid contains mildly poisonous levels of cyanide. After draining and fermenting, the dried meal is placed in large bowls to be fire-roasted into the starchy, nutritious *gari*.

When not making *gari*, Ken often drove his moped around Bori to gather political gossip.[25] Before long, gossip became talk of war. Finally, at the end of May 1967, the Igbos of eastern Nigeria seceded. The Ogonis found themselves within the borders of the newly created Republic of Biafra, in the midst of full-scale civil war.

|||||||

At the start of the Biafran War, life in Ogoniland seemed to go on as usual. It was a situation Ken Saro-Wiwa found frustrating in the extreme. There was little for him to do but listen to the radio, or try to find newspaper reports of the war's progress. Ken continued to help his mother, wife, and siblings with the cassava, but it was only writing that truly gave him a sense of fulfilment. Although he considered his strengths as a writer to lie in short stories and drama, his thoughts during the time of war began coming to him in verse.[26] His 1985 book *Songs in a Time of War* begins with his June 1967 poem "Voices," written about the Igbo secession:

They speak of taxes
Of oil and power

They speak of honour
And pride of tribe

They speak of war
Of bows and arrows

They speak of tanks
And putrid human flesh

I sing of my love
For Maria[27]

By the end of summer 1967, federal forces were gaining against the Igbos. The Biafran rebels' morale remained high, however, and Igbo militiamen began browbeating the inhabitants of the Niger delta to contribute to their care and feeding. As reward for their coerced support, the Ogonis were subjected to raids on their personal property, the looting of their money, and the rape of wives and daughters. Husbands and fathers who protested were beaten, often fatally. Since it was illegal for any non-Igbo to have a gun, the Ogonis were defenceless.[28]

Young Owens noticed that many Bori male teenagers and adults – Igbos and Ogonis alike – had started wearing Biafran uniforms, mounting roadblocks, and trying to convince more men to join the Biafran army or Civil Defence. Soon throngs of new soldiers in threadbare uniforms were marching up and down the streets of Bori singing war songs. To make matters worse, Pa Wiwa was forced to leave Ogoniland when the Igbos transferred him to Umuahia, a hundred kilometres north.

After federal forces captured nearby Bonny Island, Ken hoped it would only be a matter of time before Nigerian soldiers marched into Ogoniland and liberated the area. But the Biafran rebels countered their defeat by invading the Mid West, turning the outlook for the liberation of Ogoniland decidedly gloomy. Ken could take no more. Since it was impossible for him to oppose the Biafran nightmare from within, he felt his only answer lay in fighting it from without.

In September 1967, the Wiwa family's routines changed. After Pa's transfer to Umuahia, Ma had continued to run her market stall in Bori during the week as usual and move the family back to Bane on weekends. On Sunday evening, September 24, Ma Wiwa surprised her children by announcing that the family would now remain in Bane.

That night, around 2 a.m., Owens was awakened by unfamiliar activity. Creeping to the door of his room, he saw his mother moving through the house with her lantern. Out the window, he saw people moving between the buildings, in and out of the compound, helping Ken with a few large sacks. Owens recognized Ken's wife, Maria, and two of his uncles from his mother's side. In the morning, Owens learned that Ken and Maria had fled Ogoniland. In a hired canoe, under cover of darkness, they had manoeuvred through the maze of delta creeks to cross the Biafran front lines. At the federally held port of Bonny Island, they boarded a ship for Lagos.[29]

Trouble soon arrived for the family Ken and Maria left behind. The Biafrans set up a police command post at Kono, on the Imo River about seven kilometres from Bane. Biafran soldiers came to Bane every day to terrorize the villagers, and seize, slaughter, and eat the village livestock and ravage the community's supply of plantains, breadfruit, and coconuts. Although most Ogonis did not support the war, the soldiers persuaded many village males to join the army and the Civil Defence. Those who resisted were evicted from their homes and transported into the heart of Igboland. The deportees were invariably poor farmers and fishermen who spoke only Khana, Gokana, or Eleme; some had never before left their ancestral village.[30] Few ever returned.

Months passed with no news about Chief Jim Wiwa. Word did, however, come about Ken. Three months after Ken and Maria's escape from Ogoniland, the Nigerian National Broadcasting Service reported that the federal government had appointed Ken Saro-Wiwa the new administrator (i.e., the governor) of Bonny Province.

In Lagos, Ken had found a position as a lecturer at Lagos University, and he and Maria had moved into a university-owned flat in the suburb of Surulere. More importantly, Ken managed to get himself appointed to serve on the Rivers State Interim Advisory Council,[31] an inner circle of advisors to the current Rivers State governor who was in exile in Lagos.[32] Through his work on the council, Ken soon found himself being offered an appointment as administrator for Bonny. Fearing that once the Igbos learned of her husband's appointment they might exact revenge on their relatives, Maria implored Ken not to accept. Although he was not oblivious to the dangers, he opted to accept the position and entrust the safety of his family and of Ogoniland to God.[33]

||||||

Maria's fears proved well founded. After Ken's appointment, Jessica's neighbours and friends became convinced that Igbo soldiers would brand anyone having contact with the Wiwa family as saboteurs. Worse, the Wiwas began noticing that whenever members of the Civil Defence led the Biafran army through the village, they would point out the Wiwa compound.

One afternoon, Owens and Letam were playing at home while Jessica and her teenage children were working in the fields. Hearing raised voices, Owens looked out his mother's front door and saw the Civil Defence boys leading a squadron of Biafran soldiers into the Wiwa compound. Grabbing Letam, Owens ran to hide in his father's house. The children watched from an upstairs window as soldiers entered their mother's house with gallon jugs of kerosene. The soldiers splashed fuel over the walls, floor, and furniture, then stepped outside and set the building alight.

With Ma's house in flames, the soldiers turned their attention to Pa's house. Seeing the soldiers coming, Owens and Letam fled the building and ran unnoticed to the neighbouring Anglican church. They escaped just in time as the soldiers burst into Pa's house and went directly to Ken's room upstairs and tossed all his books, files, and clothes out the window into the courtyard. The soldiers appropriated the items that caught their fancy and made a bonfire of the rest. Watching from the church, Owens wept. Ken had always kept meticulous care of his volumes of great literature; to him, books were sacred. Ken was also an obsessive record-keeper and the family's archivist; into the fire went everyone's birth certificates, report cards, medical records, family photos, personal letters, award certificates, and diplomas.

The soldiers ransacked Pa's house, smashing windows and carting away the furniture. With the building reduced to a shell of concrete and cinder block, they spilled kerosene over the downstairs floors and set a match to the liquid. The kerosene flared up, burned brightly, and went out.

Puzzled, the soldiers poured more kerosene and struck more matches, but still the house refused to burn. Deeply superstitious (and naive about modern building materials), the soldiers became frightened.

"The father must have buried a powerful thing in the house!" said one.

"A *juju*! This is a house so pure it cannot be burned!"

Spooked, the soldiers halted their destruction and focused on the search for Jessica and her children. Owens and Letam were quickly discovered at the church. The soldiers then went door to door in the village rounding up anyone they could find, demanding to know if they were related to the Wiwas. When word reached Jessica in the fields that her house was on fire, she and her children raced home, where they were captured.

The soldiers lined up the family, as if to be shot. Instead of killing them, however, the soldiers beat them, forced them into a military vehicle, and set off for Kono. En route the soldiers taunted Jessica, saying that her son Ken had betrayed Biafra.

At Kono, Jessica and the five children were locked in a tiny mud and stick hut with a thatched roof and dirt floor. Once a primary school class-room, the hut was now an aboveground dungeon, its windows boarded and its door locked and guarded. A few beams of sunlight filtered through the thatch. Rain came in the same way.

Occasionally, armed soldiers escorted the Wiwas into the bush to attend to sanitary matters. At other times, the family was allowed to sit at a table in another former classroom to watch the soldiers bring in other prisoners, who were tortured, frogmarched, and beaten with bullwhips.

Holding Owens, Jessica would say, "Don't worry, Mon. One day, he will come and see that we will all survive and they will get their punishment." Owens did not know if his mother was speaking of Ken or Jesus Christ.

After a three-week detention at Kono, the Wiwas were released. They walked back to Bane, although there wasn't much to come back to. For a few days, Jessica and the children lived in the charred shell of Pa's house. Each day, the sound of gunfire grew louder, and Jessica took the children to hide in the forest, daring to return only after dark. The other Bane families did the same. Tales had reached the village about the reactions of the Biafran soldiers to the approaching federal bombing raids. Facing defeat, the Biafrans had become vicious, making daily forays into Ogoni villages, seizing animals, raiding houses, and raping every woman they came across.

Although the Ogonis by and large approved of Ken's decision to leave Biafra for Nigeria, the villagers of Bane began avoiding all contact with the Wiwas. This was understandable: to be associated with Ken's family now was to risk arrest. Feeling no longer welcome in the community, Ma gathered the children and left Bane on foot in search of anonymity. By now, however, the family of the market master and their connection to Ken Saro-Wiwa were so well known that Ma and the children were shunned at every turn. Fortunately, in Lubara, the ancestral village of Jessica's grandparents, one uncle was brave enough to let the Wiwas stay at his house.

One afternoon in Lubara, Biafran soldiers arrived to arrest Jessica and her five children again. This time, the Wiwa family was detained at the abandoned Teachers Training College in Bori. The war had brought an influx of people to Bori, and the Biafran soldiers, frightened of losing control of the town, had turned the small college into a holding area for Ogoni refugees fleeing the Nigerian army shelling.

As days passed and the sounds of explosions and gunfire came closer, the ranks of Biafran soldiers at the Teachers College thinned. Soon the guards were gone, leaving Jessica and her children free to begin another trek away from the advancing federal forces. On the way out of Bori, they passed their town house, now ransacked and desolate.

The Wiwas marched northward, stopping only briefly in villages where Ma's former customers dared to offer them food or water. They kept walking until they arrived at a village called Bangoi, the farthest outpost in Ogoniland. Beyond lay Igbo territory. Here Ma stopped, sat down, and declared she would go no farther. "I will not leave Ogoni," she told her children. "If we die, we die right here."

Moved by the Wiwas' plight, a Bangoi villager offered them nighttime shelter with his family in their thatched-roof house. Because the approaching battle made staying in the house far too dangerous during the daytime, the villager also offered to share the primitive bomb shelter he had dug for his family at the edge of the forest.

Though crude, the shelter proved effective. Getting to it safely, however, was another matter. A week after their arrival in Bangoi, Owens was walking with his mother and siblings to the shelter one morning when they heard a high-pitched whoosh followed by a sharp smack. Bullets were whizzing by their heads and striking tree trunks. The Wiwas were caught in a crossfire, the Nigerians behind them and the Biafrans in front. Luckily, everyone made it to the shelter, but the Wiwas and their hosts had to stay there for three days and nights as the firefight raged.

On the third day, the sounds of fighting grew intense, with long volleys of shooting punctuated by explosions. Suddenly, the shooting stopped and there was an eerie silence – then a melody. Owens recognized the song from Ken's transistor radio. The federal soldiers were singing in Hausa the Nigerian army war victory song:

Mu je Mu Kerkeshe Su,
Tu Tatara Kayan Su,
Mu Ber Su Suna Kukan Banza

("We go, we slaughter them / We ravish their precious wares / We abandon them crying useless tears.")[34]

The two families remained in hiding for several hours more. After what seemed an eternity, the Bangoi villager timidly opened the door and stepped out to see if everything was safe. Nigerian soldiers spotted him at once and apprehended him.

Hearing the soldiers cart the man away, Owens's sisters, Barine and Comfort, believing they might be able to use their command of English to help, left the shelter. The girls returned several hours later, excited and happy. Instead of arresting the brave Bangoi man, the soldiers were treating him as a refugee. When Barine and Comfort found the soldiers and their protector, they introduced themselves. The soldiers said they had been instructed to look everywhere for the Wiwa family and take them to safety.

The Nigerian soldiers brought the Wiwas tinned fish, rice, and other food, and drove them back to Bane. On the trip home, Owens stared at the corpses lining the road. By this time, more than thirty thousand Ogoni lives had been claimed by the Biafran war,[35] but many of the bodies Owens now saw belonged to Biafran soldiers.

||||||

A few weeks after the Wiwas' return to Bane, news of a grand arrival spread throughout the village. People spoke of a black man wearing a "French suit," a type of stylish safari outfit usually worn by Europeans in Africa, arriving by water. Owens and Letam raced down to see what was happening. A short, dapper young man holding a walking stick and surrounded by army officers and private guards stepped off the boat. It was Ken.

The new administrator for Bonny and his entourage marched into the village surrounded by the remaining residents of Bane. At the crook of the road behind the Anglican church, he caught sight of his family compound. His mother's house was in ashes and his father's house reduced to an empty shell. The house of one of Pa's other wives was a rain-filled crater. Seeing the destruction, Ken burst into tears, but when he caught sight of Ma and the children, his despondency turned to joy: at least his family was alive.

Jessica looked a mess in one of Ken's torn and filthy T-shirts. Ten-year-old Owens and little Letam came to be hugged and kissed, but his sister Barine stayed in her aunt's house, afraid to venture out.

Like their Biafran predecessors, the federal troops billeted at Bane had become a menace. Not only did they rape village girls and steal from the houses and farms, they laid siege to the town, imposing a dusk-to-dawn curfew.[36] Ken pleaded with the superior officers stationed at Bori and at the headquarters of the 14th Brigade to discipline the rapacious soldiers. The young officers promised to do what they could. "But you have to understand," one officer said to Ken, "to control all the men under our command is not an easy task in the circumstances."[37]

Realizing Bane was not safe, Ken decided to take everyone back to Bonny Island. Jessica would not hear of it. "I don't need a holiday," she said. "I have to stay and wait for my husband to come back. And look at my house. I have to rebuild!"

Anxious to return to school, Ken's brothers and sisters welcomed the opportunity to leave their war-torn home.[38] Ken said goodbye to his mother and took his siblings to catch a boat for the short trip to Bonny.

||||||

The voyage to Bonny was Owens's first time in a motorboat. It was also his first glimpse of the ocean. Salt water fascinated him. The only water he had ever known ran clear in the fresh streams or swirled lazily in brackish swamps. Here were green-blue water and crashing waves as far as the eye could see. When the boat docked in Bonny, Owens jumped ashore to taste the sea. The only thing he could compare it to was tears. He was having problems with his eyesight and reasoned that this miraculous water, this ocean of tears, must have curative powers. He scooped water into his cupped hands and splashed it into his eyes. The reward for his faith was searing pain.

Less than a month after Ken left Bane with the children, the federal soldiers began pulling out of Ogoniland. Goods unavailable during the federal blockade of Eastern Nigeria began to arrive; things were returning to normal.[39] The Bane villagers warmed to Jessica again, engaging her in trade, seeking her advice, coming to her to borrow money. Still, Ken felt it best that the children not return home. Pa Wiwa was still missing, and perhaps had not survived the war. There were no schools left in the Niger delta. Now the de facto head of clan, Ken saw it as his duty to ensure that his brothers and sisters continued their schooling. Since there were no schools in Bonny either, Ken decided to take the children to live in Lagos. Owens's first ride in a speedboat was nothing compared to his first ride in a seaplane. He and his brothers found the experience of flight astonishing; his sisters spent the ride vomiting.

Lagos of the late 1960s was unlike any other city in Nigeria. Both the national capital and commercial nerve centre, it was the home of the president, government ministries, and many national institutions. Lagos was cosmopolitan, colourful, crowded, and loud. Tall corporate buildings rose out of the tangle of elevated downtown freeways, mansions rested serenely in elegant neighbourhoods, while nightclubs pulsed to the beat of a thriving music scene.

Ken had rented a house in the lower-middle-class suburb of Surulere. He lived modestly – far beneath his means as a provincial administrator – retaining only one servant and purchasing a ten-year-old Volkswagen. One thing he did spend money on was his siblings' education. Ken dispatched Jim and the girls to boarding schools right away. Letam was easily placed in a local elementary school. Eleven-year-old Owens was the only remaining problem.

One morning, Ken came into Owens's room and asked him to "please dress smart." (When Owens was young, Ken was extremely polite with him, as Owens was the clear favourite among his siblings.) With his little brother looking his best, Ken drove to the Ministry of Education, where he conferred with officials about the best possible school. The ministry staff instructed the boy to be delivered to the Surulere Baptist School.

At Surulere Baptist, the principal greeted the Wiwa brothers cordially, then turned to Owens and asked, "What is seven times eight?"

Owens stared at the man, saying nothing. The principal asked again. Still Owens did not respond.

"I don't think we can take him," said the principal.

"Why?" snapped Ken. "Because he doesn't know seven times eight? Do you know what this child's been through?"

The principal apologized but remained adamant: his school was a little too advanced for Owens. Back in the car, Ken turned his full fury on his junior brother.

"What is seven times eight?" he demanded.

"Fifty-six," Owens replied without hesitation.

"Why didn't you tell the principal that?"

Owens said nothing. How could he tell Ken he had been too nervous to speak? Or that when the men began arguing, he was even more afraid?

Ken drove back to the Ministry of Education, where he read the riot act to the staff. "This is a boy who has just come from the war front and the principal is asking him these questions, telling him to do this on the spot, and when he doesn't say an answer, he says he won't take him in his school!"

The ministry official tried to calm Ken down and said, "Let's go back to the school together." The official followed Ken's car back to Surulere Baptist, had a talk with the principal, and made the arrangements for Owens's registration himself.

Although named Surulere Baptist, the school was a secular institution run by the government. Compared to the small brick schoolhouse in Ogoniland, the three-storey concrete building in Lagos seemed almost futuristic. The classrooms had glass windows. Doorways sported doors. Back home, Khana County elementary school had large fields to play in and a working farm where the children learned agriculture and composting. The school in Lagos didn't have any of these things, but it did feed the children lunch, something unheard of back home.

The first day at Surulere Baptist proved particularly hard for Owens. While sitting in class on the second floor, Owens heard students in the room upstairs get up from their desks at break time. Their metal chairs scraping the cement floor made a sound like that of approaching aircraft on bombing runs over Ogoniland. Hearing the sound, Owens panicked and ran out into the hallway. When the noise stopped and he noticed that nobody else had tried to "take cover," he slunk back into the class, embarrassed. Although peeved that his fellow students were laughing at him, it took Owens several months to control these panic attacks.

Distances in the big city also took some getting used to. In Ogoniland, Owens lived a five-minute walk from school. In Lagos, the school was five kilometres away through a maze of confusing streets and crowded alleyways choked with people and cars. It was also Owens's job to take Letam to school each day. Luckily, Letam's school was not far from Surulere Baptist and he was good company. One morning, the brothers' route was blocked by panicked throngs. Terrified people were packing belongings into lorries and cars or racing down the streets with basins of personal items balanced precariously on their heads.

"What is wrong with these people?" Owens asked a shopkeeper who was closing down his stall.

"There was a bomb somewhere in the city!" said the man.

In the waning days of their fight for independence, the Biafrans had tried to conduct a bombing raid on Lagos, but managed to detonate only one bomb. Owens burst out laughing. These people who had never known war were now scared out of their wits simply at the news of a bomb exploding elsewhere in their vast metropolis. At eleven, Owens had been exposed to much worse.

Owens was beginning to feel cut off from his family. His older brother Jim and his sisters, Barine and Comfort, were away in boarding school. Ken's young wife, Maria, had just given birth to their first child, Ken Junior, and was devoting her attention to the baby. Ken himself was hardly ever around. As administrator for Bonny Island, he spent most of his time at his new homes in Bonny and in Port Harcourt. Owens had gone from being part of a crowded, polygynous rural household to living in a small city house with his younger brother, a sister-in-law who was almost a stranger, and his new baby nephew. It was a lonely time.

|||||||

During the Biafran rebellion, Ken published a pamphlet entitled *The Ogoni Nation Today and Tomorrow*,[40] a twenty-five-page booklet meant to introduce the little-known Ogoni people to the nation and to themselves. Up until this point, the Khana, Gokana, and Eleme tribes did not share a pan-Ogoni sense of self. On the contrary, the very name "Ogoni" was offensive to many people in these groups. (Ogoni is a corruption of the Ibani word *igoni*, or "strangers.") In *Today and Tomorrow*, Ken used geographical, statistical, historical, and cultural evidence to convince three groups spread across six kingdoms to unite under one ethnic identity in a region called Ogoniland – or simply "Ogoni." ("The land and the people are one," Ken would insist testily whenever he heard the term "Ogoniland" or "Ogoni People" bandied about.) This call for the Khana, Gokana, and Eleme people to embrace some sort of amalgamated cultural pride worked better – and more quickly – than even the self-assured Ken Saro-Wiwa could have anticipated.

In 1969, trudging with his stack of booklets through more than one hundred rural Ogoni villages, Ken discovered that the common Ogoni man was ready to follow a dedicated leadership.[41] Certainly one beguiling aspect of his campaign for Ogoni solidarity was his call for a share of the oil revenues that the government and Shell-BP were dividing between them.

"We find it most intolerable that we who sit on oil should be one of the poorest, if not the poorest, people in the country," Ken wrote in the booklet. "The common man in the Western State benefits directly from cocoa; so does the man in Kano State benefit from groundnut. We, on the contrary, are chased out of our land by oil-prospecting companies and left to languish in poverty. We find it untenable that the oil industry should bring prosperity to everyone but us. We refuse to accept that the only responsibility which Shell-BP owes our nation is the spoliation of our lands to satisfy the company's needs for the silly sum of five shillings per acre in an area where 1,200 people live on each square mile of land, and the only source of income as well as subsistence is agriculture. Shell-BP can earn the goodwill of our people. She can do for us what she knows is right."[42]

His criticism of Shell won Ken the confidence of the Ogonis. As his popularity rose, so did his political profile. By the start of 1969, in addition to being administrator of Bonny Province, he was a member of the Interim Advisory Council of Rivers State and the Rivers State commissioner for Works, Land and Transport.[43] As a Rivers State Cabinet member, he worked to rehabilitate the Ogonis and other ethnic groups who had been the main victims of the civil war.[44]

Owens, meanwhile, was flourishing at school. He advanced from Grade 4 to 5, and his teachers felt that his performance was good enough for him to skip the sixth grade and take the entrance exams for secondary school. In Bori, he had learned to speak Igbo. Now, in Lagos, he was picking up Yoruba and interacting with children from that cultural group.

Ken's house in Lagos began to fill up. Although Ken was rarely there, many Ogonis from Biafra who had come to Lagos to look for work used the residence as a sort of transit house. With familiar Ogoni faces and language surrounding him once more, Owens's loneliness began to abate.

|||||||

On January 15, 1970, Biafra surrendered and the Nigerian Civil War officially ended. At the end of the war, Ken found Pa Wiwa at an Igboland refugee camp in Ikot Ekpene and delivered his father back to Bane for an emotional homecoming.[45]

Ken moved his wife, newborn son, and two little brothers out of Lagos to Port Harcourt where the government allocated him a new house. Number 11 Nzimiro Street was a large, two-storey building of white cement block with a corrugated zinc roof, porches flanking the front door, and upper balconies screened with a railing of lattice brick. The grounds were spacious, covered in well-tended grass and surrounded by trees. Out back was the "boys' quarters," a small detached apartment for Ken's house staff and night watchmen. The house filled up as quickly as the Lagos house had. A few of Maria's relatives came to stay and Ken billeted some of the Ogonis he regularly took under his wing. Ken also brought his half-brother Harry Wiwa, the son of Pa's fourth wife, Tapwa, to live at Nzimiro Street.

This bursting house wasn't entirely the product of altruism. In Ogoni culture, elder siblings become responsible for their junior bothers and sisters as soon as they move out of their father's house and build a home for themselves. Any younger family member who has not yet built his own house or is unmarried and wants to live away from his father's house can expect to live with his older brother.

Because Harry was about the same age as Owens, Ken enrolled them both at Port Harcourt's Baptist High School. A boarding school, Baptist High also had provisions for day students. Owens wanted to live at school with the other children – just as Jim, Comfort, and Barine were allowed to do – but Ken did not want to pay to send him to a school only eight kilometres from home. Eight kilometres was, however, a big distance to a twelve-year-old boy. In Lagos, Owens had gotten used to walking five kilometres to and from school every day. Now, in Port Harcourt, he and Harry had to get up at four every morning to get to school on time.

Besides the frustrating commute, eating lunch was another difficult thing to get used to at Baptist High School. Because of the early seventies American charity campaign featuring photos of starving Biafran orphans, Nigeria had received great quantities of humanitarian aid, much of it food, from all over the world. After the war (and after graft and theft had taken their share), the dregs of this charitable aid filtered into the school system as "lunch." The children were fed dried salt cod and a baffling substance called cheese. They were also the beneficiaries of mountains of tinned meat, a product of suspicious provenance. Owens and his schoolmates suspected it was dog food.

The long walk to and from school offered Owens plenty of time to think each day. Many of the other students in his class were there on scholarship. He reasoned that, if he applied himself to his studies, he might earn one as well. By the end of the school year, his grades were good enough to earn him a scholarship for his entire board and tuition.

Thrilled, he raced home to tell Ken. "Look, I've got a scholarship!"

"How do you think you got the scholarship?" Ken asked.

"I'm the best in my class," said Owens. "In all my subjects I'm always among the first three – and there's more than thirty of us!"

"I am the commissioner for education in Rivers State," said Ken. "I am the person in charge of all state-funded scholarships. One of the first things I did was to guarantee that every Ogoni student who desired it received a government scholarship. All Ogonis are given scholarships no matter how well or poorly they do in school – all Ogonis except you and Harry."

Ken explained that, to avoid any taint of nepotism, he made sure that his policies did not extend to his own family. He forbade Owens to accept the scholarship.

Owens did not understand. Neither did the other students. "How come your brother is commissioner for education and he has two official cars but doesn't let the driver take you to school?" they'd ask. "Are you really his brother?" But it was just Ken's way. It was important that Owens and Harry not be seen to have special privileges.

In the end, Owens won the day. After he proved to everyone that he had won his scholarship on merit, Ken relented. At the start of the next year, Owens moved out of the congested house on Nzimiro Street to savour his first taste of independence.

CHAPTER**TWO**

On April 25, 1970, a group of Ogoni chiefs handed the military governor of Rivers State a document entitled "Humble Petition of Complaint on Shell-BP Operations in Ogoni Division." The petition argued that Shell-BP was "seriously threatening the well-being and even the very lives of the people of Ogoni Division.

"About two decades ago," the chiefs wrote, "agriculture was the mainstay of the economy of Ogoni Division. But to-day, the entire economy of our people has been completely disrupted through the connivance of a nation which seems to have allowed the Shell-BP, a purely commercial organization to enter upon and seize the people's land at will. So long as the nation gets her royalties, nobody bothers what happens to the poor rural farmer whose land has been expropriated." The chiefs described the way Shell-BP bulldozed cash crops and trees, and said that crude oil spills had killed mangrove swamps along with the "fishes and crabs, mudskippers, oysters, shell-fishes, etc. on which the livelihood of the poorer people depends."[1]

The petition was, in truth, the work of Ken Saro-Wiwa. As a member of the very government he was petitioning, he understandably opted to use the chiefs as a front.

Two months later, Shell-BP responded. "As you know, the main aim and purpose of an oil Company must be to find and produce hydrocarbons as

efficiently as possible," wrote J. Spinks, Shell-BP's manager of eastern operations. "This is the area in which it makes a very significant contribution
to the overall economic development of any country in which it operates. As
is the case with the other oil Companies operating in Nigeria our obligations and responsibilities are clearly delineated in the agreements made with
the Federal government and by the Laws and Regulations relating to the oil
industry in Nigeria. These have always been meticulously observed by this
Company. We have, however, been extremely careful to ensure that our
operations cause minimal disturbance to the people in the areas in which we
operate. . . . There can be no doubt, however, that the incidental benefits
accruing to the Ogoni Division from Shell-BP's presence there greatly out-
weighs any disadvantages."[2]

Spinks's assertion that Shell-BP had made a significant contribution to
the national purse was correct. Petroleum Decree No. 5 of 1969 and other
revenue-allocation measures had effectively transferred all the oil resources
of the Niger delta into the coffers of the federal government, giving the
Gowon dictatorship more funds than any previous Nigerian government
had ever received.[3]

As for the "incidental benefits accruing to the Ogoni Division," these were
incidental indeed. Ogoniland remained impoverished, with no electricity,
telephone service, sewage treatment, or running water. Ogoni farmers whose
fields had disappeared under pipelines were left with no means of earning an
income and were given no special consideration for jobs in Shell-BP. At the
time of the petition, only one Ogoni man had ever held a senior staff position
at Shell-BP, and fewer than a dozen had secured junior staff assignments.[4]

Interestingly, Ken Saro-Wiwa had once applied for a job with Shell-BP.[5]
During an interview in 1965, Shell manager Bobby Reid was impressed with
Ken yet mystified. "I don't quite see why you want to work at Shell-BP," he
said. When it became clear that the twenty-four-year-old was simply looking
for any halfway lucrative employment, Reid offered him a Shell scholarship to
any university in the United Kingdom. Thrilled, Ken accepted, and Reid (who
as a child had lost his right arm in his father's butcher shop) warmly shook
Ken by the left hand.[6] Five years later – a few months after the petition –
whatever warmth Ken may have still felt for Shell evaporated with the first
major oil-related ecological disaster in Ogoniland.

In July 1970, an oil well at the Bomu/Shell-BP oilfield exploded, causing widespread pollution and outrage. (It should be noted that the name of this oilfield was a misnomer on Shell's part as the oilfield was actually located near the Gokana village of Dere – quite far away from the village of Bomu.) "Our rivers, rivulets and creeks are all covered with crude oil," wrote the Dere Youths Association in a protest letter to Shell-BP. "We no longer breathe the natural oxygen, rather we inhale lethal and ghastly gases. Our water can no longer be drunk unless one wants to test the effects of crude oil on the body. We no longer use vegetables, they are all polluted."[7]

Four days after the explosion, several Ogonis came to Ken complaining that villages were "drowning in oil." Ken called for his driver and, with twelve-year-old Owens by his side, went to see for himself.

From a distance, Ken and Owens could see a grey-white column of smoke rising over the kingdom of Gokana. As they got nearer, the smoke became a roiling black pillar and the air took on a sharp petrochemical tang. Fireballs the size of buildings erupted from the smoke with a low roar. On the outskirts of Dere village, Ken's car became so enveloped in the inky cloud that his driver was forced to retreat upwind.

On the roads near the Bomu/Shell-BP oilfield, twenty thousand Dere inhabitants had gathered to watch the fire. Many people recognized Ken and swarmed around him. They told him their farms and houses were gone, their streams were filled with oil, there was no water or food. Even if food were available, it could not be cooked because Shell had warned them that it was dangerous to light fires. Some villagers reported coughing up blood. The foul miasma made Owens cough and his eyes water.

This was Owens's first exposure to the consequences of oil production. Ken, on the other hand, had been noting the impact of the oil industry on the Ogoni landscape for almost a decade. Oil fires and direct pollution had been an occasional problem in the past. More serious were the profound effects from the infrastructure required to support oil exploration and production. In an area lacking paved roads, the few modern thoroughfares the oil company built in Ogoniland tended to bypass villages and head directly to wellheads and flow stations. The new roads opened up pristine forests to hunting, commercial logging, and cultivation. The road construction was often slipshod, restricting drainage across flood plains, causing flooding on one side of the road and excessive dryness on the other. The death of trees and disruption of seasonal fishing were devastating the income of local communities.[8]

The oil companies were also bulldozing seismic lines across Ogoniland. These cleared lanes were used for laying pipeline or conducting seismic surveys, a method of looking for oil without drilling. To do a seismic survey, an oil company must first clear vegetation in a grid pattern. Along this grid, dynamite is placed into holes dug a few metres deep; a small artificial earthquake is created by detonation. Computers record the time it takes for shock waves to be reflected by subsurface layers, giving geologists a sophisticated three-dimensional image of what lies below.[9]

According to Shell, dynamite is used only in remote regions. "In densely populated or environmentally sensitive areas, where explosions are not practical, vibrator trucks are used," said the company.[10] Shell's definition of "remote region" was rather relaxed. Seismic dynamiting had been documented in many locations close to human habitation. Pa Wiwa's concrete house in Bane still bears cracks in the walls from these man-made earthquakes. Other Bane homes of mud and thatch have suffered total collapse.

Along seismic lines cut through dry land and fresh water, the tropical plant life of the delta often regenerates quickly. Mangrove forests in brackish swamps have, however, proven more ecologically sensitive. Because of their slow regeneration rate, mangrove forests bear the scars of seismic lines for decades. Despite this, in 1993, Shell estimated that it had cut 39,000 kilometres of seismic lines through mangrove forests in the delta and said it planned to cut 17,400 kilometres more.

Excessive flaring was the worst consequence of all. Volatile and highly toxic, natural gas is a by-product of oil extraction. To deal with it, a company has three options: trap the gas and sell it; pump it back into the earth; or ignite it and let it burn off into the atmosphere. Capturing the gas and pumping it back into the well increases underground pressure and oil yields, but it's a complicated and expensive process. The cheapest solution is to burn it off. The World Bank estimates that as much as 76 per cent of all natural gas from petroleum production in Nigeria is flared (compared to 0.6 per cent in the United States, 4.3 per cent in the United Kingdom, and 21 per cent in Libya).[11] Natural gas flaring creates heat, air, and noise pollution. In a deafening roar, the flames (with temperatures as high as 1,400 degrees centigrade) produce carbon dioxide, carbon monoxide, volatile organic compounds, and nitrogen oxides. In some Ogoni locations (for example, in the Yorla and Korokoro oilfields) Shell has flared gas dangerously close to human habitation. In Dere, Gokana – the town closest to the Bomu/Shell-BP oilfield blowout – a flare was installed right in the middle of the village.[12,13]

Over the years, when Ogonis complained to Shell about houses and farmlands being destroyed, Shell frequently offered compensation. Unfortunately, Shell's arbitrary, non-transparent compensation schemes often caused confusion and friction. Much of the time, the compensation offered was either insufficient to cover the damage or was unevenly and unfairly distributed. Villagers who'd suffered little damage sometimes received more than those who'd suffered greatly, causing bitter squabbles and divisions.

|||||||

At Dere, Ken listened to people's complaints as he toured the area with Owens. Vegetation around Dere was dying and oil was everywhere. Ken talked to Dere's chiefs and elders and attended a meeting with some of the educated villagers. Then he began mobilizing.

He invited General Yakubu Gowon to visit the disaster site, and the Nigerian dictator toured the area two weeks later. Ken issued the same invitation to the region's state governors and began phoning reporters. He wrote letters to the editor and op-ed pieces in the country's prominent newspapers. He wrote letters to Shell, expressing his dismay over the disaster. And he organized the students of Dere to write letters and hold peaceful demonstrations during Gowon's tour of the affected area.

Seeing Ken pour his energy into the Dere campaign both excited and confused Owens. Ken had raised a fuss over things in Ogoniland before, but nothing like this. When Owens asked him what was so special about what had happened at Dere, his brother replied that he was laying the groundwork to take Shell and the government to court. "I want to make this a test case that will make the company and the government look at Ogoni in a different light."

Some Ogoni elites had a similar idea – except that their plans focused on how to turn the Dere tragedy to their own advantage. Before Ken could launch a legal action against Shell, a lawyer went around to the chiefs of

Dere and negotiated an out-of-court settlement with the oil company. Although the chiefs were well compensated, the government and Shell did not offer nearly enough relief to the people of Dere, nor did they begin to clean up the spilled oil in a timely manner. The chiefs' settlement negated Ken Saro-Wiwa's legal strategy, leaving him frustrated and furious.

||||||

During the 1971–72 school year, Owens excelled academically, remaining at the top of his class in Port Harcourt. Naturally enough, this led Ken to suspect that Owens's school was not challenging enough. In truth, Baptist High had been greatly affected by the recent war and was not offering sufficient academic competition or enough extracurricular activities for a complete education. Therefore, at the end of the term, Ken packed off both Owens and half-brother Harry to join their brother Jim at Edo Government College in Benin City.

Edo College was very different from Baptist High School. The principal, a white Englishman, ran the institution like an English public school and ensured his school's top spot in athletics by actively recruiting the best student athletes from around the country. At Edo, Owens learned to play cricket and soccer and was on the 100-metre relay team for his house.

At the same time, Ken's political career continued to thrive. In 1972, he was appointed commissioner of the Rivers State Ministry of Information and Home Affairs. On breaks from school, Owens stayed with Ken in Port Harcourt, not with his parents in Ogoniland – Ken's age and status made him more a father figure to Owens than an elder brother. And, like a father, Ken was always concerned with Owens's academic performance.

Although Owens did well at Edo College, the teachers reported he was "erratic" in his behaviour, noting that he was hanging out with a bad crowd. Even though Owens himself never got into trouble, the boys he befriended were sometimes suspended for serious infractions, and this gave Ken cause for concern. During Owens's holiday visits, Ken noted Owens spent a great deal of time attending parties or playing with Ken's small sons, Junior and Gian, or with Ken's houseboys. What could not be denied, though, was that Owens was also an avid reader. From the age of twelve, Owens regularly read *Newsweek*, *Time*, and *West African Magazine*, as well as every newspaper Ken brought into the house. Ken also supplied Owens with a variety of abridged

British and American classics, which Owens read voraciously. Pleased, Ken once boasted to a visiting writer friend that his junior brother "read everything in black and white." When the writer asked Owens if he had read one particular book, Owens said yes, and excitedly recounted part of the story.

"And do you remember who wrote this book?" asked Ken.

Owens thought a moment, then shook his head. "I don't remember," he said.

The writer and Ken let out a laugh. "I did. That was my book," said the visiting writer.

Included in Ken's household at that time was a young medical student named Bennett Birabi whom Ken considered something of a son or younger brother. Bennett was the child of the most prominent person to have come out of Ogoniland: Timothy Naaku Paul Birabi.[14] To the Ogonis, the elder Birabi was a hero. As the first Ogoni ever to graduate secondary school, he had furthered his studies at Achimota College in Ghana and graduated in mathematics from Britain's Southampton University in 1948. He was elected to the Eastern Nigerian House of Assembly and nominated to the House of Representatives in Lagos. Around 1950, Birabi began to put forward the idea that the Ogonis needed not only good administration but autonomy as well. He inaugurated the Ogoni State Representative Assembly with the idea that the Ogonis, as well as the many other ethnic groups, should have self-determination within the upcoming Nigerian Federation.[15] Unfortunately, Birabi met an untimely death on his return from the Nigerian Constitutional Conference in London in 1953. The Birabi family tragedy was compounded by the death of his wife, the elegant Victoria Maah, a few days later.[16]

The orphaned Birabi children – three-year-old Bennett and his baby sister, Ruth – were turned over to the care of a grandmother. Because of their shared admiration for Birabi, Ken and his close friend Edward Kobani made it their mission to enrol Bennett in school and pay his fees through a bursary from the Khana County Council. During the civil war years, the Birabi children lost their grandmother in a refugee camp in the Igbo heartland. Ken Saro-Wiwa tangentially came to young Bennett's aid again in 1969 when, as commissioner for education, he embarked on a scheme to send all senior secondary school boys in the state to the Baptist High School in Port Harcourt. Bennett was one of the young people who benefited from the scheme, but without family or fortune, his life remained difficult. From school, Bennett wrote to Ken, outlining his troubles. Ken asked Bennett to consider himself a member of the Saro-Wiwa household. During the next school break, he came to stay at Ken's house, shared a room and a bed with Owens, and remained an honorary member of the family until he graduated from medical school, married, and established a practice in Port Harcourt.[17]

Instead of gratitude, Bennett developed an intense resentment toward Ken. Bennett felt it was his birthright to step into his father's shoes and become the next "Great Ogoni Man." It appeared to many, however, that Ken was assuming that mantle for himself. Accepting charity from the man who was "stealing" his birthright was a bitter pill for Bennett, whose hatred of his benefactor would haunt Ken Saro-Wiwa in the years to come.

Like young Birabi, several others who had been liberated or had escaped from the war front came to stay with Ken while they sought employment or education. Many were Ogoni elites who had joined the Biafran secessionist government only to find themselves unemployed, internal refugees at the war's end. After the surrender of Biafra, many of these elites beat a path to Ken's door in Port Harcourt.[18] They came because, at the collapse of the Biafran rebellion, Ken was the only Ogoni who had openly sided with Nigeria from the outset. In an administration that discriminated against the Ogonis, Ken represented the only chance of access to governmental appointments. Though they had been on the opposite side of the conflict, Ken treated these educated Ogoni men with courtesy and respect, gave them food and temporary lodging, and instructed Owens to polish their shoes. As a result of Ken's interventions, those who wanted to continue their careers in politics received positions in the state government, and those who wanted to further their studies were accepted back to school with full scholarships.

As Owens shined the shoes of these Ogoni elites, he listened to their stories about the war and their terrible treatment by both the Igbos and the federal soldiers, who didn't know which side the Ogonis were on. Because Ken was Ogoni, and because he had famously escaped Biafra to fight for the federal government, the Nigerian soldiers were apt to treat the Ogonis they found with more mercy than they bestowed on the Igbos. Although they recognized this, although they actively sought out Ken's help after the war, and although they profited from his influence and assistance, the Ogoni elites, like young Bennett Birabi, contrarily grew to despise Ken Saro-Wiwa.

To be a Nigerian man (and an Ogoni in particular) is to be inordinately concerned with one's place in the hierarchy of society and family. Unquestioning obeisance is expected from one's social or familial inferiors. Although in Ogoniland's particular caste system Ken was considered an elite, he was nonetheless seen as a threat to his elite peers. Certainly Ken was younger, richer, and more politically powerful than they. But worse, Ken's writing was making him well known and exceedingly popular. This, more than anything, infuriated the other elites. If the proud Ogoni elders saw themselves as the sun, Ken was their solar eclipse.

In the midst of the Biafran War, a wise friend, Dr. Obi Wali, had cautioned Ken, saying, "If you're not careful, the great unanimity you are achieving among the Ogonis will be thwarted and brought to a standstill once the educated Ogonis return from the rebel area." Prophetic words. As soon as they were able, the elites, coveting a niche in the Port Harcourt business community and in the Rivers State government, promptly set themselves to the task of undoing everything Ken had achieved for the Ogoni people.[19]

Ken maintained good relationships with officers in the Nigerian Army. As administrator of Bonny Island, he had needed military cooperation to aid Ogoni refugees and to see that relief such as food and clothing could get to Ogoniland. Ken kept up these important relationships by throwing regular parties at number 11 Nzimiro Street.

One evening's guest list included the commander of the Nigerian Army Engineering Corps, Olusegun Obasanjo;[20] the army chief of staff under Murtala Mohammed, General T.Y. Danjuma; and Ken's next-door neighbour, a short, quiet soldier named Sani Abacha. Because Ken's wife, Maria, was away at boarding school finishing her education, Ken had arranged the party himself and seconded Owens to act as butler. Dressed in his finest clothes, Owens milled about with a tray of hors d'oeuvres trying to figure out who were the most important men in the room in order to serve them first. He had approached only one or two guests before Ken took the tray away and said, "No, that's not the way. Go and sit down and watch me."

Ken started by serving all the women at the party first. He then served the men by their order of seating and not by their importance or rank. In Ogoni society, this was shocking – not only was this order of service unheard of, but the act of service itself was considered degrading, something only servants or junior relations performed. Open-mouthed, Owens watched his brother move about the room with the tray until finally it was Owens's turn to be served.

Proffering the plate of hors d'oeuvres, Ken looked down at Owens and explained quietly, "A good leader must first learn to serve." As Owens accepted a skewer of roast goat meat, Ken leaned forward and continued, "I will serve you because I will be your leader."

|||||||

By 1973, Ken had become more critical than ever of the government he was a part of. He was disgusted with the corruption in Nigerian politics and the lack of government accountability. The plight of the Ogonis bedevilled him as well. He was certain the Ogonis needed dedicated leadership, and he had tried to provide this by setting up a formalized structure to the Ogoni Development Association. It was a failure. At the first meeting, Dr. Garrick Leton was elected president of the ODA – and the association never met again.

Ken's every effort to instil the elites with his personal vision for the Ogonis served merely to engender jealousies among them as they scrambled for personal advantage in the state government. In March 1973, Owens was home on a school break when he heard a radio bulletin announcing that the government had fired Ken Saro-Wiwa. Ken was on his way home from Lagos; the moment he came in the door, his family rushed to him.

"You've been sacked!" said Owens. "We heard about it on the radio."

"Yes, I know," said Ken. "I heard it in the car as I was coming from the plane."

"But why were you sacked?"

"I don't know."

The government had given him no reason; it simply issued a press release. The next morning, government trucks and uniformed officers arrived at 11 Nzimiro Street to remove the Saro-Wiwa family from their house.

Ken came out onto a front balcony and sternly addressed the officers below. "Get out!" he shouted. "My family and I are not leaving this house."

"But sir," the officer in charge called up to him, "we have our orders from the state governor for you to move."

"Where does the state governor expect me to go?" said Ken. "I have nowhere to go. And I am not going to sacrifice myself after seeing that this very state, Rivers State, was created and developed. And I will not sacrifice myself after fighting a war for this government. No, my good fellow, my family and I are not leaving."

Ken's service record to the government was the least of his reasons for resisting eviction. In truth, this was the worst possible time for him to lose his job; he was nearly broke. Although he lived in a grand house and was chauffeured around, his salary was modest. Besides being the father of two and supporting his wife at boarding school, Ken was the primary caretaker of all five of his siblings and a couple of half-siblings as well, not to mention the scores of Ogoni people unrelated to him who were using his home and hospitality.

After the government trucks left, Ken went into the living room to smoke his pipe and reflect. Cautiously, Owens approached him. "Dede, what are you going to do?"

Ken chewed on the stem of his pipe. "We may have to move."

"Where will we go?"

"Bane. If it is the government's insistence that I leave this house, we may have to move back, all of us, to our father's house."

"Bane?" said Owens, disappointed.

"I have thirty-six pounds in the bank," Ken explained. "I have to think about how we will eat. If we go to Bane, Mama is there. She has enough farms, and the produce from there will feed us."

"What will you do for work?"

Ken exhaled smoke in a slow, thoughtful stream. "Trade," he finally replied. "I will go back into trade."

||||||

Back in the mid-1960s, Ken had purchased a small rattletrap bus that he used to start a passenger shuttle service covering the thirty-six-mile route between Bori and Port Harcourt. After the war ended and he started working for the state government, Ken decided to find out what had happened to his old bus. Though it had fallen into disrepair during the war years, it was salvageable. During his tenure as a Rivers State commissioner, he'd had the bus repaired and restarted his shuttle service. Before long, the old bus had earned Ken enough money to allow him to invest in a larger bus, which he used to start long-distance service between Port Harcourt and Lagos. Both buses were being run from a company Ken had named Khana and Sons in deference to his home kingdom in Ogoni. The office was situated in a flat he rented at 24 Aggrey Road in Port Harcourt.

Now that the government had fired him, the income from Khana and Sons was all he had. Unfortunately, the profits were slim, not nearly enough to sustain his extended family. If he wanted to earn money, Ken would have to think of something bigger – and quickly.

After losing his job, Ken approached all the banks in Port Harcourt, trying to borrow enough capital to start a new enterprise. When asked what type of business he would like to go into, Ken said that he was opening a grocery store. He had, after all, learned from his mother something about buying and selling food products. After being turned down by a half-dozen banks, Ken had some luck: one bank manager turned out to be an old school chum, who authorized a small loan. Ken immediately started looking up other old friends, in particular those working for grocery wholesalers and distributors. He made deals and arranged delivery for cases of canned milk and other food items. Within a week, he'd transformed the dingy little Khana and Sons office at 24 Aggrey Road into a tidy food market. With characteristic optimism, he adopted a plural name for the business: "Gold Coast Stores."

As the paint was drying and the last of the shelves were being installed, two lorries pulled up. Walking outside, Ken called to Owens, who was helping him, "They are here!"

Not knowing who "they" were, Owens followed his elder brother to the curbside as the drivers flung open their trucks to reveal a mountain of groceries. After Ken signed the waybills, the drivers started handing him cases of canned goods.

A former state commissioner lifting goods off the back of a truck was something the Aggrey Road neighbours didn't see every day. Few Ogoni men in Ken's position would have publicly exposed themselves this way, but it didn't seem to bother Ken. Owens pitched in, and the brothers offloaded groceries until the little Gold Coast Stores was bursting with merchandise.

The following day, Ken called Owens into his office. "I want you to do something for me," he said.

Ken took out a sheet of lined paper and wrote a list of basic grocery items with brand names and sizes. He divided the rest of the page into columns and handed it to his brother.

"Go to all the shops on the next several streets and get me the prices of each of these items."

Confused, Owens looked over the list. "You don't want me to buy these things?"

"No, just write down the prices."

"Well, if I ask them their prices and I say I'm not buying, they're going to be angry with me."

"Just go. Tell them you're from school, doing research for a project."

Why was Ken asking him to do this? Nigerian shopkeepers charged whatever they wanted. If a customer thought a price was too high, she haggled. What was the point of knowing how much the shop down the street was charging?

Owens took his list and visited all the grocery stores in the area. As he predicted, some grocers took angry exception to his queries. Before long, though, he stumbled on a trick. If there were boys his own age in the store (or if the shopkeeper had a pretty daughter), Owens would strike up a conversation, tell them he needed the prices of certain items for a school project, and have his new mates or the pretty girl get him the information. He also found a way to inquire about the store's hours in a casual manner. Over the next couple of days, he compiled a scribbled database of the business practices of most of Ken's competitors.

Owens had no idea that Ken was about to introduce the radical concept of price competition to Port Harcourt.[21] Using the information his little brother had gathered, Ken undercut his competitors and provided more convenient hours. He opened at seven in the morning and closed at ten in the evening; the other shops opened at nine and closed at five or six.

Ken made sure that Gold Coast Stores was also cleaner and brighter than his competitors' dingy little markets. But advertising was his biggest innovation. No grocery store had ever advertised on a billboard or on radio. Now the people were hearing a radio jingle: "Where in the world can you buy / Peak Milk at ten kobo? / At Gold Coast Stores / 24 Aggrey Road!"

Peak was a popular but scarce and expensive brand of imported canned milk. Ken used his company Khana and Sons to become a bulk distributor of Peak, then – using a classic "loss leader" retail strategy – offered the milk to the public at the wholesale price. Thus Gold Coast shoppers were able to purchase Peak milk at ten kobo (Nigerian cents) a can while his competitors were charging thirty to fifty kobo.

The day after the first airing of the radio jingle, Owens was walking to work early in the morning when he saw a huge queue, hundreds of people long, beginning at 24 Aggrey Road and stretching far down the block. It was as if the whole city were coming to that one small store. Once they started letting people inside, they couldn't keep enough Peak milk on the shelves.

Before long, Ken called Owens upstairs to issue new orders: from now on, shoppers could buy Peak milk only if they also purchased other items, and there was a limit of three cans of milk per person. The demand remained so intense that, by afternoon, Ken reduced the limit to one can. The shoppers also bought more groceries than expected. Ken had to arrange for an emergency delivery. The next day the crowds were even larger.

Within a month, everyone in Port Harcourt knew the Gold Coast Stores. Unlike other grocery stores, Ken's stocked packaged versions of native foods such as *egusi, gari,* and pounded yam along with bottles of palm oil and other staples found at outdoor village markets. Profiting from the knowledge gleaned from his trader mother and market master father, Ken had, in a matter of four weeks, transformed himself from a broke, unemployed government cast-off to a successful retailer.

As for the house, the government tried several more times to evict the Saro-Wiwas, but Ken remained adamant that the government provide him another house before he would move. Every day, Ken wondered if the government would haul away his furniture. Every night, he returned to find everything just as he had left it. Inexplicably, the Saro-Wiwas continued living at 11 Nzimiro Street for the next three years with little further incident.

|||||||

Before long, Ken started importing kitchen stoves, sacks of corn and flour from the United States, and crates of European liqueurs and spirits. The purchase of a printing press allowed him to start his own small publishing house as well.

In the spring of 1975, seventeen-year-old Owens graduated from secondary school in Benin City and returned to Port Harcourt to work in Ken's store. The business had greatly expanded, and Owens was no longer relegated to stacking cans. Instead, he was made manager of Khanatek, the new technical division of Saros, selling cooking gas and kitchen stoves.

Life as a stove salesman quickly lost its charm, and Owens began making plans to achieve his lifelong dream – becoming a doctor. In anticipation of medical school, Owens opted to do his A levels at Port Harcourt's College of Science and Technology, throwing himself fully into the study of physics, chemistry, and biology. At the college, he shared a hostel room with a couple of students from South Africa, ANC activists, who had been given scholarships by the Nigerian government. Through them, Owens became aware of the human rights abuses caused by apartheid. He ended up being more concerned about events at the southern tip of the continent than the upheavals at home.

Since 1966, Nigeria had been under the shaky leadership of General Yakubu Gowon. Gowon's real achievements included having kept the country united during the Biafran rebellion and having treated the Igbos in a statesmanlike fashion after the civil war. Gowon had also been the guiding force in creating the twelve Nigerian states and the trading bloc known as the Economic Community of West African States (ECOWAS). His regime was, however, undisciplined and corrupt, and despite his enormous power he seemed an inept dictator. Gowon vacillated over the reorganization of the armed forces and failed to rescind the wartime ban on political activities. No political parties were formed during his administration and no elections were held. He could not control his corrupt governors, commissioners, and bureaucrats; neither was he able to manage the enormous wealth generated by the 1970s oil boom. Instead of investing in education, health, agriculture, and sorely needed infrastructure, Gowon boasted that Nigeria's problem was too much money with no idea how to spend it. He granted stupendous salary increases to public servants and engaged in profligate spending on relatively frivolous projects such as the World Black and African Festival of Arts and Culture. Agriculture, the mainstay of Nigeria's economy before the oil boom, fell into decline as farmers left their deteriorating villages to start new lives in the cities. Because a significant proportion of the country's food suddenly had to be imported, inflation increased sharply.[22]

In July 1975, a coup led by a group of army officers deposed Gowon and brought General Murtala Mohammed to power. Fed up with the Gowon regime, the people welcomed the coup and hailed its bloodless execution. Mirroring the mood of the public, the newspaper *New Nigerian* rhapsodized over this turn of events, saying, "The swiftness of the latest change in our government gives us cause to be proud."

The coup was ostensibly carried out to restore the "dignity" of the military, which had been compromised under the Gowon regime. The new president and his deputy, General Olusegun Obasanjo, compulsorily retired Gowon and the other generals, the inspector general of police, the inspector general's deputy, and all the military governors and federal and state civilian commissioners.[23] (Had Ken Saro-Wiwa not been fired in 1973, the coup would probably have ended his political career anyway.)

The Mohammed regime attempted structural reform. To centralize power in the federal government, it assumed control of all education from primary schools to universities. All television and radio stations were brought under federal control, and the government's grip on the press was tightened with the establishment of the News Agency of Nigeria. Mohammed's government also became the major shareholder of the country's two largest newspapers, the *New Nigerian* and *Daily Times*.[24]

Next, Mohammed and Obasanjo began a campaign of "cleansing" society by creating anti-corruption and public morality institutions such as the Public Complaints Commission, the Assets Investigation Panel, and the Corrupt Practices Investigation Bureau. Such measures did little to prevent corruption, as most of the anti-corruption policies were easily circumvented. The country's new leaders went out of their way to create the image of a financially responsible government, the antithesis of their spendthrift predecessors. As part of his new austerity plan, Mohammed cut much of his predecessor's elaborate security systems and reduced the number of presidential bodyguards. It was an ill-conceived decision.[25]

On February 13, 1976, after only seven months in power, Mohammed was killed in an unsuccessful coup attempt. The motives behind the attempted coup were said to have involved grudges held by army officers upset about being passed over for promotions as well as a wish for the restoration of business as usual. It was rumoured that the United States and Britain had backed the attempted coup because they disapproved of Mohammed's "communist" inclinations and were outraged that Nigeria supported the Marxist MPLA party in Angola.[26]

Ken Saro-Wiwa's old friend, General Obasanjo, became the new head of state and General T.Y. Danjuma was named chief of army staff. Obasanjo was Yoruba, Danjuma was from a minority tribe, and both men were Christian. To lend ethnic and religious balance, Obasanjo chose the highest-ranking northern Muslim officer, Shehu Musa Yar'Adua, to be his deputy.[27]

Because Murtala Mohammed was widely perceived as the best leader in the country's history, the Nigerian people strongly condemned his death and the attempted coup. Even before the outcome was known, university students took to the streets in solidarity with the Mohammed regime.[28] More concerned with his medical studies than with Nigerian politics, Owens stayed out of the struggle, though he could hear the mobs of protesting students outside his dorm windows screaming as they ran from police firing bullets and tear gas into the crowds.

||||||

During the Obasanjo regime, thirty-three-year-old Ken began writing his own widely read political newspaper column in *The Punch* and through it became a familiar name across the country. When General Obasanjo decided that Nigeria needed a new constitution, Ken saw it as an opportunity to correct some of the wrongs suffered by the Ogonis at the hands of government and the oil industry. In speeches and in his column, Ken began agitating for the political restructuring of Nigeria, allowing for the creation of Port Harcourt State, a political entity consisting of Ogoniland and the neighbouring oil metropolis of Port Harcourt. Obasanjo's new constitution was to be drafted by a constitutional assembly; Ken began a campaign to have himself elected as a delegate.

Ken's campaign had scarcely begun when trouble found him: he was arrested for smuggling whisky.

At the time, Ken was importing goods of all sorts, including whisky, through a wholesale trading company he called Dorbi and Sons, named in honour of Ken's great-grandfather Gbenedorbi. Dorbi and Sons, along with Khana and Sons and the growing Gold Coast Stores, had been subsumed under a newly created private corporation Ken called Saros International.

Owens was at school when he got the news of Ken's arrest. Panic-stricken, he raced to the jailhouse by the wharfs. He found Ken alone in a small detention cell, sitting calmly reading an old copy of the 1959 book

A Ladder of Bones: The Birth of Modern Nigeria from 1853 to Independence by
Ellen Thorp.

"What's the matter?" asked Owens. "Why are you here?"

"You shouldn't worry," Ken said. "This is just my political enemies trying
to smear my reputation." He explained that the candidates running against
him for the constitutional assembly had jointly filed a formal complaint
accusing Ken of smuggling liquor. "There's nothing to it," he said.

The following day, the police raided Ken's warehouse in Port Harcourt
but found no trace of illegal whisky. Ken had proper bills of lading for every
item in storage. After three days, the authorities satisfied themselves that
everything was correct and accounted for. They apologized and let Ken go.

But the damage had been done. Ken's arrest and detention were front-
page news across the country. Editorials and letters to the editor expressed
shock and dismay about the popular political columnist, war hero, and
respected businessman's descent into a life of crime. Ken's opponents collected the worst of the tabloid clippings and posted them to President
Obasanjo with a formal letter asserting that Ken's alleged crime made any
candidacy for election to the constitutional assembly inappropriate.

Ken was disqualified from the race, but soon after his release he succeeded in having the ruling overturned. He then won the local election for
the delegate seat. At the national caucus held to winnow down the number
of delegates, his political opponents (including old friend Edward Kobani)
formed a bloc and "convinced" many pro-Ken delegates through bribes and
intimidation to vote against Saro-Wiwa. In the end, he lost the election to
the constitutional assembly by a single vote.[29]

Ken took the loss very hard. Watching the election results on television
at home, he broke into tears. The next day, he declared that perhaps he had
done enough for the Ogoni. It was time to think about himself.

||||||

In September 1978, Nigeria's new constitution was signed into law; the state
of emergency in place since May 1967 was lifted. The next summer, Alhaji
Shehu Shagari was elected president, the supposed end to thirteen years of
military rule in Nigeria.[30]

Saros International was flourishing. The trade in kitchen stoves,
propane, and maize brought in money hand over fist. Ken progressed from

being comfortable to being wealthy. His immediate goal was to give his children – sons Junior and Gian and two-year-old twin daughters, Zina and Noo – the best education possible.

After his election defeat, Ken washed his hands of Nigerian politics and moved from his government-provided house at 11 Nzimiro Street to a new home at number 9 Rumuibekwe Road. The house was in a pleasant middle-class neighbourhood north of Port Harcourt, opposite the posh Shell staff residential area. A few months later, he bought another home as well, this one in England. In order to give his children the best possible educational opportunity, he moved his immediate family permanently to the United Kingdom where Maria soon gave birth to Ken's third son, Tedum. After installing his two boys, ages nine and seven, at boarding school in Derbyshire,[31] Ken stayed in London as much as he could. Because his business was still in Nigeria, however, he made frequent trips to Africa to support his growing family's new bi-continental lifestyle.

Owens, meanwhile, after finishing his A levels at the College of Science and Technology, stayed on in Port Harcourt and worked as a supervisor in Ken's store. During a year away from school, he spent five months working with a scientific team from UNESCO conducting research on the Ogoni people. Owens accompanied the team of British and American anthropologists as they travelled throughout Ogoniland, measuring the sizes of farms and interviewing people about the food they ate. Working with the UNESCO team exposed Owens to much more of Ogoniland than he had known. To his surprise, villagers who heard his name often connected him with his elder brother and told him how much respect they had for the famous Ken Saro-Wiwa.

They also confided their problems to Owens. Shell was destroying their farms, they said, and passing pipelines through fields and villages without asking permission or paying compensation. Nearly everyone made mention of the terrible blowout at Bomu a few years earlier. Although Owens had witnessed the devastation and subsequent demonstrations at Bomu, he hadn't, at the time, grasped the level of anger the local people felt toward the company.

After his stint with UNESCO, Owens resumed his job at Ken's Khanatek division selling gas cookers, refrigerators, air conditioners, and pressurized canisters of gas. Because there is no infrastructure in Nigeria to allow for natural gas to be piped into homes, all gas-powered appliances have to be fuelled by small, pressurized containers. Owens had an idea. Instead of having customers bring their empty propane bottles back to the store to be refilled, it might be better to take the gas to the customer. With Ken's blessing, he set up twenty distribution nodes around Port Harcourt – small shacks manned by "distributors" who collected empties and swapped them for full tanks. Sales increased once again.

Ironically, the area of town in which Shell company staff lived was one of the Wiwa brothers' most lucrative distribution nodes. "Little London," in Port Harcourt's north end, was a place where the basic public services worked, and people cooked with clean-burning cooking gas rather than over open fires. Inside the fortified perimeter of the faux European suburb, hidden from the gaze of the locals, Shell staffers lived in well-appointed houses on paved streets lit by street lights powered by uninterrupted electrical service. Little London had a clubhouse, a swimming pool, a cricket pitch, and a golf course. In such a place, one could almost forget that jolly old England was half a world away.

IIIIIII

University admission in Nigeria is highly competitive. In 1979, the year Owens applied to medical school, approximately 145,000 university students vied for 20,000 spaces. On the day the admissions lists were posted in the newspapers, Owens anxiously scanned the roster of successful applicants. To his great relief, he read that he had been admitted to medical school at the University of Calabar.

"Wow, Calabar," Owens said to Ken. "I didn't apply to Calabar."

"I didn't know there was a medical school at Calabar," Ken replied.

"Yes, it's strange. Calabar didn't have a medical school when I applied. Oh well, the newspaper says I'm going to Calabar. We'll see what happens!"

Owens was fortunate. Calabar's school of medicine was, in fact, brand new, and, as a member of its first class, Owens profited from air-conditioned classrooms, new laboratory equipment, new skeletons, and a cache of new cadavers. The provost of the medical school, a pathologist from the University of Ibadan, had recruited an impressive staff of visiting lecturers from the United Kingdom. Expectations ran high at the new school, and the students were expected to help Calabar achieve a stellar reputation. During his opening address, the provost joked that the incoming freshmen had just been sentenced to five years' imprisonment with hard labour.

Taking the challenge to heart, Owens immersed himself in physiology, biochemistry, anatomy, and neuroscience. His first year's grades were so good that Ken rewarded his younger brother with a trip to Europe.

||||||||

Once enrolled in medical school, Owens became something of a Nigerian frequent flyer. His trips to and from Calabar were almost always by air, and, thanks to a friend who was an air hostess, were often free or at reduced fares. Still, he had never in his life left the country of Nigeria – much less the continent of Africa – and the prospect of intercontinental travel was thrilling.

At Ken's instruction, Owens flew to London, then continued to Paris via ferry and train in order to meet up with Ken and his sons, Junior and Gian, at the Hotel Meridian Étoile just opposite the Palais des Congrès. In Paris, Ken had rented the cheapest possible room for the four of them (one bed, no view, next to the service elevators) and was bursting with pride in his parsimony. To be sure, Owens and the kids would have to sleep on the floor; according to Ken, the money saved would be better spent soaking up culture.

For Ken, Paris wasn't so much to be enjoyed as to be crammed like a textbook. At Napoleon's tomb in Les Invalides, Ken made sure his brother and young sons learned about the Napoleonic Wars. At the Place de la Bastille, he quizzed them about the French Revolution. At the Louvre, it was art history and architecture. The two little boys were often bored senseless, but Owens was transfixed. He had cared little for art, but once he found himself standing before the *Mona Lisa*, he could hardly be torn away.

In Paris, Owens climbed to the top of the Eiffel Tower and prowled through Notre Dame, but the most amazing thing he saw was at the end of a meal when Ken handed the waiter a small piece of plastic. Although he had read about such things in *Time* and *Newsweek*, Owens had never actually seen a credit card before.

The waiter behaved as if he hadn't either and did not want to accept Ken's card for payment. Only after the restaurant supervisor was summoned did it become clear to Owens that the problem was not with the credit card. Apparently, one did not accept credit from Nigerians. Without making a scene, Ken took out a pipe with a gold stem from his pocket and lit some tobacco.

"*Est-ce qu'il y a une problème?*" he asked.

"We have to telephone," the supervisor replied in English.

After a lengthy delay, the waiter returned and offered Ken a suspicious glare and a receipt. Noting the stricken expression on Owens's face, Ken said, "Ignore them, Mon." Ken signed the paperwork with a nonchalant flourish, as if it were beneath his dignity to acknowledge the slight.

|||||||

Owens graduated from medical school in 1985 after five years of intensive study and rewarded himself by taking a couple of months off to spend time with his parents in Ogoniland.

He had not stayed at the house in Bori since he was a child and much had changed. For one thing, there were no longer many Igbos around. The most striking difference, however, was in the vegetation. The grass had a yellowish hue. The guava trees, the oranges, the papayas, and the wild pears were withering. It had been fifteen years since the war had ended, and the natural world had not rebounded. The air reeked of sulphur, and the water holes and creeks had an oily iridescence. Even the pitch blackness of night had vanished. Lit by the macabre orange flames of Shell's gas flares, the nighttime sky had become perpetual twilight, a scene out of Dante.

It became clear to Owens that Ogoniland was in serious trouble. He vowed to do whatever he could to convince the Ministry of Health to post him permanently to Bori once he completed his fast-approaching internship.

At the University of Port Harcourt Teaching Hospital, Owens went through the usual internship rotations of surgery, pediatrics, obstetrics, gynaecology, and emergency medicine. When he was not on call, he lived at Ken's house on Rumuibekwe Road. Except for the domestics, he and Ken were the only two living in the house, and now that Owens was a doctor, he and Ken were on a more equal footing. Alone together, the two ate at the same table (something a junior would normally never do with a respected elder); Ken started asking for Owens's views on personal problems and social issues (also unheard of in their former lives).

One evening, Ken discussed his plans for the future with Owens. Ken loved five-year plans. In 1980, he had resolved to devote five years to business and, accordingly, had succeeded in becoming wealthy. Now he announced to Owens that he intended to spend 1985 to 1990 writing and publishing books under his own imprint, Saros Publishing. After establishing himself as a major literary figure, he intended to shift focus; from 1990 to 1995 he would re-dedicate himself to the Ogoni struggle and environmental issues. For this work, he said he expected to receive the Nobel Prize for either Peace or Literature. As for the five years after that, Ken considered the possibility of returning to academia, but he couldn't really envision his future much past 1995.

Just as he loved to lay out and realize his five-year plans, Ken also followed a rigid daily routine. Each morning he arose at four and turned on the radio, as he used to in Bori before the war. After listening to the BBC, the French news, and the Nigerian report, he would lock himself in the bathroom and remain there for at least an hour and a half. Once his lengthy and fastidious ablutions were complete, he would join Owens for breakfast, then head off to his office at 24 Aggrey Road. Returning home at 5:30, he would spend the rest of the evening and long into the night writing in his study.

Ken's confidence in his literary abilities was well founded. With astonishing speed, he established himself as a major voice in African poetry with the 1985 publication *Songs in a Time of War*, a collection of poems capturing his personal response to the Biafran conflict. In *Songs*, he reflected in verse on the dilemma of loving Nigeria while longing to be freed from both the Biafran menace and the corrupt federal government, and lamented the befouling of the Ogoni environment through years of oil spills and gas flaring.

Hot on the heels of *Songs*, he released his first work of fiction, *Sozaboy: A Novel in Rotten English*, which William Boyd in the *New Yorker* acclaimed as "not simply a great African novel but also a great antiwar novel – among the very best of the twentieth century."[32]

Written in a quirky mélange of pidgin English and lyrical prose, *Sozaboy* ("soldier boy" in pidgin) is the tale of a naive youth who joins the Biafran army for economic reasons only to become disillusioned and embittered. The young protagonist narrates the book in a uniquely Nigerian patois – the "rotten English" of the title. Through its use of the common speech of uneducated Nigerians, *Sozaboy* transformed itself into a biting social commentary and a sort of beat poem of epic proportions.

Sozaboy met with mixed critical reaction. Some praised it for its innovative style and antiwar message. Others were appalled by Ken's extensive use of pidgin, claiming that it projected an image of African cultural inferiority. The controversy sold books, and *Sozaboy* was shortlisted for the prestigious Noma Awards for Publishing in Africa.

In 1985 Ken became a columnist for the Nigerian newspaper *Vanguard* and signed a contract with the Nigerian Television Authority (NTA) to produce a weekly comedy television series called *Basi and Company*.[33] The plot and characters of *Basi* were inspired by his 1972 radio drama *The Transistor Radio*. Set on a fictionalized Adetola Street in Lagos, *Basi* followed the exploits of its eponymous hero and his feckless friends as they hatched an endless series of unsuccessful get-rich-quick schemes. The public adored *Basi and Company* and turned the show into a "street sweeper," a program not to be missed.[34]

The *New York Times* reported: "'Basi' is Nigeria's hottest comedy show. Every Wednesday night at 8:30, an estimated 30 million Nigerians, or almost one third of Africa's most populous nation, gather around television sets to see their national foibles skewered. . . . 'Basi' seems to have struck a chord because it lampoons modern Nigeria's get-rich-quick mentality."

Over the next four years, Ken rode a crest of popularity. He left the *Vanguard* to begin writing a column entitled "Similia" for the government-owned *Sunday Times*. He also produced an impressive number of books, including *A Forest of Flowers*, *Prisoners of Jebs*, *On a Darkling Plain*, *The Singing Anthill*, *Adaku and Other Stories*, *Pita Dumbrok's Prison*, and *Nigeria: The Brink of Disaster*, and a number of works related to *Basi and Company*. Episode scripts went on sale (*Basi & Company – Four Television Plays* and *Four Farcical Plays*), as did a novelization of the sitcom (*Basi and Company: A Modern African Folktale*). For young *Basi* fans he created "The Adventures of Mr B," a children's book series.

By 1990, Ken's writing career had gone as he'd predicted. The only part of his five-year plan he did not achieve was the Nobel Prize. Although *A Forest of Flowers* was shortlisted for the Commonwealth Writers Prize, the 1986 Nobel Prize had gone to his friend Wole Soyinka. Ken's Nobel dream was dead. According to prize committee politics, it would be unlikely for another Nigerian writer to win a Nobel for decades.

||||||

In 1983, Nigeria's presidential elections had resulted in a landslide for incumbent Alhaji Shehu Shagari – a victory due less to Shagari's popularity than to collusion among the National Party of Nigeria (NPN), the Federal Electoral Commission (FEDECO), and the armed forces.[35] Though it wrapped itself in the mantle of democracy, the Shagari administration was among the most corrupt in the nation's history. As in the days of military dictators, Nigeria's politicians used the public purse to enrich themselves while the masses spiralled into poverty. The economy drifted into recession.

On December 31, 1983, Shagari was overthrown and Nigeria once more became a dictatorship. The new despot, General Mohamadu Buhari, became perhaps Nigeria's most brutal ruler to date, turning the country into a virtual police state.[36] A bloodless coup that ousted Buhari in 1985 seemed justified, but it brought to power Buhari's chief of army staff, General Ibrahim Babangida, a man who once referred to himself as an "evil genius."[37]

During the early days of the Babangida regime, Owens completed his internship at the University of Port Harcourt Teaching Hospital and left to join the National Youth Service Corps. In Nigeria, Youth Corps service is mandatory for all young adults. As a licensed physician, Owens was posted to the Command and Staff College in the city of Jaji, far north in Kaduna State. Though he spoke no Hausa, the regional language, and had not spent a great deal of time in the Muslim north, Owens welcomed the posting because, oddly enough, it reunited him with much of his family. His sister Comfort had graduated from law school and was now living in the nearby city of Zaria, where she had set herself up in private practice. His other sister, Barine, had completed her schooling in London and moved back to the northern city of Kano, where she owned and operated a catering school. Junior brother Letam, having recently joined the Nigerian army, was stationed in Kaduna State.

The Command and Staff College was the training facility for all Nigerian military officers – army, navy, and air force – between the rank of major and lieutenant colonel. Along with a permanent military doctor, Owens was the physician for the college. During his tenure he made friends with many of the officers in training. Parties at the young doctor's house were a big attraction because there was usually a bevy of Youth Corps girls for the soldiers to chat up.

In 1985, after finishing his one-year Youth Corps service, Owens applied to the Rivers State Ministry of Health for a posting near Ogoniland. He was assigned to residency in obstetrics and gynaecology at Port Harcourt's Braithwaite Memorial Hospital. Moving back into Ken's house on Rumuibekwe Road, Owens assumed life would be as it was during his internship. Ken, however, had different ideas.

In 1986, just before Ken was about to leave for a three-month trip to England, he came into Owens's room and shook his head in wonder. "Well, well, well," he said. "Look at how you live. You own a radio."

At such an odd comment, Owens laughed and agreed. "You're right. That is my radio."

"But everything else in this room belongs to me. You don't have your own bed or your own table. You don't have a pot. What is your budget for this year?"

"I don't keep a budget."

"What do you mean? Even this corrupt military government has a budget."

"Well, the concept of budgeting never occurred to me."

Ken stared at his younger brother. "And why is that? Because you are in this house. Six months of working, you should be on your own. You are twenty-nine years old. I think that within three weeks you should get your own house."

Ken's tone was unemotional, firm, and businesslike. Owens wasn't surprised at his request. In fact, he had expected this day would come sooner or later.

"That is not a problem," said Owens.

"Three weeks," Ken repeated, and left to pack.

Owens wasn't daunted by the prospect of moving out. He had a good job and a little money put aside. Although finding a decent place in three weeks would be difficult, Ken was going to be away for three months, so there seemed no rush.

To save for the move, Owens put in long hours at the hospital and moonlighted at a clinic. He hadn't even begun his apartment search when, three weeks after Ken's departure, he arrived home from work to find that all the locks at Ken's house had been changed. Owens found a phone and called Ken's manager, Apollos. "Was there a problem?" Owens asked. "Did somebody spoil the door?"

"No," Apollos said. "Ken told me to change the locks if you are not out of the house in three weeks."

"What is all this? Come and open the door."

"Your brother said that I should open the door only when you show me your new house. He said you should also show me your bed, your cooker, your pots, and your fridge. When I see all this and they are new, that is the time I can let you take your clothes and things. These are the instructions. Your brother pays me, and if I disobey him I might lose my job."

Owens was furious. "Apollos, come and let me in! Ken is not here. He won't be back for months. Nobody will tell him."

Apollos stood firm. He did not even let Owens reclaim his chequebook so that he might begin buying things. That night, Owens was forced to sleep at a friend's home, and the following day he procured a temporary chequebook from his bank. Within a week, with the help of an uncle and a girlfriend, he installed himself in a bare-bones flat with the essential items that Ken had demanded. After inspecting the flat, Apollos allowed Owens into Ken's house to collect his clothing and books.

Owens remained livid. Ken, on the other hand, returned from England, invited Owens over for dinner, and received him in a particularly jovial mood: "I am so happy."

"What have you to be happy about?" Owens growled.

"Mon, you're now a man of your own."

|||||||

After Babangida came into power, Ken stepped up his criticism of the government in his columns and media interviews. He knew that if he were to present himself as an advocate of the suffering masses, it was important to avoid association with any political party or government office. Nevertheless, when Ken was offered an appointment by the new military dictator, he considered it carefully. Babangida asked Ken to become executive director of Mass Mobilization for Self-Reliance, Social Justice, and Economic Recovery (MAMSER), one of the government's new (ostensibly) anti-corruption measures. The mandate was to awaken Nigerians to their rights and obligations as citizens; to create consciousness about the proper role of the government; and to propagate the virtues of hard work, honesty, self-reliance, and patriotism. Although Ken detested the government that created MAMSER, the job description was enticing and he decided to give it a one-year trial.

It took Ken only a few weeks to discover how naive he'd been. None of the programs he tried to initiate were implemented. None of the suggestions he made to Babangida were considered. Instead of admitting defeat, Ken decided to maximize his time at MAMSER by increasing his contacts with the media and government and increasing his own knowledge. He read voraciously on the topics of public relations, communications, mass mobilization, and Marxist theory.

By the time he resigned from MAMSER, in October 1988, Ken Saro-Wiwa had become a fervent believer in non-violent struggle as a tool of the dispossessed.

As for Owens, he spent barely at year at Braithwaite Memorial before learning that he was to be transferred to a small clinic in the Ijawland area of the delta. He did not report for the posting, refusing to go because he was preparing for post-graduate exams in Port Harcourt. When summoned to the board for a disciplinary hearing, Owens was given an ultimatum; he still refused to leave his Ogoni patients in Port Harcourt. Arguing that he didn't speak Ijaw, Owens convinced the state to extend his posting to Port Harcourt. The minute his second year at Braithwaite was completed, another letter came announcing that he was being sent to an even more remote area, in the country's interior.

Appearing before the board once more, Owens asked, "Why am I being posted to this area? I can't speak the language. The majority of the people there will not be able to communicate with me in English. How can I treat them?"

The board replied that the transfer was mandatory; other doctors had managed to cope, and interpreters would be available. It was clear that Owens was being punished for objecting to the earlier Ijawland transfer. Now he would have to take his lumps.

Owens went to the rural posting, but found the language and cultural barriers intolerable. After pleading with the Health Management Board once more, his posting was changed to Bori General Hospital.

Owens was thrilled. At last, he was going home to Ogoni.

CHAPTER**THREE**

Bori General Hospital represented a home-coming of sorts for Owens. On his first day of work at the fifty-six-bed facility, which had neither electricity nor running water, one of the matrons proudly pointed out bed number three in the maternity ward. It was the bed in which Owens had been born. In the thirty years since the hospital had been built, nothing had changed, not even the bed numbers.

Owens and his three physician co-workers all commuted from Port Harcourt. When Owens learned that if a medical emergency arose at night, no doctor was available in Ogoniland, he felt he had no choice but to move back to Bori. The new posting brought with it new challenges. In Calabar, Jaji, and Port Harcourt, Owens had treated a wide spectrum of diseases. In Ogoniland, however, asthma and bronchitis were unusually prevalent. Standing in the fierce orange glow of a Bori night, it wasn't difficult to imagine why the rate of respiratory illness was so high.

After Owens had spent only two years at Bori General Hospital, the state government sought to transfer him out of Ogoniland, though he was the only physician in the area who could speak Khana. He refused to go. Threatened with dismissal, Owens stalled for time by using four years' worth of accumulated vacation to go on paid leave, infuriating the Health Board.

Owens's "vacation" was hardly restful. At all hours of the day and night people came to his house begging for treatment. He was soon seeing so many patients privately that he began to think of opening his own clinic.

The idea of a private practice was shelved, at least temporarily, when a letter from the Health Board arrived, telling Owens that he was being posted to Taabaa General Hospital. Although he had never heard of such a facility, Owens accepted the transfer because it was in Ogoniland.

When he arrived in the town of Taabaa, in Nyo-Khana Kingdom, Owens realized that the Health Board had won this round. The cottage hospital, built around an existing maternity health centre, was only half completed. Empty of furniture and devoid of medical supplies, Taabaa General Hospital would have to be created from scratch.

Owens began by submitting a list of items necessary to staff and supply a small rural hospital and pharmacy. The government gave him little of what he requested. As equipment and drugs trickled in, Owens started accepting a few outpatients. When word spread that Dr. Wiwa had moved his practice to Taabaa, a flood of Ogoni patients threatened to overwhelm the meagre facilities. His inability to handle the needs of his seriously ill Ogoni patients kindled his frustration, which revived his ideas about a clinic of his own.

According to regulations, new clinics had to be registered with the state. Hoping to bypass this expensive, time-consuming formality, Owens talked to a friend who owned the Princess Medical Centre, a small private clinic in Port Harcourt. Owens presented his friend with the idea of establishing an annex to the Princess Medical Centre in Bori, in essence giving Owens a franchise. Drawing up the paperwork to create the Princess Medical Centre Annex – Owens's own Ogoni clinic – was relatively easy. The only snag was that he had no money for staff, supplies, furniture, or rent. He went to his older brother with his vision.

"It's a very good idea," said Ken.

"I will need a small loan to start."

"I'm not a bank," Ken replied flatly.

Undeterred, Owens used his next paycheque to lease a vacant second-floor flat above a warehouse in central Bori and to pay a local cabinetmaker to make a padded examination table. For a desk, Owens took the dining table and kitchen chair from his own house.

A nurse named Zor who had recently left the employ of another doctor came to see Owens about a job. Impressed by her experience and enthusiasm, he hired her immediately. "I don't have any money right now," he said, "and I don't know how this will go, but I will pay you somehow." Zor set to work, scrounging another table and chair for the reception area as well as a little bench for waiting patients. The best sign painter in Bori demanded 500 naira in advance (about $125) to produce the largest allowable sign for a private medical practice. Since Owens now did not have a penny left, he borrowed the money from one of his uncles. The Princess Medical Centre Annex was in business.

Owens worked in Taabaa each day until two in the afternoon, then spent the rest of the day and far into the night at his clinic in Bori. There was no lack of patients, especially once word got out that Dr. Wiwa was charging only five naira ($1.25) a visit.

Within two weeks, Owens earned enough money to pay Zor her full wages and to pay back his uncle. By the sixth week, he'd transformed the clinic. There was a proper doctor's desk and chair in the examining room, a sturdy new desk and a full-time employee in the reception area. Owens let the adjoining upstairs flat, where he set up a small ward for inpatients and hired a nursing assistant to help Zor. Six months after opening, he took over the entire building and installed an operating theatre. A half-dozen nurses now kept Princess Medical Annex going twenty-four hours a day.

By early 1989, the Princess Medical Centre Annex required all Owens's time and attention. Without regret, he resigned from the service of the Rivers State Ministry of Health.

||||||

Ken, meanwhile, after five years and more than 150 episodes, found it impossible to keep *Basi and Company* running. Besides writing and directing every episode, he was also organizing, managing, and marketing the show, with Saros International handling the production, shooting, and editing. Everything was funded out of Ken's own pocket.[1] For the first six episodes, *Basi* had not a single sponsor. By the seventh episode, Ken had one sponsor, a newsprint factory. (Not coincidentally, Ken was serving as director of the Nigerian Newsprint Manufacturing Company.) Despite its enormous popularity, the show never paid for itself. The only way *Basi and Company* managed to run for so long was through the sale of *Basi* scripts and books, and the newspaper serialization of *Basi* stories.

One reason the show lacked sponsors was that most Nigerian businesses wanted to distance themselves from the opinionated, controversial, and disconcertingly messianic Ken Saro-Wiwa. Even Ken's newspaper column "Similia" was axed by the *Daily Times* after he ran a strident piece entitled "The Coming War in the Delta."[2]

Shortly after both *Basi* and "Similia" ended, Ken stopped by the clinic to see his brother. As usual, the conversation turned toward Ogoni issues, the environment, and Shell. Ken railed against the oil company and the discrimination suffered by the Ogonis. Then, unexpectedly, he informed Owens that he'd made an important decision.

"The time has come to put myself, my abilities, my resources, at the foot of the Ogoni people," he said.

Annoyed, Owens replied, "Dede, do you remember what happened before? Remember in the civil war when you went off to the federal side to try and save the Ogoni people? Remember what happened when all the elites that you rehabilitated after the war went to have you removed as a commissioner? Remember when you wanted to go to the constitutional assembly to defend the issues you are talking about now and you were betrayed and defeated by these elites? These people are going to betray you again. The Ogonis are not ready for this."

"They are ready," Ken insisted. "Ogoni schools have not been open for six months because the teachers have not been paid. Bori General has been closed down for three months because the doctors and nurses have not been paid. The Ogonis are faced with environmental degradation, political marginalization, economic strangulation, slavery, and possible extinction. There is stupendous oil and gas wealth on our land, yet we have no social amenities. Worse, our environment has been devastated. No, Mon, the Ogonis are ready. I can feel it from the rallies I've been going to. I can feel it from what they have been saying to me. I can feel it in the air."

Owens simply sighed and shook his head.

|||||||

Ken's first step was to resurrect the moribund Ogoni Central Union, an organization he had created in the 1960s as a forum to explore Ogoni development and education. As the OCU's newly elected president, he organized a seminar on Ogoni culture, education, agriculture, history, economy, and women's issues. The seminar's participants concluded that the Ogonis needed to take responsibility for their own political existence.[3] A new organization would, however, be pointless without the support of the Ogoni elites.

With the exception of Buhari, every Nigerian military dictator since Ironsi had declared support for the goal of demilitarization. After every *coup d'état*, the military was expected to perform its rescue operation and then restore democracy to Nigeria. Since independence from Britain, only the Mohammed and Obasanjo governments had been successful in executing such a program. Despite this history of failed intentions, Babangida announced his intention to restore civilian rule. One of his first efforts was to expel many of the political appointees from previous administrations, stripping each of the Ogoni elites of a cushy patronage position. Resentful over being banned from government, the elites were hungry for something – anything – that would restore them to relevance.

To woo these disaffected elites, Ken attended meeting after meeting of Kagote, a club of Ogoni elders (KAGOTE being an acronym of Khana-Gokana-Tai-Eleme), and repeatedly pounded out the need to form a mass organization. His proposal to create an Ogoni Bill of Rights was enthusiastically received, and he applied himself to drafting the bill. On August 26, 1990, Ken called a meeting in Bori, inviting influential Ogonis from all six kingdoms to sign the Ogoni Bill of Rights. In the carnival-like atmosphere, nearly every Ogoni elite volunteered to put his signature on the historic bill; however, for political reasons, signatories were limited to six per kingdom.

In twenty numbered points, the Ogoni Bill of Rights traced the history of the Ogoni people, the discovery of oil on their lands, the estimated $30 billion in revenue the oil had provided, the dearth of return on such revenue, the damage done by the government and Shell Petroleum Development Company to Ogoni culture and environment, and the desire of the Ogoni people to run their own affairs. The bill ended with a demand that the Ogonis be granted political autonomy and the right to control the use of their natural resources.[4]

The Ogoni Bill of Rights was sent to Babangida's Armed Forces Ruling Council and published in one of the national newspapers. When Ken toured Ogoniland to present and explain this new document to the people, the Bill of Rights generated intense excitement.

The next logical step was for the Ogonis to create a structured protest organization to agitate for their clearly stated goals. At one of the Ogoni Central Union's meetings, Ken's long-time friend Edward Kobani suggested a name for this new organization: The Movement for the Survival of the Ogoni People, or MOSOP. Something about the acronym caught Ken's fancy, even after someone pointed out its uncomfortable similarity to Mossad, the Israeli secret service organization.

The Ogoni Central Union created a steering committee to choose the MOSOP leadership. In early 1991, it named Dr. Garrick Leton president and Lekue Laa Loolo vice-president. Ken dissuaded the committee from choosing him to be president; he preferred a rather loosely defined role as MOSOP "spokesman," which gave him the authorization to speak not only on behalf of MOSOP but of all Ogoni people.[5]

Ken was adamant that MOSOP be committed to non-violence. "*Alo be, iko be, nale begin*" was his slogan. "We will fight with our brains, not with a knife."

At first, any Ogoni with a distinctive achievement to his name was invited to join MOSOP's steering committee. As a doctor, Owens received a ready welcome to these meetings. In the hope of dispelling government fears about the organization, MOSOP was eager to recruit lawyers and high-ranking civil servants as members of the steering committee.

MOSOP seemed unable to function without Ken, however. Owens noticed that, in steering committee meetings, if Ken were not around nothing substantive would be discussed.[6] During such meetings, the members, mostly older Ogoni elites, preferred to eat goat, chew corn, and talk about girlfriends from their younger days. One evening, as Ken arrived, Dr. Leton's servant started to serve drinks, goat meat, and pepper soup. Doing his best to suppress his irritation, Ken chided them, "Gentlemen, we are not joking here! We should be discussing serious things. Is it possible that we eat only when the meeting is over so that we can at least get things done?"

Ken's rebuke of the elders was a relief to some, but to others it was simply a sign of disrespect. Going further, Ken proposed a change of venue for the meetings. Instead of meeting in Dr. Leton's living room, he said that in the future he would like MOSOP to convene in the more businesslike atmosphere of the Saros offices in Port Harcourt.

To further upset matters, Ken began promoting the idea that MOSOP should create community subgroups, following the model of the ANC in South Africa. His vision was of an umbrella organization of many grassroots bodies so that decisions would come directly from the Ogoni people themselves. The younger members of the constitutional committee were in favour, but the senior members – notably Leton and Kobani – were decidedly against it. The elders wanted MOSOP to be an organization with the elites at the top dispensing trickle-down edicts.

In addition to joining the MOSOP steering committee, Owens had become a member of the Council of Ogoni Autonomy (COA), a group made up of about a dozen young Ogoni professionals concerned about Ogoni home rule, democratic rights, and the environment. After MOSOP was firmly established, the COA decided to change its name to NYCOP, the National Youth Council of Ogoni People. This new organization proved so popular that before long NYCOP meetings were filling church halls.

When NYCOP applied to become a subgroup of MOSOP, its application was accepted despite the discomfort this action caused Leton and Kobani. NYCOP's president and several of the organization's officers then joined MOSOP's steering committee.

Soon after, a second MOSOP unit came into being. For as long as anyone could remember, Ogoni women had been meeting to discuss community affairs. Jessica Wiwa had been leading groups for decades. Noting Ken's interest, she introduced her son to some of the informal groups' leaders. Ken was also interested in a group of educated Ogoni women who had started meeting after hours at his Port Harcourt office. He encouraged the two groups to merge, creating FOWA, the Federation of Ogoni Women's Association. The new organization began sending representatives to the MOSOP steering committee.

COTRA, the Council of Ogoni Traditional Rulers, became yet another unit of MOSOP, which was fast turning into an umbrella organization for various Ogoni groups. During the colonial period, the Ogoni chiefs had steadfastly refused to sign treaties with the British, and the British had simply replaced these non-cooperative chiefs with government-appointed stooges. Following independence, the successive military regimes had continued to recognize only the appointed chiefs and even to create new ones. These ersatz chiefs enjoyed a lucrative relationship with the government, as well as profitable contracts with Shell. The descendants of the traditional rulers were nevertheless still regarded by many Ogonis as the legitimate village chiefs. These symbolic leaders had banded together to form COTRA.

In quick succession, the Council of Ogoni Churches and the National Union of Ogoni Students formed their own organizations and also became units of MOSOP. Ogonis who had emigrated to the United States had previously formed a group called LOOP, the League of Ogoni People; when MOSOP was formed, they renamed their organization MOSOP-USA. Ogoni expatriates in England formed the Ogoni Community Association of the United Kingdom, or OCA-UK.

Whether the MOSOP leaders liked it or not, Ken's vision of MOSOP as a multi-group organization with decision-making power percolating up from the grassroots was becoming a reality. Dr. Leton and Edward Kobani could do little but accept the situation.

In early 1992, an addendum was added to the Ogoni Bill of Rights. Because the Nigerian government and Shell had largely ignored the OBR, the addendum resolved to give Ken Saro-Wiwa the authority to take the Ogonis' case to the international community. Ken wasted little time putting the addendum to action.

|||||||

"The Ogoni are being consigned to slavery and extinction," Ken Saro-Wiwa told the United Nations' Working Group on Indigenous Populations in Geneva.[7] In July 1992, at the Palais des Nations, he spelled out the Ogoni predicament: "Studies have indicated that more Ogoni people are dying now than are being born. The extermination of the Ogoni people appears to be policy. The Ogoni have suffered at the hands of the military dictatorships which have ruled Nigeria over the past decades. The Ogoni are faced by a powerful combination of titanic forces from far and near, driven by greed and cold statistics. Only the international community acting with compassion and a sense of responsibility to the human race can avert the catastrophe which is about to overtake the Ogoni."[8]

Ken's speech aroused intense media interest in Nigeria and abroad.[9] His
UN appearance also led to MOSOP's joining UNPO, the Unrepresented
Nations and Peoples Organization, a small non-governmental organization
(NGO) based in The Hague. UNPO was founded in 1991 to promote the interests of minority groups who believed themselves to be unrepresented in
national or international political forums.[10] MOSOP and UNPO were an ideal
match, and UNPO soon became one of the Ogoni's most influential international champions.

Ken's bold campaign against the oil companies was becoming a role
model for other minority groups in the Niger delta. In October 1992, the Ijaw
nation launched the Movement for the Survival of Izon Ethnic Nationality in
the Niger Delta (MOSIEND). The same year, the Ogbia community in Oloibiri
(the site of the first oil discovery in Nigeria, in 1956) established the Movement
for Reparation to Ogbia (MORETO); and in 1993, the Ikwerri established the
Council for Ikwerre Nationality. The Saro-Wiwa "ripple effect"[11,12] began to
seriously affect oil production in the Niger delta. According to Shell estimates,
losses from community uprisings and "sabotage of oil installations" amounted
to nearly US$1.4 million in 1992.[13] Nigerian authorities frequently met these
protests with swift and sometimes deadly force.

On November 8, 1992, Britain's Channel Four aired *Heat of the Moment*,
a television documentary focusing on Shell's destructive practices in Nigeria,
Gabon, Ecuador, and Papua New Guinea.[14,15] The program provoked a
strong anti-Shell public reaction. At the same time, the release of the report
on Ken's United Nations trip added to the growing global awareness of Shell's
practices in Ogoniland.

Ken began to notice many younger faces at MOSOP meetings,[16] which
pleased him. MOSOP had been primarily represented by passionless, intransigent men of advanced age, and Ken felt it was time to bring in young
blood. MOSOP's affiliation with NYCOP had been a good beginning. To continue courting Ogoni people under forty, Ken organized a two-day tour of
all six Ogoni kingdoms in November 1992.[17]

The public enthusiasm during the tour took even Ken by surprise.
Thousands of angry Ogoni youths turned out to hear what the men from
MOSOP had to say – and they liked what they heard. Each rally began with an
introductory speech by MOSOP president Dr. Leton, which ended with his
calling for a few words from Ken Saro-Wiwa, "the spokesman of the Ogoni

people." Ken would then step forward and, in one of the three Ogoni languages, ignite the crowd by proclaiming, "Great Ogoni people! We must do something to extricate ourselves from our cruel fate."

With the audience hanging on his every word, Ken proposed a motion calling for the three oil companies operating in Ogoni – Shell, Chevron, and the Nigerian National Petroleum Corporation (NNPC) – to pay the Ogoni people $4 billion for destroying their environment and another $6 billion for unpaid rents and royalties: "They must pay this all within thirty days, or it will be assumed that they have decided to quit the land." Joyous crowds ran screaming after Ken's car when he left the rally.

After the tour, MOSOP asked Ken to draft a document entitled "Demand Notice." He posted it to the three oil companies on December 3. Not surprisingly, none of the oil companies bothered to reply.

|||||||

On December 10, 1992, the United Nations General Assembly proclaimed 1993 the International Year of the World's Indigenous People. The goal was "to strengthen international cooperation for the solution of problems faced by indigenous communities in areas such as human rights, the environment, development, education and health."[18] The UN declaration seemed tailor-made for the Ogoni cause and Ken Saro-Wiwa. It accorded well with an event he had recently conceived – a massive non-violent protest march to be held on the coming January 4, to be known as "Ogoni Day."

The decision to create Ogoni Day was pure Ken Saro-Wiwa, for better or worse. He had hatched his plan without bothering to consult the Ogoni people and had chosen January 4, 1993, because it was two days after Babangida had pledged to hand over power to a civilian government. If the dictator did so, Ken's enormous Ogoni rally would be a signal to the new administration that the Ogoni people would no longer accept exploitation. If Babangida failed to relinquish power, the protest would have even greater impact.[19]

Early January marked the midpoint of Zua, an ancient Ogoni pre-planting festival that spanned the period December 28 to January 14. The birth of a new agricultural season was marked by sacrifices to the spirits of the land, and singing and dancing to promote fertility. Zua is also a period of peace and social harmony. To help convince the gods to smile on the people and allow their crops to flourish, quarrelling with one's neighbours and excessive noise are forbidden. In Ken's mind, there could be no better time for a non-violent protest.

After submitting his Demand Notice to the oil companies, Ken had gone to London intending to catch a flight to New York to attend the inauguration of the United Nations International Year of the World's Indigenous People. Visa problems in the United Kingdom prevented him from flying to the States, so Ken used his time in England trying to convince Amnesty International and Greenpeace to send observers to his Ogoni Day march. Amnesty declined, but Greenpeace agreed to send a photographer if Ken would cover airfare and lodging. Shelly Braithwaite, a representative from the London-based Rainforest Action Group, agreed to attend as well.[20] On his return to Lagos, Ken asked the major Nigerian newspaper publishers to send reporters to cover the upcoming protest.

While Ken was busy jetting around, raising awareness in the media and among human rights groups, Owens contacted an acquaintance in the army to find out what the military thought of the impending event. Owens also stopped in to see an old friend from his Youth Corps days in Jaji who was now a naval commander in Port Harcourt and a high-ranking member of the Rivers State security council. According to these sources, security agents had been instructed that if a problem arose during the Ogoni Day rally, they were to shoot three times into the air. If the people did not disperse, the agents were then to shoot into the crowd.

On December 25, 1992, Dr. Leton phoned Ken to say that the State Security Service (sss) had been pestering him about Ogoni Day and that someone named Mr. Egwi had summoned him to the sss offices that day.

"On Christmas Day?" asked Ken.

"Yes."

"What could they possibly want that couldn't wait until tomorrow or the day after?"

"God alone knows."[21]

Ken offered to accompany Leton to the meeting. When the two arrived at the Port Harcourt offices of the sss, they asked to see Egwi's superior and were told that the State Director, Mr. Terebor, was at home – it was

Christmas morning, after all. Ken and Leton went to Terebor's house, where the director welcomed them amicably.

"Why are we invited on Christmas Day to a security chat and not allowed to enjoy the festival with our family?" asked Ken.

Terebor, a homely little man from Ondo State, apologized in Yoruba-accented English and offered Ken and Leton a drink. Terebor then telephoned his subordinates, and within minutes Egwi and another sss operative showed up. Egwi was annoyed that Ken and Leton had gone to his boss instead of reporting directly to him – precisely the reaction Ken had hoped for.

Egwi informed Ken and Leton that MOSOP would have to cancel Ogoni Day. Babangida had decided not to hand over power and had prohibited mass gathering or protests. The military was ordered to shoot anyone found carrying a placard in a march.

"I assure you that the Ogoni people and their leaders are not up to mischief," said Ken. "We are only celebrating the end of the year in our usual way." He offered full cooperation with the authorities in order to ensure a peaceful Ogoni Day.[22]

Although the tenor of the meeting had remained friendly, Leton was unnerved and proposed cancelling Ogoni Day. He pleaded with Ken to call off the demonstrations, confiding that he had recently had a dream in which he was shot while carrying a placard.[23] Ken found Leton's cowardice disturbing, especially for a man who was supposed to be leading a resistance movement. Realizing that Leton did not have the resilience to withstand further sss harassment, Ken assured him that he would accept responsibility for the protest himself.

Ken had barely returned home before the governor phoned to summon him and Leton to lunch. At Government House, among a full complement of Rivers State cabinet ministers, the two MOSOP representatives were treated to a sumptuous Christmas feast with Ken seated at the governor's right hand.

Governor Rufus Ada-George was a large, pious man with glasses and a gap-toothed smile. As always, he wore an oversized white Victorian pyjama shirt over white trousers, an affectation meant to advertise his membership in the Pentecostal church.[24] When Ken asked why he and Leton had been summoned by the sss on Christmas morning, the governor claimed to know nothing of the matter.

No pressure of any kind was put on Ken or Leton during the Government House banquet. The following week, however, Ada-George and the SSS contacted Ken to demand the cancellation of the Ogoni Day march.

||||||

During this time, Owens had been coming under intense pressure to "settle down." His mother, his nurses, his patients, and even his father (who didn't normally interfere in such matters) were urging him to get married. Nearly all of Owens's friends and siblings had married, and it was becoming socially inconvenient for him to continue as a bachelor.

That is not to say that he was ever lonely. If Bori was a small pond, Owens was a big, eligible fish. Every unmarried female over twelve and under a hundred had her eye on him. He dated many young women, and this bothered Ma Wiwa. Every time she saw her son, he seemed to have a different girl on his arm.

One afternoon, Jessica took him aside to say that she did not like the way he was handling his love life and was concerned about his health.

"Don't you know about Eight Days?" she asked pointedly.

"Eight Days?" he said. "What is Eight Days?"

"Have you not heard about this thing, Eight Days? If you get sick with it, you have only eight days and you will die! Look, they tell us at the women's meeting and they're all talking about it: Eight Days."

Owens had never heard of the mysterious ailment. When he questioned his mother further and found that it was spread through casual sex, he realized she and her friends had been mishearing it. "Eight Days" was AIDS. He assured his mother that as a doctor he knew how to protect himself and promised to be careful.

Even Ken tried to talk his brother into starting a family. At first he would casually ask about the women Owens was dating. Before long, he would lecture about the importance of marriage. Owens balked at the intrusion. Ken was hardly a role model when it came to matters of the heart. His marriage to Maria was in ruins, and he was keeping mistresses in both Lagos and Port Harcourt. In fact, none of the Wiwa siblings' marriages had turned out well. By 1992, Barine, Comfort, and Jim were all either divorced or separated. After witnessing all this unhappiness, Owens had decided it would be better to devote his life to medicine and to never marry. By the end of the year, however, the pressure from family and friends had become intense.

IIIIIII

The beach at Kono Waterside on the eastern edge of Ogoniland was filled with couples celebrating the last day of 1992. Although the water of the Imo River was too polluted for swimming, university students and their dates flocked to this little Ogoni resort village known for its fresh fish, pepper soup, and "western-style" amenities – a single motel and a beachfront restaurant.

On this New Year's Eve afternoon, twenty-year-old Diana Barikor was sitting in the sand with her boyfriend and several other couples. As the friends chatted, they noticed the well-known Dr. Monday Owens Wiwa and his date walk by, and the young people asked them to join their group. Owens accepted and helped his girlfriend sit on the ground, though he himself remained standing.

"I can't sit in the sand," he said. "My trousers will get stained on the beach."

Diana whispered to her boyfriend, "Who does he think he is?" The doctor's expensive gabardine pants were clearly inappropriate for an afternoon at the beach. And he was so haughty, she thought, going to sit on a nearby swing while she and the others sat on the sand. Still, Diana did her best to engage the doctor in polite conversation. After a while, she found him bearable.

Party music was blaring from a hut down the beach. Owens asked his girlfriend to dance, but she declined, saying that she didn't want to crash someone else's party. Owens looked down at Diana, sitting in the sand and gently bopping to the distant music.

"Diana, would you like to dance?" he asked.

"Why not?" Even if he was a bit annoying, her own boyfriend wasn't paying her much attention and she loved to dance. To the displeasure of their respective dates, Owens and Diana went off to join the other beach party.

Owens didn't tell Diana that this was not the first time she had caught his eye. He had seen her on another evening at a restaurant with a young man. He had also watched her at MOSOP rallies and church services whenever she appeared in her role as organizing secretary of MOSOP's youth wing, NYCOP. An articulate Ogoni beauty from Gokana, Diana Barikor was a political science major at the University of Port Harcourt and the eldest daughter of one of the professors there. Beginning in the late 1970s, she had spent five years in the United States while her father attended graduate school in Michigan. This experience left her with an American accent, which magnified her worldly charm.

"Would you like to come to my village and see the fireworks tonight?" Owens asked as they finished dancing.

Diana's friends had been talking about the New Year's Eve fireworks that the village of Bane put on once every seven years but she had never seen it herself. "Yes, I would," she answered, "but I don't have a ride. My boyfriend said he doesn't want to see the fireworks."

"It's no problem. I can take you in my ambulance."

"You want me to ride in an ambulance?"

Owens shrugged sheepishly. "My other car has disappointed me."

Diana laughed, but she did not go to the fireworks that evening. Her boyfriend saw to that. And Owens did not have much time to press the point. It was New Year's Eve, and he was expected at home.

For decades, Pa Wiwa had been performing a family New Year's Eve ceremony called the Impregnable Circle. He would call his children together into a downstairs room in his house. As evening fell, he would sacrifice either a goat or a chicken and drain the blood into a hole dug in the corner of the room. Then, as the wives and daughters came to take away the slaughtered animal, he'd pull out his large old Bible and lead his sons in an excruciatingly long prayer. The prayer was the same every year. Pa would pray to the Angel Gabriel (and to as many other archangels as he could remember) to guide and protect the Wiwa family and the Ogoni people. Then he asked each of his sons to pray aloud his resolutions for the coming year and to state the offering he would make during the next Impregnable Circle if God were to allow this goal to be realized. After the prayers, the women would return, bearing bowls of cooked goat meat, chicken, fish, and rice, and the family would enjoy their holiday dinner.

On this New Year's Eve, Owens announced the three goals he had set for himself. The first two were financial: he planned to build a second clinic in

Ogoniland, and to buy real estate in Bori and Port Harcourt. His third was to get married.

The choice of wife would be a difficult one, however. Owens did not know exactly whom he would marry. At the time, he was dating three women. Would Diana Barikor make it four?

CHAPTER**FOUR**

Owens Wiwa tossed and turned, worrying. Would anyone show up for Ogoni Day? Would people be too frightened to march? Was his brother to face disgrace again? MOSOP was still debating whether to cancel the march, although loath to do so after so much work. The biggest problem had been coordinating such a huge event. Without modern communications, how could they get half a million people spread over six kingdoms and 120 villages to show up in the same place at the same time?

Ken felt the key to success lay in the Ogonis' sizeable young adult population. He had written a seventeen-page booklet entitled *Second Letter to Ogoni Youth*[1] and had his publishing company, Saros International, print up thousands of copies. He then arranged for another grand tour of the kingdoms of Babbe, Gokana, and Tai to mobilize community leaders and hand out his inspirational booklet to the young people.

To spread the word of the Ogoni Day rally, Ken developed a "four-by-six" plan. Six coordinators were selected from each Ogoni kingdom and four volunteers were recruited in each village.[2] Of the four villagers, at least one had to be female (a radical concept in Ogoni society). Above all, the village representatives were to stress the non-violent nature of the event.

The plan was that in every community, on Ogoni Day, people would come to their village square at dawn to watch local cultural groups sing, dance, and drum. Entertainment, Ken hoped, would eliminate the possibility of violence.[3] After the performance, a representative from the village, either a MOSOP official or an elite, was to deliver an Ogoni Day address before leading a march to the headquarters of the kingdom. There, Ken and the other MOSOP leaders would direct everyone to converge for a massive, multi-kingdom rally in Bori.

By January 3, the planning committee had finished installing PA systems at bandstands in the six kingdom headquarters. A team of Ogoni youths had created and stockpiled hundreds of protest placards, banners, and posters. It was assumed that only the bravest Ogonis would dare carry placards, for Babangida's order was still in effect for the military to shoot any protester caught brandishing a sign. But would the Ogonis march at all?

A sympathetic soldier covertly informed Ken about the troops stationed in the area. Yes, the army and the police were massed and standing by, he said, but had been ordered not to shoot into the crowd unless they had first given three unheeded warning shots into the air.

On the morning before the march, a religious service was held at 10 a.m. at St. Peter's Church in Yeghe, just outside Bori. Yeghe was the birthplace of T.N. Paul Birabi and St. Peter's was the site of his tomb. Ken arrived early for the service accompanied by one of his mistresses, Hauwa Madugu, and Alfred Ilenre, the general secretary of the Ethnic Minority Rights Organization of Africa (EMIROAF). Garrick Leton, Edward Kobani, and Owens were also on hand. Conspicuously missing, however, was Bennett Birabi. Ken considered his former ward's absence a personal insult and feared the Ogoni people would be offended at Birabi's slight as well.

At 10 a.m., St. Peter's was nearly empty. The bishop delayed the start of the service for some time but eventually got it underway. Ken was crestfallen. Were all his plans coming to nothing?

Inexplicably, the congregation arrived in full force halfway through the service. By the time Ledum Mitee, who played a prominent role in MOSOP, read a stirring lesson from the book of Lamentations,[4] the church was packed. After the bishop's sermon, the service moved outside to Birabi's tomb, where Garrick Leton gave a short speech. Leton, shy by nature and with a slight stammer, failed to move the crowd, but the next speaker, Edward Kobani, spoke with passion.

"We have come to this sacred ground to reaffirm our faith in Ogoniland, to reaffirm our determination to fight with every drop of blood in our veins that the Ogoni man cannot be a slave in our God's given country. We have come here to get inspirations to carry us through the days ahead."[5] Kobani implored the crowd, "Do not be afraid! Nothing will happen to you on this land which God gave us!"[6]

Kobani's oration thrilled the congregation. It also inspired Ken Saro-Wiwa, the next speaker, to give a splendid extemporaneous speech.

"The United Nations recognizes the rights of all the world's indigenous people," he said. "The indigenous people have been cheated through laws such as are present in Nigeria today. Through political marginalization, they have driven certain people to death. That happened in America and in Australia. They are trying to repeat it in Nigeria – and we do not want it. In recovering the money that has been stolen from us, I do not want any blood spilled – not of an Ogoni man, not of any strangers amongst us. We are going to demand our rights peacefully, non-violently, and we shall win!"

After the service, Owens spent the day meeting with his staff about a potential mass medical emergency. He had stockpiled the clinic with first aid kits, syringes, IV needles, sterile saline solution, antiseptics, and drugs. All the beds in the clinic had been emptied, many patients having been taken home by their families in preparation for Ogoni Day. The tension in Ogoniland was palpable.

||||||

On January 4, Owens rose from his fitful sleep at five o'clock and took his ambulance for a drive around Bori. The town was nearly deserted; the streets, eerily quiet. He noticed that a few shops were opening, stores owned by some of the remaining Igbos.

There seemed nothing for Owens to do but go home. By seven, there was still no sign of activity in Bori, but Owens swore he could hear the sound of drumming far off in the distance. Returning to his clinic, he put on a lab coat and stethoscope, and asked his ambulance driver to take him to Port Harcourt. There, he planned to join Ken and any journalists who showed up.

Five minutes out of town, as the ambulance passed St. Peter's Church in nearby Yeghe, Owens noticed that the sound of drumming was growing louder. He could hear singing. It seemed to be coming from somewhere over in Gokana.

"Something must be happening there," Owens said. "Let's go and see before we continue to Port Harcourt."

By the time they reached the first village in Gokana, there were so many people in the road it was impossible to proceed. Everyone was singing, dancing, and waving palm fronds and small branches fluttering with leaves. It looked as though a forest had come to life and begun to dance.

Owens leaned out the window and asked what was going on.

"He is here!"

"Ken is here!"

"Ken!"

People were pointing and shouting. To catch up with Ken's entourage, Owens had the driver put on the siren and emergency lights; the crowd parted to let them pass. A woman handed Owens a branch and he waved it out the window in solidarity.

Driving to Mogho Junction in Gokana, the ambulance encountered an impenetrable crowd, but Owens spotted Ken's car in the crush ahead. To catch the attention of Ken's contingent, he had his driver sound the siren and flash the lights once more. Ken's handlers muscled their way back through the throng. Owens stepped out of the ambulance and was escorted straight to the rostrum where Shelly Braithwaite from the Rainforest Action Group, Greenpeace cameraman Tim Lambon,[7] and a group of Nigerian reporters were already stationed.

The stage for the Mogho rally had been constructed on the soccer field. The crowd would not allow Ken to walk to the speaker's platform; people insisted on carrying him shoulder-high. Dressed in a loose-fitting, burgundy red African shirt, his reading glasses hanging on a chain around his neck, Ken stood at the microphone and looked out on the largest crowd he had ever seen. Tears welled in his eyes.

He held up his hands to silence the crowd. "*M kana mon Gokana!*" he shouted: I salute all Gokana!

"*E zira!*" responded the people: We return your salute!

Ken spoke in the Gokana language. "Today we are taking over the oilfield. This is now ours. We have been liberated!" In English he proclaimed, "Shell, go to hell!" and the crowd exploded with cheering.

As Tim Lambon snapped pictures, Owens turned to him and asked, "What do you think?"

"This is marvellous," said Tim. "I've never seen such a huge environmental protest in my life."

Owens, imagining stories in *Newsweek* and *Time* and all the European papers, asked, "So is it going to make the news in the west?"

"I don't think so."

"You mean you have this sort of crowd, these sorts of pictures, and it won't make the news over there?"

"No," he said matter-of-factly. "Not unless someone is shot."

The next stop was the rally in Kira, Tai, five kilometres from Mogho, Gokana. Because of the crowds, it took Ken's entourage thirty minutes to drive there. The gathering in Tai was even larger because people from the previous rally followed on foot. Leton joined the presenters and gave a short speech; this was followed by an address from Noble Obani-Nwibari, the foremost MOSOP activist in Tai Kingdom.[8]

When it was Ken's turn to speak, Owens noticed a disabled vehicle being towed by a truck bearing a Shell logo. The tow truck, accompanied by two armed policemen, was finding it slow going extracting the stalled vehicle through the sea of people. People turned to look at the Shell trucks trying to pass. Owens feared people would react in anger at the sight of the yellow scallop logo, but nothing happened. The Ogonis simply turned their attention back to Ken. A dance group broke into song in the truck's wake.

As Ken spoke, a Nigerian journalist nervously approached Owens. "Dr. Wiwa, you people have to go to Bori right now."

"Why? What is the problem?"

"I have just come from Bori and a huge crowd has already gathered there. They are becoming restless waiting for Ken Saro-Wiwa. Some people are shouting and cursing."

The journalist feared that if Ken didn't show up soon, mayhem would erupt. "I have been following politicians and political rallies in Nigeria since 1960. Even at the rallies of the most popular politicians, I have never seen that sort of crowd. I don't know what's going to happen next."

"We have to move," Owens told his brother as soon as Ken relinquished the microphone.

"What for?"

Owens related what the journalist had said, and Ken agreed. He and his entourage fought their way back to their vehicles. As they departed, Owens's ambulance was joined in the lead by more than a hundred Ogoni youths on motorcycles, flashing their headlamps and revving their engines to push Ken's motorcade through the crowds.

Although the road into Bori was wide and well-paved, it was invisible under the flowing river of humanity. The dust cloud raised by the marchers was stifling. Ken and the others sweltered all the way to the rally with their windows rolled up, to keep the dust out.

Owens noticed a huge white banner hanging over the main street: "1993 OGONI DAY International Year of the World's Indigenous People." Beneath it, thousands of revellers were singing and jogging in place while waving tree branches. Many people brandished white cardboard with slogans painted in oddly elegant calligraphy:

"Save Ogoni environment"

"Give Ogoni oil money today"

"No oil right, no peace"

"UN Save Ogoni"

"Shell leave Ogoni"

"Assassins go home"[9]

As the VIP convoy got closer to the stage, women in identical dresses shimmied and sang about Ken Saro-Wiwa's heroic exploits while male drummers beat out a furious, pulsating rhythm. Colourful wooden head-dresses of mythical man-beasts and sardonic human forms bounced above the heads of the crowd.

The marchers let out a roar as Ken stepped from his car. In front of the rostrum sat the cream of Ogoni society (minus the government-paid chiefs and party politicians), fanning themselves on plastic lawn chairs. As Ken's contingent took their places, the crowd pressed forward so intently that Ken feared the rostrum would collapse. He motioned for the master of cere-monies to begin.

Three traditional chiefs invoked the gods of the land to bless their endeavours, then Leton gave the first address. Ken had written Leton's speech for him – as indeed, he had written or edited every speech given by the presenters. Joy Nunieh, representing the women of Ogoni, spoke next. Then Ledum Mitee said a few words on behalf of the Ogoni youth.

When it came time for Ken to speak, he chose to deliver his remarks in Khana, not in English as the previous speakers had done. "Your Royal Highnesses, Respected Chiefs and Elders, President of the Movement for the Survival of the Ogoni People, ladies and gentlemen," he said. "The year 1993 has been formally declared the International Year of the World's Indigenous People as directed by the General Assembly of the United Nations. The opening ceremonies took place in New York on 10 December 1992, International Human Rights Day. The declaration of the year signified the interest which the fate of indigenous people is receiving in the international community. Although the case of indigenous people in America, Australia, and New Zealand is well known, indigenous people in Africa have received scant attention.

"The fate of such groups as the Zangon Kataf and Ogoni in Nigeria are, in essence, no different from those of the Aborigines of Australia, the Maori of New Zealand, and the Indians of North and South America. Their common history is of the usurpation of their land and resources, the destruction of their culture and the eventual decimation of the people. Indigenous people often do not realize what is happening to them until it is too late. More often than not, they are the victims of the actions of greedy outsiders. It is in this regard that we have undertaken to publicize the fate of the Ogoni people in Nigeria."

Ken went on to say that Shell had taken over $30 billion from the Ogoni people and put nothing back but degradation and death. As for the Nigerian government, it condoned and legalized this theft so it could usurp the Ogoni's wealth and resources.

"I call upon you, my brothers and sisters, to fight relentlessly for your rights. As our cause is just, and God being our helper, we shall emerge victorious over the forces of greed, wickedness and obduracy. God bless you all."[10]

The reaction was overwhelming. Ken led everyone in an impromptu chorus of "Arise, Arise, Ogoni People Arise!", a song he had written in 1968 during the civil war, then launched into a second, extemporaneous speech in which he called on the Ogonis to face the Nigerian government and Shell Oil. He challenged these forces to kill off all Ogoni men, women, and children before taking more oil from Ogonis.

"From today onwards, Shell is declared *persona non grata* in Ogoni!"[11] he cried.

Amid cheers, many in the multitude began chanting, "No to Shell! No to Shell!" and punching the air with their fists.

"No to Shell! No to Shell!"

||||||

At the end of his speech, Ken nearly fainted from exhaustion and dehydration. Making his way back to his car, he announced he was going home.

"But Dede," said Owens, "we still have to go to Khana and Nyo-Khana."

"Why?" said Ken. "Everybody is here in Bori. Why should I go there?"

"Because you said you were going. You have to. There are other people at those rallies, especially the women and the old people who cannot make the trek all the way to Bori."

"I can't make it," said Ken, climbing into the back seat. "Have Dr. Leton go." He closed the door and instructed the driver to take him back to Port Harcourt.

Leton was exhausted as well, not having slept the night before. The shy MOSOP president had been overwhelmed at having to speak in front of so many people, and it was difficult for Owens to convince him to continue on to Nyo-Khana and Ken-Khana, but eventually Leton capitulated.

Owens was correct that the people who had gathered in the two unvisited kingdoms would be awaiting Ken Saro-Wiwa. Late in the day, when Owens arrived in Baen, Ken-Khana, on his own, they let out a cheer.

"Ken will come!"

Owens stood before the assembly and shook his head. "No, he won't come," he said. "Ken is tired. He has gone back to Port Harcourt to rest."

"No! This is his kingdom!" a man shouted. "Look at all of us here!"

"Look, he's gone," said Owens apologetically. "Dr. Leton is coming."

But the people of Baen had to wait a good while longer before Garrick Leton showed up. Owens could see the disappointment when they understood that Ken was really not going to appear at their rallies – and no one was very enthusiastic about hearing what Leton had to say. The rally came to an awkward close. The one in Nyo-Khana also sputtered out, a disappointing end to an otherwise overwhelmingly successful day.

More than three hundred thousand people had marched that day – nearly two-thirds of the Ogoni population. Those who could not go to the central rallies at Bori or their kingdom headquarters had marched in their village squares. There had been no violence, no arrests. Uniformed police and army personnel had been at the rallies, and Ken had instructed people to be kind to them and offer them food and drink. Most of the soldiers seemed to enjoy the event, ogling the pretty girls, smiling and waving at the marchers from their transport wagons. A few soldiers danced with the Ogonis in the streets.

When Owens returned to his clinic, the nurses reported that no one had come in. If an Ogoni had so much as stubbed a toe, they hadn't heard about it. All the drugs Owens had so carefully stockpiled remained unused. Even the aspirins had gone untouched.

||||||

The next day, January 5, the MOSOP steering committee convened a meeting at Chief Edward Kobani's house in Bodo. Kobani opened by saying, "We should congratulate ourselves for embarking on a demonstration devoid of violence." He said the feat was achieved in the face of stiff opposition from some "unpatriotic Ogoni sons," referring to the pro-government Ogoni elites who had opposed the event. It was agreed that there should be some sort of follow-up. Proud of themselves, the MOSOP steering committee closed the meeting with a short prayer[12] and feasted on roasted goat.

In the days after the momentous demonstration, Shell vehicles arrived in Ogoniland in greater numbers than ever before. Shell went about its business as if no protest had happened. The Ogonis interpreted this as defiance and became increasingly angry.

Five days after the march, Ken was summoned to the Port Harcourt office of the deputy inspector general of police, a courteous man named Malherbe. He informed Ken that, next morning, the presidential jet would be coming to take him, Leton, and Kobani to a meeting with senior government officials at the capital. Ken read this as a good sign – the government was willing to start negotiating. When Bennett Birabi heard about the trip, he decided to accompany them as well.

Birabi, who'd spent seven years in Ken's house sharing a bedroom with Owens, had trained as a doctor and practised medicine for a couple of years before going into politics. Ken did not approve of Birabi's political aspirations – and for good reason. When Ken had been canvassing for the creation of a Port Harcourt state, Birabi betrayed Ken's confidence and relayed his strategies to government officials who adamantly did not want Port Harcourt State to be formed. Ken's campaign failed. Birabi had become Nigeria's Senate minority leader and one of the Ogoni elites categorically opposed to the massive rally. Despite their troubled history and personal differences, for political reasons, Ken thought it wise for Birabi to join them on the trip.

Even after the presidential jet took off from Port Harcourt, Ken had no idea where they were headed. At the time, Babangida had decided to move the capital of Nigeria from Lagos to Abuja, and the federal government was still transferring civil servants and government functionaries. Shortly before arrival, the jet's four passengers were informed that the former capital, Lagos, was their destination.

Ken and the others were escorted to the headquarters of the Nigerian police. There, the inspector general of police, Aliyu Atta, received them in unexpectedly hospitable fashion. It was evident that the inspector general was a good friend of Bennett Birabi. Atta said he would like to know exactly what the recent Ogoni Day protest had been all about. After listening to the four men describe the agony of the Ogonis, Inspector General Atta thanked them and asked Ken to prepare a memorandum and deliver the document to him in the morning.

The memorandum Ken prepared that evening was essentially a restatement of the Ogoni Bill of Rights. In particular, the memo asked the Nigerian government for an Ogoni state.[13] Ken included a copy of the "demand notice" that had been sent to Shell. The next morning, Atta assured Ken that the government would reply within ten days.

No response ever came.

|||||||

While Ken was in Lagos, six Ogoni friends of Governor Ada-George held a meeting at the beach at Kono Waterside. There, this gaggle of government-appointed chiefs drafted a communiqué in which they took the side of the government against the Ogoni people.

The communiqué stated that "at a fully representative meeting of the Ogoni people," the events of January 4 were reviewed in detail and that the Ogonis reaffirmed their faith in the Republic of Nigeria. The pro-government chiefs announced that the Ogoni people were aware of the recent encouraging actions of the federal government and that "no further demonstration will be held in expectations of the positive reaction of government to their cry."[14]

In Lagos, Ken was furious when he saw this communiqué printed as a full-page advertisement in the government newspaper, *Nigerian Tide*. Particularly galling to him was the chiefs' assertion that they represented the Ogoni people. As far as Ken was concerned, MOSOP was the only organization authorized to speak on the Ogonis' behalf.

A couple of weeks later, the MOSOP leadership met with the six chiefs who had signed the communiqué. Asked what they found troublesome about a non-violent rally, the chiefs gave no specific grievances but indicated that MOSOP's actions had embarrassed them with their colleagues in government. When it was pointed out that they could not have afforded a full-page ad in a national paper, the chiefs admitted that the state government had picked up the tab.

||||||||

The next day, a handful of Ogonis came into Owens's clinic saying that a Shell employee was stranded with car trouble just outside Bori. "What should we do?" asked one of the men.

"Go help him, of course. See the Shell man gets some transport back to Port Harcourt," Owens advised.

Later that day, a patient mentioned to Owens that a Shell vehicle was overturned and burning. Some Ogonis who witnessed the event explained what had happened. They confirmed that the Shell vehicle that had broken down in Bori had been repaired. As the Shell employee drove at high speed back to Port Harcourt, he lost control of the car and it overturned. Fortunately, the driver managed to escape unharmed and found a ride back to Port Harcourt. When the local people heard the crash, they came out to see the wreckage and noticed the Shell logo on the side of the car. After the driver left, a few village men set fire to the vehicle.

The incident was reported widely in the local news. According to the reports in the media (based primarily on statements by Shell and third-hand accounts), a MOSOP mob caused the accident, then beat the driver and burned his car. Disturbed by the inaccuracies of the stories, Ken took out a newspaper advertisement letting the people of Nigeria know that MOSOP was a non-violent organization, that MOSOP had nothing to do with the Shell vehicle being burnt, and that no one was injured.

Nevertheless, Shell was sufficiently spooked by the event that it released a statement saying it would immediately withdraw its staff from Ogoniland and close oil operations in the area.[15] In a briefing paper drafted later entitled "The Ogoni Issue," Shell reiterated clearly that Shell Petroleum Development Company (SPDC) "withdrew all its staff from Ogoni in January 1993."

The truth was there were no Shell staff in Ogoniland to withdraw. Shell's offices were in Port Harcourt, well out of Ogoniland, and (with rare exception) all Shell employees lived in Port Harcourt. Before and after the incident with the Shell vehicle, Shell staff continued to come and go at will while subcontractors worked in Ogoniland on Shell's behalf. In this way, despite the company's assertions to the contrary, Shell continued its operations in Ogoniland for most of 1993.[16]

Shell had good reason to promote the idea that it was no longer operating in Ogoniland. After the Ogoni Day demonstrations, Shell was concerned about how events in the Niger delta were harming the company's reputation internationally. At meetings at Shell's central office in London on February 15 and 16, 1993, and in The Hague on February 18, Shell executives discussed the Ogoni problem. Ken Saro-Wiwa was a key topic of their discussions.

"The overriding PA issue facing Shell in Nigeria comes under the overall umbrella of community relations," recounted Shell general manager Nnaemeka Achebe in the draft minutes of the February meetings. "It is commonly felt that other parts of the nation are benefiting from oil more than those who live in the oil-producing states, and this manifests itself in community attacks on installations, disrupting operations. . . . International networking, most prominently so far involving the Ogoni tribe and Ken Saro-Wiwa, is at work and gives rise to the possibility that internationally organised protest could develop. Ken Saro-Wiwa is using his influence at a number of meetings, last year in Geneva at the UN Commission on Human Rights and most recently one organised in the Netherlands by the Unrepresented Nations and Peoples Organisation (despite its acronym, UNPO, is not a UN offshoot). . . . Ken Saro-Wiwa/UNPO will be using every opportunity made available by 1993 being the UN's declared Year for Indigenous Peoples.[17] . . . SPDC [Shell Petroleum Development Corporation] and SIPC PA [Shell International Petroleum Company Public Affairs] departments to keep each other more closely informed to ensure that movements of key players, what they say and to whom is more effectively monitored to avoid unpleasant surprises and adversely affect the reputation of the Group as a whole."[18,19]

Shell started monitoring Ken Saro-Wiwa's every move. Ken had indeed recently been to the Netherlands; the day after MOSOP's meeting with the pro-government chiefs, he had flown to an UNPO meeting at The Hague. There, he attended a workshop on non-violent struggle and was elected vice-chairman of UNPO's Third General Assembly. At this time, UNPO also officially admitted the Ogoni people into its organization. CNN covered the ceremony and broadcast scenes of the Ogonis' January 4 anti-Shell demonstrations, and a story about the Ogoni issue appeared in *Time* magazine.[20]

In short, MOSOP was becoming a PR nightmare for Shell. Affiliated organizations of MOSOP – NYCOP, FOWA, COTRA, and others – had started organizing rallies nearly every weekend in villages throughout the six Ogoni kingdoms. Some weeks there were five or six anti-Shell rallies. All stressed the same themes: the rights of the Ogonis must be recognized, the environment must be cleaned up, and above all – unless it acceded to the Ogonis' demands – Shell must be driven out of Ogoniland.

Ken was adamant about maintaining non-violence. Of course, MOSOP being unarmed and impoverished had no course other than non-violent protest. Ken had no choice but to have faith that piety and PR would eventually win the day. As for Shell, he suspected that the company would fall back on its usual modus operandi of relying on the Nigerian regime to resolve its problems while claiming to be unaware of what was happening.

Ken began telling people that he believed his life was at risk. Preparing his will, he informed his family that they should brace themselves for the worst. In his public speeches, he told the Ogoni people to expect the same.[21]

||||||

To continue the momentum of the Ogoni Day rally, MOSOP planned follow-up events: a pan-Ogoni night vigil/candlelight march, a series of church services to dedicate the movement to God, and the launching of the One Naira Ogoni Survival Fund (ONOSUF).

On behalf of ONOSUF, Ken asked that every Ogoni person contribute one naira as a symbol of commitment to the movement. Ken expected that the half-million Ogonis would raise at most 500,000 naira (about US$23,000 at the time). By asking for a token sum, Ken felt it would solidify the Ogonis' commitment, provide a sort of census,[22] and would gauge just how acceptable the idea of MOSOP was to the Ogoni.

MOSOP units went from village to village, collecting one naira from each person and recording names in school composition books. The enthusiasm was so great that pregnant women came forward to donate one naira for themselves and another for their unborn child, deciding on the baby's name at that moment. When the campaign was completed, each kingdom held a public counting ceremony. In each Ogoni kingdom, dance groups performed and then either a MOSOP representative or a chief from each little village came forward to give an emotional speech announcing the total contribution and turning over the cash and composition books to the president of that kingdom's MOSOP chapter.

On February 27, Ken Saro-Wiwa officiated at the final pan-Ogoni ONOSUF ceremony in Bori. In an address to the thousands gathered that day, he said, "Why one naira, we might ask? The Ogoni people are an extremely rich but dispossessed people. In establishing this fund, we want to emphasize not money but the symbols of togetherness, of comradeship, of unity of endeavour, of the total commitment of young and old. Money cannot win the war of genocide against the Ogoni people. God Himself will win the war for us. But all Ogoni men, women, and children, including newborn babies, will contribute to ONOSUF as a statement of their will to survive as individuals and as one indivisible nation."[23]

At the end of his speech, Ken held up a wad of bills. "This is my own donation. I donate this, and my life."

|||||||

At the rally, young Diana Barikor was furious with NYCOP president Goodluck Diigbo. In organizing the day's event, Goodluck had set up only a single table for community representatives to present their contributions to the ONOSUF. Goodluck's plan was that the contributors would, one by one, have their pictures taken while turning over their cash and ledgers. Before long, the line became so long it threatened to take all afternoon and most of the evening just to collect the contributions.

"Are you crazy?" an angry Diana asked Goodluck. "Set up several tables to collect all this money."

"There is only one photographer," he said, "so there can only be one table. One table is enough."

"But you can't ask them to line up forever. It is beyond sense!"

As they were arguing, Owens approached Goodluck to reprimand him as well for not putting out more tables.

"That's what I was just telling him!" said Diana.

After they finished ganging up on Goodluck Diigbo, Owens asked Diana to come to his clinic, explaining, "I want to discuss something with you."

A couple of hours later, Diana took a break to get out of the sun and walked over to the Bori clinic. Stepping inside, she told the receptionist that Dr. Wiwa wanted to see her.

"He's with a patient," she said. Diana started to leave, but the receptionist motioned for her to have a seat. "Don't worry," she said. "I'll tell him you are here. When the patient comes out, Doctor will see you."

Diana wondered what Dr. Wiwa wanted to talk about. Probably something about NYCOP or about finding a way to work out the ONOSUF money collection.

Owens, in his white lab coat with a stethoscope around his neck, soon called Diana into his office and offered her a seat. He kept the conversation going, but was taking a long time getting around to saying why he had asked to see her. Diana found their chat somewhat tedious, though it is not unusual in Nigeria for two people to talk about everything under the sun before getting to the point.

Finally, he said, "Come have dinner with me and we can continue our talk."

At his house, Owens offered Diana some red wine in a cut-glass goblet. She stared at the glass a moment, hesitating. He had no idea she was a Seventh-day Adventist and had never touched alcohol. Diana wondered whether to tell him it was forbidden by her religion, but the sparkle of the glass tempted her. In any event, she told herself, unlike her family, she was not personally opposed to trying wine.

"Thank you," she said, accepting the goblet. She took a tentative sip and thought it tasted awful.

"It is not good?" he asked. "You would prefer something else?"

"No, it's fine," she said, smiling. "I'm the sort of person who will finish everything she starts."

As they talked, the evening grew late and Diana's glass grew empty. The one small serving of wine had made her head swim. "If I'm going to keep sitting here talking to you, you're going to have to take me back to Port Harcourt tonight," she said.

"My car and my driver are here. Don't worry."

Owens's valet came in to announce that a staff member from the clinic had just arrived to summon Owens back for an emergency. Owens excused himself, saying that he would return as soon as he could. As he departed, Diana stood up to have a look around. To her surprise, the room was spinning. Noticing a bed in the adjoining room, she went in and lay down.

After performing an emergency caesarean section, Owens returned home to find Diana asleep in his guest room. He gently awakened her.

"What time is it?" she asked.

"It's after ten."

"What do you mean, it's after ten? I have to go home! Go tell your driver."

"He's gone for the night," said Owens.

"What am I going to do?"

"Don't get agitated," he said.

Diana wasn't worried so much about what she'd tell her parents – she was used to concocting one lie or another to mollify her strict father – but there was no public transport back to Port Harcourt at that hour and she had a class in the morning. She had an uncle who lived in Bori, but it would cause gossip if she were to drop in at such a late hour. Owens seemed unfazed by her predicament. No matter how she impressed on him her need to get back to Port Harcourt, he told her she needn't worry because he would drive her there himself.

Diana retired to the guest room. When the sun rose, Owens kept his promise.

|||||||

Over the following weeks, Diana and Owens's friendship grew into a roman-tic relationship. One night at a NYCOP gathering, the friend Diana was with pointed at Owens across the meeting hall and asked, "Isn't that your boyfriend? He's Dr. Wiwa, right?"

"Yes, he is," she said.

"Isn't he related to Ken Saro-Wiwa?"

Diana looked at him askance. "No, they can't be related. If they were, I would have known."

The following evening at Owens's house in Bori, Diana was taken aback at the sight of Ken Saro-Wiwa letting himself in the front door.

In amazement, Diana whispered to Owens, "Ken knows you so well that he would come to your house?"

"What do you mean?" said Owens. "He's my brother."

Diana felt foolish. During the time she and Owens had been dating, he had never once mentioned his relationship to Ken. To her, fifty-one-year-old Ken was just a famous little old man. Now, seeing the two brothers side by side, she wondered how she could have missed the resemblance.

|||||||

The next big MOSOP event was a night vigil to be held in March. The Ogoni people had been flocking to Christianity in greater numbers than ever before, feeling that if Jesus could not solve their problems, at least he could grant them solace. Ken, who had little religious feeling, did not like this development; the Ogonis, he felt, needed more than divine intervention to see them through the coming months. Still, he believed the church was an important instrument to sell MOSOP to the masses, so he approved the plan for a night of dedicating the movement to God. To keep the crowds small, MOSOP instructed the Ogoni villages to hold separate vigils simultaneously.

At sunset on March 13, the citizens of Bori lit candles and marched single-file from the grammar school to the conference hall at the other end of the town. "Go down, go down, go down to Abuja," they sang. "And tell the government, government let Ogoni go!"[24]

All night, Ogoni bands played and church choirs sang Christian songs. Most of the elites stole away to their beds at midnight, but the entire Wiwa family marched and sang until the break of day. At four in the morning, Ken nudged Owens and said, "Look, Papa is still around." Even after many had gone home, eighty-nine-year-old Chief Jim Wiwa was still there with his candle, watching the dancers.

Ken's joy was quickly followed by great sorrow. Two days after the night vigil, he received terrible news from England about his youngest son.[25] On the playing fields of Eton College, fourteen-year-old Tedum had complained of feeling dizzy and thirsty. A few minutes later, the teen died of a heart attack resulting from a congenital defect. As Ken rushed to London for the funeral, his grief was compounded with feelings of anger and guilt. During this difficult time, his relationship with Maria and the children was already at an all-time low. It didn't help when he wondered aloud on the way back from the mortuary if his anti-Shell campaign had something to do with Tedum's death. Hearing this, Ken Junior, Gian, and the twins flew into a rage. How could their father be so narcissistic? Above all, how could he be so insensitive as to put into words the very thoughts they were trying to suppress?[26]

After Tedum's funeral, Ken returned to Nigeria broken, bitter, and wracked with guilt. His only relief came from throwing himself even more obsessively into the Ogoni struggle. He travelled to the Delta State city of Warri to deliver a lecture. The Delta State authorities, however, wanted no part of Ken Saro-Wiwa. Twenty policemen were waiting to intercept him. After briefly detaining him, the police deported him, escorting him to the Patani Bridge on the Niger River and watching him cross out of their jurisdiction.[27]

|||||||

One day in March 1993, Owens said to Diana, "I want to ask you something, but I would not want you to laugh."

"Okay," she said, hesitantly. "You're a sensible man. What could you possibly say that I would laugh at?"

"You should not laugh."

"Well, Mon, if it's funny, I will laugh. Are you going to be funny?"

"No," he said, "this is not funny."

"Okay, then. I will not laugh."

"What if I asked you to marry me?"

Diana gave a shriek.

Like Owens, Diana had questioned if she ever wanted to marry, although she did wish to have both children and a career. As Owens grimaced, Diana continued to laugh. She never gave him an answer and he never asked again, yet from that day, it was understood they would wed.

||||||

At this time, there were two main political parties in Nigeria – the Social Democratic Party (SDP) and the National Party of Nigeria (NPN). In January 1993, Babangida lifted his ban on politicians from previous administrations, and most of the Ogoni elites rushed back into politics. In a series of elections, the Ogoni members of the NPN triumphed, as did many NPN candidates across the country. Those Ogoni politicians belonging to the SDP, having lost their elections, filtered back into MOSOP.

At the onset, Ken had made it clear that MOSOP would be non-partisan. "We are not going to influence the Ogoni people to vote for one party or the other," he said.

Among the elites now able to resume their political careers were Garrick Leton and Edward Kobani. In a steering committee meeting, Ken pointedly reaffirmed MOSOP's non-partisan nature; no MOSOP official would hold elected or appointed political office. The steering committee again agreed.

Although Kobani had assumed the role of MOSOP vice-president, he decided to defy MOSOP policy by running for chairman of the Rivers State branch of the SDP. To MOSOP's satisfaction, Kobani suffered a humiliating defeat. Shortly thereafter, MOSOP president Dr. Leton secretly attended as an official delegate the SDP political convention in Jos in central Nigeria.[28] At the next steering committee meeting, Leton was castigated for disgracing the movement and for destroying what MOSOP was trying to build. Those in the NPN complained that Leton's selfish actions gave the appearance that MOSOP was now officially backing the SDP. Even though MOSOP remained officially apolitical, partisan politics had created the first irreparable crack in the organization.

To compound matters, the country was heading into its first democratic presidential election in many years. To ensure that MOSOP remained neutral, Ken Saro-Wiwa rashly announced during a televised interview with CNN that the Ogoni people would boycott the upcoming presidential election.

||||||

In the spring of 1993, the State Security Service began to target Ken Saro-Wiwa. On April 18, at the airport in Port Harcourt, the state police ushered him into a back room where he was strip-searched and held for sixteen hours before being released.[29] Five days later,[30] he was sitting in his office at 24 Aggrey Road talking to Owens when three SSS officers barged in to conduct a search.

"What do you want to search the office for?" asked Ken.

"We know what we're looking for."

"Just tell me. If I have it, I will give it to you."

Ignoring Ken, the police started going through his shelves, tossing books to the floor. Owens expected his brother to fly into a rage. Ken was an ardent bibliophile who treated books like sacred objects, yet to Owens's amazement his brother remained impassive. Owens, who knew some of the officers personally, sprang from his chair to stop the vandalism, but Ken motioned for him to sit back down.

"They have guns," Ken warned him softly in Khana. "If you do anything, they will use the opportunity. Just sit down." When Owens reluctantly resumed his place, Ken asked the men again, "What do you want?" One mumbled something about a flag. Ken said, "Well, if you want a flag, I'll give you a flag."

"You have it? You have a flag?" said one of the policemen with accusatorial glee.

"Of course we have a flag. The Ogoni flag. I have the prototype around here somewhere."

Among the criteria for UNPO membership was a commitment by the unrecognized nation to non-violence and the existence of a flag. To this end, Ken had arranged for the design and manufacture of the first Ogoni flag – a circle of six red stars representing the six Ogoni kingdoms on a field of blue, gold, and green bars.

"Is this the flag you're talking about?" Ken asked.

"Yes, yes, yes. Now you are under arrest. We are taking you down to headquarters."

"For having a flag?" exclaimed Owens.

"Okay, fine, let's go," said Ken with equanimity. "But first, look out the window. What do you see? Across the street, Coca-Cola has a flag. Down the street, there is a flag flying for the NNPC. Shell has a flag. Are you going to arrest them as well? Whenever we fly the Ogoni flag, we hang it under the flag of Nigeria. There's nothing illegal about it."

The police took Ken to the sss detention centre in Port Harcourt. At this deceptively pleasant-looking building surrounded by high whitewashed walls, Ken was placed in a bungalow in a small room with a desk and a chair. During his interrogation, Ken tried to make it clear to the sss that MOSOP was a peaceful organization and that this harassment was only going to bring more attention to the cause. Unmoved, the Secret Service agents grilled Ken until he collapsed from hunger and exhaustion twelve hours later and was released.

||||||

During the time Ken was being targeted by the sss, Owens began to record incidents of petroleum-related human rights abuses. The first incident that Owens recorded in his clinic occurred in February 1993. A pregnant woman and her husband came in for treatment. Covered in cuts and bruises, they told Owens they owned a farm on which Shell had decided to place an oil well. At daybreak, when they'd gone to their field to harvest crops, a van of mobile police pulled up, beat them, and drove them off their own land.

A week later, two Ogoni men were brought into the clinic with gaping wounds. They said that army personnel had pulled up in front of their shop in Bori demanding fuel for their vehicle. When the shopkeepers said they were out of gasoline that day, the soldiers roughed them up. This event particularly upset Owens, for the two injured men worked for Owens's uncle, the man who had loaned him money to buy the sign for his clinic.

At the beginning of April, Owens read a newspaper account of an incident in the Ikwerri region in which soldiers had arrested several chiefs protesting the passage of the Trans-Niger pipeline over their land. The chiefs had become upset after an oil company ran a section of the pipeline through their sacred forest without permission. Nor did the company allow the Ikwerri the opportunity to perform the necessary rituals and sacrifices to purify the area. Shortly after reading this article, Owens learned that Shell had halted work in the Ikwerri region and planned to begin construction of the continuation of the same pipeline slated to bisect the Ogoni kingdom of Eleme.

At the time, Shell was expanding the Trans-Niger pipeline, which carried oil from most parts of the delta through Ogoni territory to the export terminal at Bonny Island. An earlier version of this pipeline had been laid in the 1960s, but Nigeria increased its oil production dramatically in the late 1980s. In preparation for the expansion of the pipeline through Ogoni territory, Shell Petroleum Development Company failed to carry out an environmental impact assessment study and made no effort to negotiate with the landlords whose land they were about to cross. To avoid a repeat of the Ikwerri problem, the company arranged for soldiers of the Nigerian army to accompany its work crews on the Rumuekpe-Bomu section of the pipeline.[31]

On Thursday, April 29, nearly a dozen people from Biara came into the Bori clinic with gunshot wounds. Owens asked what had happened.

"Soldiers shoot us," one young man said.

"You're telling me Nigerian army soldiers shot you?"

"Yes."

"This I have to see for myself."

After making sure that the wounded were being properly attended, Owens had his driver take him to the place the villagers described, a field located near the twenty-two-kilometre marker on the Bori-Port Harcourt Road. There he came on a noisy gathering of Ogonis, a crowd so large it spilled back into the bush. On the other side of the road were army personnel with guns drawn, guarding two jeeps, two bulldozers, and a Willbros Company bus; Willbros West Africa, Inc., of Tulsa, Oklahoma, was the contracting firm hired by SPDC to construct the pipeline.

Stepping out of his car, Owens went over to the crowd. "Sit down!" he shouted. "Sit down! Just sit down!"

Like his brother, Owens had been studying the tactics of non-violent protest. To avoid further shootings, he knew the people had to protest but not confront the soldiers.

"Sit down!" he repeated until a large section of the crowd obeyed.

The Willbros construction workers – a group of Caucasian Americans in work clothes – were leaning against the bulldozers and taking snapshots of the protesters with Instamatic cameras. Beside one of the jeeps were two white men dressed in business attire. Recognizing one of the army captains from his days in Jaji, Owens walked over to the officer and struck up a conversation. As they spoke, Owens noticed a group of women crying over another woman who was lying on the ground. Trying not to show his anger,

Owens asked the army captain if it would be all right if he spoke with the man in charge of the Willbros workers.

"That's the manager there," said the captain, pointing to one of the white men in a shirt and tie.

Owens introduced himself. "Tell me what's happening. I had some people who came to my clinic who said they were shot. Now I come here and I'm seeing this. What is it?"

The man, whose name was J.K. Tillery, shrugged and said, "Hey, this is none of my business."

"How do you mean?"

"These are Nigerian soldiers and these are Nigerian people. It's none of my business."

"Sir, I passed this road yesterday and your trucks were not here," said Owens. "These soldiers were not here and these people were not here. Nobody was shot. You brought the soldiers to this area. So now this is your business."

As they talked, a commotion arose from the crowd. To the delight of the protesters, a couple of small Ogoni boys had sneaked up behind the Willbros crew and dropped sand into the fuel tanks of both bulldozers. When Owens tried to settle the crowd, Tillery gave the order for all the Willbros workers to climb onto the bus. He instructed a couple of his employees to put the injured Ogoni woman on the bus as well. He and the other Willbros manager then got into the jeeps and drove off, leaving the bulldozers behind.

As the workers left the scene, the army also packed up to go. Climbing into their transport vehicles, the soldiers fired several rounds into the air and sped away.

After the soldiers had gone, Owens asked what had happened. He was told that a Biara man and his wife had gone to their farm at about five that morning and discovered the Willbros company bulldozing their crops while the army watched. When the farmer's wife asked the crew why they were doing it, the couple was assaulted by the soldiers and chased away. The farmer and his wife ran back to Biara and alerted the entire village. In no time, a spontaneous protest developed at the site of the ravaged farmlands. As soldiers kept guard over the Willbros work crew, the protesters danced, sang, and shouted to convey their displeasure over the destruction of their crops.[32,33] Suddenly, the soldiers opened fire on the crowd, wounding eleven people. Mrs. Karalolo Korgbara,[34] a mother of five trying to salvage what was left of her vegetables, was hit by bullets that shattered the bone of her upper arm and pulverized the muscle and tendon above her elbow, necessitating amputation. An Ogoni cameraman captured the shooting on film.

J.K. Tillery's account of the incident at Biara can be found in his report to Shell: "By 10.00 am a large crowd of villagers had gathered. . . . Fortunately there was a Military presence to control the situation and to offer protection to the workers and equipment. The tension developed to a level where there was real danger to personnel, and the Army were drawn into a confrontation by the hostile Villagers."[35]

The next day, thousands of unarmed Ogonis returned to the site, daring the soldiers to shoot.[36] Mass demonstrations continued through the weekend, halting construction of the pipeline. On Monday, May 3, the conflict heated up until members of the Nigerian military opened fire on the crowds again, fatally wounding an Ogoni man named Agbarator Otu.[37]

Otu's body was taken to the Princess Medical Centre Annex, where Owens logged the death as the first official fatality in MOSOP's struggle. The dead man was then transported to the University of Port Harcourt Teaching Hospital for autopsy. It was determined that he'd died from a bullet wound in the back.

All this proved too much for Willbros West Africa. The American subcontractor withdrew its workers and equipment from Ogoniland.[38] After receiving reports on the incident at Biara, Amnesty International condemned Shell and the Nigerian government[39] and issued an Urgent Action notice concerning the killing of Mr. Otu.[40] The Shell Petroleum Development Corporation contacted Governor Ada-George, requesting the "usual assistance of His Excellency to enable the project to proceed."[41]

|||||||

Around the middle of 1993, oil production in Ogoniland began grinding to a halt. Royal Dutch/Shell would have the world believe that it terminated business abruptly and withdrew its staff from within the borders of Ogoniland as a result of the January 18 incident in which one of its vehicles was set ablaze. As noted earlier, there is no evidence that Shell removed any staff from the area at all, as all its employees lived and worked in Port Harcourt and the majority of oil business in Ogoniland was performed by sub-contractors, who continued doing business as usual.

In the spring of 1993, a group of Ogoni subcontractors for Shell attended a MOSOP steering committee meeting to ascertain what they should do next. Most of these subcontractors performed simple services for the company. Some cut the grass around oil wells and pipelines; some were hired to keep the oil pumps lubricated and filled with diesel fuel; some performed janitorial services at Shell facilities in Port Harcourt.

"This has been our work," said one of the subcontractors to the steering committee. "Now with all this thing going on with the community, our dealings with Shell, we don't know what to do so that we don't starve."

The committee's response came from Dr. Leton. "As of this moment," he said, "the Ogonis are at war with Shell. If you continue to deal with Shell, it means that you are at war with the Ogoni people. Everybody sitting here is also going to suffer one way or the other because of this struggle for a short time. When we win, it will be good for everybody, so you too have to do your own part."

The Shell subcontractors excused themselves and went into an adjoining room to discuss what Leton had said. When they returned, they announced to the steering committee that they had agreed that their own contribution to the struggle would be that they would stop servicing the Shell installations. They carried out their vow, and soon the pumping stations had stopped working and the flares sputtered out.

|||||||

The Nigerian government considered these Ogoni protests a call to arms. With security reports making reference to the Ogoni territory potentially becoming "another Biafra," the entire country was put on national alert.

Babangida set his sights on exterminating the Ogoni problem at its root. The day after Agbarator Otu's killing, the government passed the Treason and Treasonable Offenses Decree, making the call for minority autonomy an offence punishable by death.[42] This decree mandated the death penalty for anyone who "conspires with himself" to so much as utter the words "ethnic autonomy." Death was also the punishment for anyone who planned secession from Nigeria or sought to alter the boundaries of any local government or state previously decreed by the military authorities.[43] The purpose of Babangida's directive was so obvious that it became known as the "Ken Saro-Wiwa Decree."

The Ogoni people and the MOSOP leadership were traumatized by the shootings at Biara. Seeking to calm the people, the MOSOP steering committee delegated Ledum Mitee, Edward Kobani, and Ken to embark on a tour of all six kingdoms. The three MOSOP leaders spoke to many large crowds, imploring the Ogoni to maintain the peace.

At the same time, a group of eleven Ogoni elites, including Chief Samuel Orage, I.S. Kogbara, and HRH G.N.K. Giniwa, presented a document to the governor apologizing for the violent lawlessness of MOSOP and suggesting that Ken Saro-Wiwa was not the legitimate spokesperson for the Ogoni people.

"We wish to state that the Ogoni people have not mandated any person or group of persons to coerce the public, cause bodily harm and threaten life and property," they wrote. "Specifically the Ogoni people have not mandated any person or group of persons to obstruct the legitimate operation of any normal business operations in the area or to interfere with Government security men who are carrying out their duty of maintaining law and order."

To legitimize the fact that troops had been sent in and one man had been killed, the chiefs ended their apologia with "Finally, may we state that no Government will allow a break down of law and order which will result to a state of anarchy. And we associate ourselves with any action by Government to protect life and property of innocent citizens."[44]

This sycophantic document was broadcast widely on the radio and printed in the newspapers. When the Ogoni people saw it, they became deeply annoyed at the eleven chiefs, and Edward Kobani flippantly referred to the small group of libellous elites as "vultures," a pernicious label that stuck for all time.

|||||||

One week after the Biara shooting, on May 7, 1993, Ken and three other MOSOP leaders were summoned to Abuja for a meeting with the country's highest-ranking security officials. Accompanying Ken to the meeting were Garrick Leton, Edward Kobani, and Albert Badey.

The son of a Methodist priest from Gokana, Badey was a career civil servant who had risen to become secretary to the Rivers State government, the highest position in the civil service. During the Biafran War, he had misguidedly cast his lot with the Igbos; after the war he required Ken's assistance to find employment. Ken admired Badey's intelligence and education, but considered him a "dyed-in-the-wool gerontocrat" – a firm believer that all of society's decisions should be made by a small coterie of elders, a view definitely at odds with Ken's vision.[45] Badey was also one of the Ogoni career politicians who had been tossed out of the civil service by Babangida and hungered to return to his former position of glory.

On first seeing Albert Badey at a MOSOP steering committee meeting, Edward Kobani had taken Ken aside and expressed his doubts about Badey's motives. Kobani told Ken that Badey had been meeting with the pro-Shell Ogoni elites who had denounced MOSOP after the shootings at Biara. Ken dismissed Kobani's concerns. It was well known that Kobani and Badey hailed from the same Gokana village and had a long-standing rivalry. Ken suspected Kobani was merely being petty, but to placate him, he assured Kobani that Badey's talents and experience in government would be useful to MOSOP.[46]

But Ken was being naive. Albert Badey's true mission was the dismantling of MOSOP from within.

At the May 7 meeting, the four MOSOP leaders met with national security advisor Major General Aliyu Mohammed Gusau, national intelligence director Brigadier General Ali Akilu, and federal secretary Alhaji Aliyu Mohammed. This top echelon of Nigerian security forces gave Ken the impression that the government wanted to help the Ogonis. As others in government had done, they asked MOSOP to prepare a paper detailing the Ogonis' demands, along with background information on how oil-producing areas in other parts of the world were treated by their governments. Oddly, the MOSOP leaders were also asked to furnish a list of all unemployed Ogoni graduates. Ken provided all the requested reports. As before, nothing more was heard from the government.[47]

Ken soon embarked on a European tour to draw world attention to the Ogoni struggle. While trying to board his plane for London,[48] he had his passport seized by the Lagos airport police. After a midnight phone call to the national security advisor, he retrieved the passport and flew to England the following night. With the Treason and Treasonable Offenses Decree hanging over his head, Ken made his objections known all over Europe to what he understandably felt was his own death warrant. He met with government officials in the Netherlands, Switzerland, and Britain, as well as with representatives from the UN Human Rights Commission and the International Commission of Jurists in Geneva. For good measure, he also arranged another appearance on CNN.[49] News of Ken's European campaign did not sit well with authorities in Abuja, but Shell responded by announcing it was "happy to discuss these matters further" with the Ogoni.

While Ken was in Europe, Albert Badey took the opportunity to engage in mischief. In Leton and Kobani, he found two willing cohorts. For some time, MOSOP president Leton and vice-president Kobani had been dissatisfied with MOSOP's Ken-dominated trajectory. They felt hamstrung by MOSOP's prohibition against their holding elected office. They were furious at Ken for announcing unilaterally on CNN that all the Ogoni people would boycott the upcoming presidential election. And they despised being overshadowed by someone whom the people had started calling "Great Ogoni Man."

In Ken's absence, Badey, Leton, and Kobani convened several public meetings in an attempt to convince the Ogoni people to allow Willbros to resume laying Shell's pipeline.[50] Reportedly, Governor Ada George had literally gone down on his knees before these senior MOSOP steering committee members, offering lucrative compensation if they agreed to allow the pipeline to continue. With Ken away, MOSOP's old guard took the bait. Understandably, the Ogonis were mystified by MOSOP's new position. When questioned by the villagers, Leton, Kobani, and Badey could not explain their sudden about-face. The people ignored MOSOP and steadfastly refused to let Shell complete the pipeline or to let Willbros onto their land.[51] Not even the offer of sizeable compensation could persuade them otherwise.

The men from MOSOP suspected that Ken had been pulling strings from abroad to convince the people to remain intransigent, but they were mistaken. It is Ogoni tradition that where blood is spilt, the site of the violence must not be touched for a mandated period of time, after which special ceremonies and sacrifices are required before the bloodied ground may be used again. To the villagers, allowing Shell to lay a pipeline across the same farms was sacrilege. As for millions of naira in compensation, the prospect didn't impress the Ogonis at all. To Leton, Kobani, and Badey's public meetings, the villagers brought copies of their previous Shell agreements. They angrily listed the promises Shell had given in the past – for things such as water, electricity, compensation, cleanup – that the company vowed to do before drilling wells, laying pipelines, and carving roads. Few had been fulfilled.

The Ogonis didn't need Ken Saro-Wiwa to tell them they were being sold a bill of goods.

||||||

Ken returned from his European campaign ten days before the scheduled June 12 presidential election. While he had been away, many of the Ogoni elites and politicians had been making deals with the presidential candidates for preferential treatment should the candidate triumph. Kobani and Leton both confidently promised to deliver the lion's share of the Ogonis' votes to presidential candidate Moshood Abiola.

A meeting of the MOSOP steering committee was convened after Ken's return, and the topic of the election soon overshadowed all. Kobani stood up and angrily said to Ken, "What is this thing we see on 'The World's News Leader'?" referring to CNN's slogan. During Ken's TV interview a week earlier, he had displayed for the cameras dramatic photographs of the shooting of Mrs. Karalolo Korgbara at Biara and had announced that the Ogoni people would not be participating in the presidential election.

"You told the world we are not going to vote in the election. Who told you that?"

"Well," said Ken, "this is something we have been discussing since November. There has been a clear understanding that some clauses in the Nigerian constitution are inimical to the Ogoni struggle. Any president elected under this constitution will have to obey these clauses. On that basis, we cannot possibly participate in this election."

He admitted, however, that it had been imprudent to announce a MOSOP decision that had not been formally made. To make things official, Ken placed a motion before the meeting that MOSOP boycott the presidential election.[52] The motion was fiercely debated. Albert Badey gave a good analytical overview, arguing that the Ogonis' most prudent strategy would be to back a candidate and hope to be rewarded. Ken reiterated the counterargument. An election boycott, he said, was one of the key strategies employed by non-violent movements. When performed by a large number of people in a concentrated location, it sends a strong message to the state and to the international community. A boycott would also bring worldwide publicity. Said Ken, "For any newly elected president, it will also give MOSOP and the Ogoni people a legitimate thing to point to, to say, 'Look, we did not elect you. You are not our president, unless you look at what we have in our demands on the Ogoni Bill of Rights and our rights as a minority group.'" The boycott, he said, would also pressure Nigeria to sign the UN protocol on the rights of minorities.

Owens took the floor in defence of the boycott. "In the past few weeks, there have been several town hall meetings in the villages, at the kingdom levels, by the teachers' union, by the Council of Traditional Rulers, by FOWA, and they have all taken their own resolutions about not participating in the election. This is not a decision we are making for the Ogoni people. The Ogoni people have already made the decision. They are not going to vote."

The debate continued, becoming quite antagonistic, until Edward Kobani, who had much to lose in the proposed boycott, pressured Leton to put the matter to a vote.

The steering committee voted to boycott the election by a count of eleven to six. Kobani was so infuriated that he threatened to resign as vice-president if the vote stood.[53] No one, however, changed his vote.

The contentious meeting ended well past midnight. When it finally broke up, Ken did not go home. He drove directly to 24 Aggrey Road to draft a press release. News of MOSOP's decision hit the papers the next day.

Two days later, Kobani, Leton, and Badey came to Ken's office. With them were MOSOP treasurer Titus Nwieke and Engineer Apenu. The five had come to pressure Ken into annulling the steering committee's boycott decision.

"A democratic decision has been taken," Ken said. "It is our responsibility to carry it out."

The five flew into a rage, but Ken would not budge. Leton and Kobani informed Ken that they would be resigning as president and vice-president of MOSOP.

Ken shrugged. "It's up to you."[54]

CHAPTER**FIVE**

Along with a handful of other Nigerians, Ken Saro-Wiwa was scheduled to attend the UN World Conference on Human Rights in Vienna on June 11, 1993. Before he left, Ken stopped at several rallies in Ogoni to talk about the election boycott. Each place he spoke, he emphasized that there must be no violence on election day and that those who wished to vote should be allowed to do so.

Ken flew from Port Harcourt to Lagos to catch his flight to Europe. At the Lagos airport, however, the Nigerian State Secret Service seized his passport and forbade him to leave the country.[1] Ken was the only Nigerian barred by the sss from attending the UN conference. Travelling separately, Ledum Mitee and a handful of other Ogoni activists were able to fly to Vienna, where they presented a photographic exhibition of the environmental degradation of Ogoniland. On this trip, Ledum had the opportunity to make the acquaintance of Body Shop founder Anita Roddick – a meeting that ignited her organization's subsequent interest in the plight of the Ogoni people.

While grounded in Lagos, Ken got a call informing him that the radio station in Port Harcourt had been broadcasting an announcement – purportedly from Ken himself – that, contrary to the decision of the MOSOP steering committee, the Ogonis should definitely vote in the presidential election. Ken issued a quick denial to the media.

Meanwhile, Owens, on hearing the same announcement, went with other Ogonis to the radio station in Port Harcourt. When they demanded to see authorization for the fraudulent announcement, a station employee produced a letter on MOSOP stationery. Ken's supposed signature bore little resemblance to his actual handwriting.

"Who brought this to you?" asked Owens.

"Bennett Birabi," said the station employee.

It was true. Dr. Leton himself later told Ken (in the presence of witnesses, including Birabi and the inspector of police) that he, Kobani, and Birabi had drafted the announcement and that Birabi had forged Ken's signature.[2]

With radio no longer a reliable transmitter of information, Owens and the others mobilized the youth of NYCOP. Diana and her fellow Ogoni university students hired buses and a public address system and started going to all the villages in Ogoni, informing people that the broadcast was not true. Ken was *not* telling them to vote in the election.

Ken's recent tour of Ogoni, the contradictory radio announcement, and then the NYCOP retraction only served to confuse and agitate the people. Before Ken left for Lagos, he had left explicit instruction for the steering committee to spread the word that MOSOP would like the Ogonis *not* to vote. Most importantly, whether they voted or not, everything must be peaceful. Because foreign observers and international monitors were expected in Ogoniland, no one should stop any Ogoni person who wanted to vote. Allowing each Ogoni to choose whether to observe the boycott would provide a further gauge of MOSOP support.

On the morning of election day, June 12, Owens was in his clinic. With no surgeries scheduled, he had time to keep an eye on the polling booth across the street. Two Caucasian election monitors and several journalists, both black and white, had come to Bori to cover the election boycott. At 10 a.m., Owens stepped outside and was recognized by a European journalist who snared him for a quick interview.

"Doctor Wiwa! Why did MOSOP decide not to participate in the election?"

"It's not that we are anti-democracy," Owens answered. "It's just that the president who is elected will obey the constitution, which, we feel, is inimical to our survival. We're just making a statement."

Around two in the afternoon, Owens heard shouting and went back outside to witness a group of people rushing toward the clinic bearing a limp, bloody body. He heard gunshots and saw three vehicles speeding away. There wasn't time to ask questions before he was presented with a moaning eighteen-year-old male bleeding from the side of the head. Owens had the patient brought inside. The boy had suffered a fractured arm and a serious wound to the left temple. After asking another doctor on duty to attend to the patient, Owens went to find out what had happened.

A small crowd was awaiting news of the lad's condition. When Owens asked what had caused his injuries, they said he had been hit by a speeding police vehicle. The police were leading a convoy of cars and buses through Ogoniland and did not even stop when it struck the teenager. The officers fired their guns out the window to frighten off witnesses and kept driving.

Many bystanders marched to the police station in Bori to file a complaint. When the Bori police barred the station doors, the townspeople battered them open. The police later claimed that when the Bori residents broke through, they looted the station of guns and ammunition. (When asked to provide an inventory of what had been stolen, the Bori police failed to comply or to offer any evidence to back up their accusations.)

The speeding police officers were escorting vehicles carrying the Andoni territory's West African School Examinations along with several ballot boxes purportedly stuffed with fraudulent Ogoni votes for the NPN presidential candidate. Despite the presence of international monitors, election rigging all across Nigeria on June 12 was pervasive. When it became apparent to the two dominant parties that the Ogonis were not going to vote, both sides took this to be a glorious opportunity for ballot box stuffing. It was reported that when the SDP heard that the NPN had beaten them to the punch, the SDP sent out thugs to prevent the NPN's ballot boxes from reaching the Electoral Commission. During clashes between supporters of the SDP and NPN, several vehicles carrying rigged ballot boxes were waylaid and the boxes' contents burned. (Unfortunately for the Andoni schoolchildren, their unmarked West African School Examination papers perished in the flames as well.) In some locations, Ogoni youths destroyed both false and legitimate election documents.[3]

In Yeghe, the hometown of Bennett Birabi, the people had resolved not to vote. Birabi called a meeting in the town square and berated them. The people, however, were not to be cowed by Birabi. Offended and angered, they asked why he had called this gathering since he hadn't bothered to speak to them since the last Senate election. And how dare he reprimand them for not voting in an election from which he would benefit and they would not? The meeting turned into a public disgrace for the Senate minority leader. Many in the crowd wondered if Birabi might be doing a little ballot box stuffing in his own house. When some of Birabi's own relations stormed the Birabi compound at Yeghe to search the premises, a fight ensued, and one of Birabi's aides sustained a minor injury.[4]

Senator Birabi wasted no time in publicly declaring that the assault was the work of Ken Saro-Wiwa, who had ordered NYCOP thugs to kill him. (Ken was a thousand kilometres away[5] in Lagos and there was no phone service in Yeghe.) Birabi's wife, Bridget, joined her husband in spreading the word that Ken wanted his one-time ward dead. The day after the election fracas at Yeghe, she sent the following letter to Ken:

> *Dede: I'm sure you'll be surprised to receive a letter from me just as I was surprised to realise that you remotely could threaten Bennett's life with the squad of hoodlums specially purposed to deal with Bennett in Yeghe. It is so embarrassing that the demon of ambition can take over situations such that life means nothing anymore. I had always thought that whatever differences in ideologies that exist between you and Bennett could be settled in a better way (at least by discussion and agreeing at some points) other than a very serious threat on Bennett's life. . . . You have never supported Bennett's political career, you know that I know at least you told me in 1983. This was supported by your non-chalant attitude towards all what we passed through during the senatorial elections of 1992. It had kept me wondering all the while that for who you've supposed to be to Bennett, for you to have been so unconcerned! Well. And now this threat on Bennett's life by your squad (MOSOP/NYCOP). . . .*
>
> *Now that your military people dealt with him on Saturday the 12th June 1993 which is in line with various death wishes, I really wonder. I've not been impressed by such wishful thinking but when it gets to a stage where Bennett's life is threatened in his own house right in Yeghe, then one must look at it seriously. I am on my way on Tuesday the 15th June to alert my people – my parents people the Asaba and the Ife-Ife people on this dangerous development, in case anything happens to Bennett. Sincerely, Bridget Birabi[6]*

Mrs. Birabi took pains to make sure copies of the letter were widely circulated.

Whether the Birabis were spreading these charges to deflect attention from Bennett's own disgrace on election day or whether they believed them is unknown. In any case, it was a major accusation that Ken Saro-Wiwa was inciting the Ogonis to violence.

||||||

The Nigerian papers were full of stories the next morning about the Ogoni boycott. CNN and the BBC also ran reports of how the Ogonis had banded together to make a statement to the world. The Ogonis had shown that they would no longer accept subjugation and exploitation and Ken couldn't have been happier. His only regret was over the actions of some Ogoni youths who stopped the movement of a few ballot boxes and harassed local politicians – including Edward Kobani, who was afraid to leave his house during the voting period.[7] The government-owned press blamed NYCOP for the altercations on election day, including the incident at Senator Birabi's house. The press also carried accounts of how NYCOP youths had ransacked the police station in Bori and stolen guns.

Ken returned to Port Harcourt two days later to find Birabi being interviewed on television saying that Ken had sent thugs to his house to kill him. As if that were not enough, the Ogoni elders had called a meeting to threaten Ken with "fire and brimstone" (as Ken put it). Naturally, Ken planned to attend, in hopes he could mollify the enraged chiefs.

Owens said to Ken, "Do not go to this meeting. Please."

"I have to go," said Ken. "I have to explain to them. These people don't understand. I'm sure that by the time I finish saying what I need to say, they will be convinced."

"No, don't go," insisted his brother. "These people already have something in mind. I heard that they have prepared a document where they will relieve you of your responsibilities as spokesman of the Ogonis and ask you not to make any further statement on the issue of autonomy or the environment."

"We'll see about that."

When Owens saw that he could not deter his brother, he decided to go with him. So did several other pro-MOSOP Ogonis.

When Owens arrived at the June 21 meeting, one of Ken's supporters approached him. "I don't think Ken will be coming," he said.

"Why not?" asked Owens.

"Somebody told me he has been arrested."

Owens drove to 24 Aggrey Road. Ken's staff confirmed that he'd left the office to attend the meeting with the elites. Hadn't he arrived?

"No. I've heard he's been arrested," said Owens. "Maybe I'll check the police station."

The Port Harcourt Central Police Station was near the University of Port Harcourt Teaching Hospital, where Owens had done his medical internship. By the time he got to the station, it was five in the evening and a crowd of Ogonis had gathered outside. Inside, Owens recognized a policeman he'd known while working at the Teaching Hospital.

"I want to see Ken. He's my brother."

The policeman led Owens upstairs to a large private office. Ken was seated on a bench opposite a senior police officer who sat behind a large desk, placidly chewing on kola nuts.[8] He was a Sokoto native, a tall, thin man in flowing cotton robes. Ken introduced Owens as his medical doctor and requested privacy. The officer withdrew, his robes dragging along the dirty floor.

Ken told Owens that he had been at 24 Aggrey Road that afternoon with some Ogoni youths, discussing the election. As they talked, five men in mufti barged in.

"I recognized one of them as a security agent," said Ken. "I greeted him and he said that I was being summoned by his boss, the state director of the sss."

Ken told the security agent that he was not available and said that if this was an arrest, there had better be a warrant. The agent said it wasn't an arrest, merely an invitation.

"I said, 'I'm sorry I can't come with you,'" Ken told his brother.

Ken said the security men left, only to return fifteen minutes later with an arrest warrant. Ken, however, was not impressed and told the men that he was late for a meeting and had to leave. He estimated that the meeting would take about four hours. "I told them, 'I'll call in at your office later today. Say, six o'clock. Only as a matter of courtesy.'"

The security men left, but then waylaid Ken a few minutes later in traffic, pointing a rifle at his chauffeur's head and ordering the frightened man to make a U-turn and head for the offices of the State Security Service. After some paperwork, the security agents transferred Ken to the Central Police Station.

"Have they told you anything?"

"No," said Ken. "They've just kept me here, and I've been chatting with anybody who comes in."

"Have you eaten?"

"Not a thing all day. Can you arrange to get me something from my place?"

"I'll buy you some food," Owens said. "I'll be right back."

On his way downstairs, Owens spoke with the assistant superintendent of police. The superintendent said that some people had come from Lagos and indicated that Ken would not remain long in Port Harcourt, and that they were planning to fly him to some distant location.

Owens went to one of the fast-food shanties across from the police station and bought Ken a meat pie and a couple of sugary drinks. He also picked up some toiletries and a refill of Ken's heart medication prescription from the Teaching Hospital pharmacy.

When he returned, he told his brother, "Someone at the reception advised me to give you enough medicine to last a long time."[9]

He and Ken laughed and then Owens reached into his pocket. Ken had climbed to that lofty social sphere where one didn't carry cash. Owens, however, was carrying quite a sum, as it was his day to make the clinic bank deposit. He handed Ken five thousand naira, which Ken politely refused.

"You've got to take it, Dede," said Owens. "I've heard that they're going to take you somewhere and nobody knows if they will give you food or you might need money."

Ken nodded and peeled off two thousand naira and handed the rest back. "I understand I'm to be taken to Lagos. There's no aircraft tonight, so it's likely to be tomorrow."

"I got some snacks on my way here," said Owens, offering Ken the drinks and the meat pie. Ken wouldn't touch the pie. A fanatic for hygiene, he was leery of the quality of Nigeria's mass-produced junk food and refused to eat even a bite.

Ken sent Owens down the hall to phone one of his lawyers. Once Owens had done so, the police ordered him to leave the building. As he stepped outside, Owens saw that the crowd had increased greatly and was becoming boisterous. The next thing he knew, the police began tear-gassing the crowd, creating pandemonium. Some protesters picked up the canisters of tear gas before they exploded and threw them back at the police. In their anger over the arrest of Ken Saro-Wiwa, the Ogonis knocked down part of the brick wall that separated the police station from the rest of the town.[10]

Owens tried to calm the Ogonis, but they would not be mollified and accused him of having taken bribes from the police. Owens had no choice but to leave them to their rioting and try once more to see his brother. Back inside, he asked again for Ken.

"That's him that's going."

In the chaos, Owens had failed to see the police whisking Ken into a waiting passenger bus. He dashed out, jumped into his car, and tried to catch the bus but the crowd blocked the way. By the time he got on the road, the bus had disappeared into the night.

Owens drove to all the police stations in Port Harcourt but Ken was nowhere to be found. He stopped off at the house of an old friend from Jaji, a naval commander who lived on the Port Harcourt naval base. The commander was also a cabinet member of the State Security Council so Owens thought he might have some answers.

The commander greeted his old friend warmly. Owens asked what was happening with his brother.

"I don't know," said the commander. "The orders appear to be coming from Lagos. They have been talking about it in the Security Council, something about a flag, you know, getting a flag."

"But they have come for him for that flag before. He has given it to them. They detained him and released him later. Is this about the flag again? Or is this about the Treason and Treasonable Offenses Decree?" The commander didn't know.

Owens went from the naval base to the home of a relative who was a policeman in Port Harcourt. He had no news. Owens returned to 24 Aggrey Road to see if the police had brought Ken there. Then back to the Central Police Station. Nothing.

Owens next tried the sss building, where the courtyard was jammed with an angry mob. By now, all the Ogonis in Port Harcourt, especially those in the waterside shanty towns called the Down Below, had heard of Ken's arrest. Along Aggrey Road, they had made flaming roadblocks of tires, stopping traffic around the police stations.

At one in the morning, Owens gave up his search and drove to Bane to tell his father and mother what had happened. Then he went to his home in Bori, packed a small suitcase, and returned to Port Harcourt to catch the first morning flight to Lagos.

|||||||

Next day, in Lagos, Owens went straight to his brother Letam's bungalow on the army base known as Dodan Barracks. Dodan Barracks also contained the official residence of the military dictator; Letam was in the presidential guard. By checking with his contacts, Letam learned that Ken was indeed in Lagos. He was being held only three hundred metres away at Alagbon Close, headquarters of Nigeria's Federal Investigation and Intelligence Bureau. Alagbon Close was notorious as the detention spot for activists and opposition leaders, and well-known for its chilling inhumanity.

Owens and Letam went to that station. In the reception area, they asked where they could find Ken; they could hear that several new "customers" were being beaten in the adjoining rooms. The laments of the prisoners sent shivers through the brothers. The officer on duty led them behind the station to a rickety wooden building that contained several woebegone offices.[11] Ken had been placed in one of the more dreadful rooms of that hovel, a space filled with cobwebs, dust, and broken furniture. The harsh light from a bare bulb emphasized the room's squalor.

Ken was being questioned by a policeman, although who was interrogating whom was not immediately clear.

"Where are you from?" Ken was asking the policeman.

"Ogoja," he said, another minority area in the delta. Ken rattled off the people he knew from Ogoja either from secondary school or university, or the Biafran War. His litany of Ogoja friends segued into a sermon on minority rights. The policeman listened and took notes.

Ken introduced Owens. "That's my brother. He's a physician. He's come all the way from Port Harcourt. People from Port Harcourt know that I'm here now."

This led Ken onto another diatribe about why the SSS felt it had to transport him by road to Lagos – a twelve-hour bus trip over dangerous and bumpy roads. If the SSS had let him, he said, he would have gladly purchased plane tickets for everybody. The policeman thanked Ken for his time and left the room.

Owens could hear other detainees screaming and pleading. Some were shouting in fear, some in pain.

"Were you manhandled?" Owens asked Ken.

"I was not beaten. The inconvenience of the trip here was torture enough. I'm feeling some pains all over. Throughout the journey I didn't sleep a wink."

Owens had thought to bring with him Ken's briefcase and shoulder bag, which he presented to his brother. Ken thanked him for being so thoughtful. He told Owens that other visitors had come by that evening, one being Dr. Olu Onagoruwa, a prominent human rights lawyer. Onagoruwa had told Ken that the police in Lagos had not yet received instruction as to what to do with him, so Ken would have to stay in custody overnight. Onagoruwa would file for Ken's release in the morning. When Ken balked at the idea of spending a night at Alagbon Close, Onagoruwa pleaded with the assistant commissioner of police for special consideration, which the commissioner granted – hence the unusual place in which Owens and Letam found themselves. To make the place halfway humane, the younger brothers set about dusting, sweeping, and clearing away cobwebs. The fastidious Ken was deeply grateful.

Ken also received a visit from his Lagos mistress, Hauwa Madugu. She brought some of Ken's clothes, a few books, and a transistor radio. As the rain fell in sheets over Lagos, they sat together as a family in the dismal little room. Ken put on a brave front and told Owens and Letam to be of good cheer. They vowed to keep Ken company through the night but Ken said no. He preferred that they return in the morning. He would have things for them to do, he said, so they needed to be rested.

||||||

With the morning light came news of possible salvation for Ken. CNN was reporting that the Nigerian presidential election results had been nullified. Although international observers had judged the election fair – Moshood Abiola had won the presidency – General Babangida deemed the process corrupt and refused to recognize the results.[12,13] By his decree, the election and all actions taken with regard to the election were declared null and void. Therefore, it was moot whether the Ogoni people had voted or not. And if Ken had been arrested for "an action taken in regard of the election," the nullification of the election also nullified the reason for his arrest.

In the morning, after securing Ken's legal representation, Owens went to the Saros International offices in the Surulere section of Lagos. Owens knew that Ken, if free, would be sending off a flurry of press releases. Owens had never written a press release, but after having read so many of Ken's, he found he was adept at mimicking the tone and style. When finished, he asked Ken's office staff to fax copies to the Nigerian media, CNN, the BBC, UNPO, Amnesty International, Greenpeace, and several other organizations listed in Ken's address book.

The staff said they wouldn't do it. "They can trace the fax."

"Who is going to trace it?" asked Owens.

"The police and the SSS have a way of knowing when a fax is sent abroad. If we send the fax, they will come here and arrest us."

"I'll give you some money to go to a business centre," said Owens.

"No, at the business centre there are spies."

There was no reasoning with the staff. They were terrified of being arrested for associating with Ken. Finally, one Saros office worker declared that the only way the faxes could be safely sent was to leave the country and do the faxing from the neighbouring Republic of Benin. He asked Owens to give him the money to make the trip. But Owens couldn't. He had spent nearly all of his clinic's bank deposit. To boost his resources, he sent word to his clinic to do whatever was necessary to assemble another bundle of cash and to give it to his fiancée, Diana, for her to bring to Lagos.

The following day, Owens borrowed Letam's car to pick Diana up at the airport. As she emerged from the terminal, Owens was both enraptured and chagrined at her appearance. Of course she was lovely. Her clothing, however, was another matter. Diana was a university student and she dressed the part in faded denims and T-shirts. In the delta, the clothing she wore hadn't mattered much, but here in the big city Owens felt she should look more adult. In the car, driving back into Lagos, he broached the subject.

"Diana, I am going to take you to meet a friend of mine. She can take you to the shops, if you get what I mean."

"No, I don't get what you mean."

"You have to buy some new clothes. You dress like a hippie."

"You don't like the way I look?"

"I just want to you look a little more presentable."

"*Presentable?*"

Before they could quarrel in earnest, Owens and Diana found themselves in one of Lagos's omnipresent traffic jams. In Lagos, impossible congestion is not surprising, but something about this traffic jam was different. Something seemed wrong.

All at once, a wave of howling, shirtless men engulfed the idling cars. Owens shouted to Diana to roll up the windows seconds before stones rained down on the automobiles. People were running in all directions.

"What's going on?" Diana cried.

"It's a riot! It must be a protest about Abiola and the cancellation of the election."

Rioters swarmed the car and pounded on the windows and doors, screaming at Owens and his terrified fiancée. Diana shrieked as the mob forced open her door and grabbed her by the arms. Owens reached over and clutched her around the waist, trying to prevent her from being dragged outside to her death. Luckily, the hooligans were satisfied with merely stealing her watch. After ripping it from her wrist, they let her go. Owens pulled the door shut and locked it.

As quickly as the wave swept over them, it moved on. Owens tried to calm Diana as she wept uncontrollably. "You are an activist, a student leader of your school and in MOSOP, and you are crying because . . ."

She waved him away. The stolen watch, he realized, was a present he had given her on her last birthday.

"Well!" he said with a smile. "Now you see what it is to be on the other side, when you are not the one doing the demonstration."

|||||||

By the time Owens and Diana returned Letam's battered car to Dodan Barracks, Ken had been transported to the Imo State Police Command in Owerri, a two-hour drive from Port Harcourt. Others had been apprehended and joined Ken in detention: MOSOP steering committee member N.G. Dube and NYCOP local village chairman Kabari Nwinee.[14] The police had relocated Ken close to the delta due to the intense local reaction to his arrest and the desire to keep the unrest from spreading.

In Port Harcourt, the Ogoni waterside shantytown residents – destitute, pitiable people who looked on Ken as their salvation – had erupted into mayhem. Seeing that the situation was out of control, MOSOP members went to Bane to fetch Pa Wiwa in the hope that Ken's elderly father might be able to calm the raging crowds. Pa came to the city and toured the pockets of Ogoni settlements down in the fetid Port Harcourt swamps – Okrika Waterside, Nembo Waterside, and Creek Road Market Waterside – speaking to the poorest of the poor, spreading the message that Ken wanted them to behave civilly. Ken's father had a profound influence on the Port Harcourt Ogonis and hastened the end of their confrontation.

In Ogoniland, meanwhile, thousands of angry youths descended on the villages of the eleven so-called vultures, the elites who had presented the apology to the governor after the Willbros incident shootings. The youths went on a rampage at the elites' homes, beating down doors and breaking windows. Priscilla Vikue, director-general of the Ministry of Education in Port Harcourt, was one of those branded a vulture. The mob descended on her house in the village of Bodo and demanded that she resign her government appointment. She refused. Vikue and several other elites later claimed that their houses were burned to the ground.[15] The claim, however, was false – throughout this period, there is no evidence that any Ogoni house was burnt as a result of Ken's arrest. What is certain, though, is that within a matter of hours, all the "vultures" had fled to Port Harcourt.

In reaction, the federal government dispatched soldiers to Bori, where they began a campaign of indiscriminate beatings and arrests.[16] When word of the tumult reached Ken, he became greatly disturbed and drafted a message to the Ogoni people, which he sent via MOSOP activist Noble Obani-Nwibari. Ken implored the Ogonis to cease all lawless activity. If the people were outraged by his arrest, he suggested they engage in sit-ins, letter-writing campaigns, and other forms of non-violent protest. Gradually, things calmed down in Ogoniland.

||||||

On June 25, Owens arrived at the dank, two-storey State Police Command building in Owerri. At the front desk, he could hear men intoning, "Praise God, hallelujah!" over and over in their despair.

"That's the guardroom," said the desk officer, nodding in the direction of the prisoners' cries.

"My brother's in there?"

"Not quite. Follow me." The officer led Owens upstairs to a sort of grimy waiting room that had been transformed into an impromptu cell. The room contained two dirty settees, an electric fan, and a bulbless light socket. Owens was shocked by Ken's physical deterioration and the ungodly stench of human waste.

"Oh my God, Dede. They are keeping you here?"

"It's an improvement, believe me, over where we have been."

On their arrival on June 23, Ken, Dube, and Nwinee had been placed in a dark chamber next to the guardroom. There, the stink had been eye-watering and the toilet facilities beyond description. After two days in this cesspit, Ken verbally assailed his captors about the black man's inhumanity to his own kind. An hour later, he, Dube, and Nwinee were moved to this comparatively plush pigsty.

These conditions had taken a toll on Ken's health. In addition to the heart problems he already had, Ken was suffering from painfully swollen feet, diarrhoea, and boils. After his constant complaints, a young doctor had been brought in, but his only gesture had been to provide Ken with a prescription that Ken had no way of filling.

Owens offered to go out and pick up Ken's medication. As he came out of the police station, he noticed Dr. Garrick Leton talking to one of Ken's arresting officers. Leton looked up and pointed at Owens: "That is Dr. Wiwa. That's him, there. That's Dr. Wiwa."

The policeman motioned for Owens to approach. "You are Dr. Wiwa?"

"Yes."

"Do you know that I can arrest you?"

Owens looked at the officer in contemptuous disbelief. "For what? I'm visiting my brother who's being kept here. And I'm trying to get medicine for him to be comfortable. Why would you arrest me?"

Brushing the men off, Owens went to find the nearest pharmacy, dismissing the encounter from his mind. Later, he learned that at the time of Ken's arrest there was a warrant out for twenty-two Ogoni individuals. After Ken, Owens was the next most-wanted. Why he was allowed to remain free was a mystery.

When Owens returned with the medicine, Ken's Port Harcourt mistress, Elfreda Jumbo, arrived with a bucket, brushes, rags, detergent, and disinfectant. She began vigorously scrubbing the little waiting room and the adjoining senior police officer's toilet, clearing away decades of cobwebs, dried spit, urine, and caked-on offal. Ken was beside himself with joy.

Over the next few days, Owens spent most of his time with Ken in the detention room, attending to his brother's health.[17] He also got in contact with Ken's lawyer, who sent one of his staff to Owerri to petition the court for Ken's release. No judge, however, was willing to sign the writs.

As for Ken, he spent his time writing and musing on the Ogoni situation. Among the works he produced at Owerri were poems about prison life and a gripping account of his detention, a manuscript he later entitled *A Month and a Day: A Detention Diary*.

|||||||

On June 30, news reached Ken that Amnesty International had issued a "Fast Action" bulletin concerning his detention. He also learned that his good friend, the British journalist William Boyd, had been writing about Ken's predicament for the London *Times*. Nigerian journalists, too, were doing their best to keep Ken on the front pages. Almost daily, reporters came to Ken's cell requesting interviews.

Most of Ken's visitors, however, were everyday Ogonis. A constant stream of friends, relatives, and well-wishers stopped by to buoy his spirits and offer solidarity. Word was circulating that the Ogoni elites were no longer comfortable returning to Ogoniland because protesters were constantly appearing at their homes, demanding that the elites publicly call for Ken's release. The news was also spreading that Ken had been interrogated over the affair in Yeghe in which Bennett Birabi's house had been violated, and it was rumoured that Ken's arrest was due to a petition drafted by Bennett Birabi. So incensed were the Ogonis by this rumour that they gave notice to Senator Birabi that he should not dare to come home to Yeghe unless he had undone whatever he had done to cause Ken to be arrested. Birabi's state of mind during this period can be clearly seen in a July 2 letter he sent to an Ogoni village chief:

> *Your Highness . . . I am going to Owerri today to see Kenule. I've been negotiating his release since the day he was arrested. I have almost succeeded but he needs to cooperate. So I am taking Dr. Leton & Mr. Idemyor to see him to sign a paper for us. I* did not *write any petition against anybody. The order for his arrest came from the Presidency. So don't be afraid. God is with me. . . .*

Birabi closed the letter with this dire postscript: *"Kenule wants to kill me and destroy my father's name. But let God's will be done."*[18]

When former MOSOP vice-president Edward Kobani visited Ken in jail, he smugly reminded him how Ken had once said that the revolution would invariably claim its victims – implying that Ken had been caught in his own trap. Kobani said, "Right now you have the knife, but the knife is sharped on two sides and can hurt both ways."

Ken glowered. "I don't know what you're talking about. I have embarked on a non-violent struggle. I urge that we all go on with what we started peacefully together."

Kobani lashed out at Ken, accusing him of using the hot-headed youths of NYCOP as his own personal army of thugs. Because of Kobani's fear of reprisals from these violent young people, his family was now no longer free to return to their home in Bodo.

"You are no longer free?" Ken bitterly replied. "Here I am, unable to take a walk outside my cell without an armed guard following me, 150 kilometres from my hometown, unsure what the next day will bring, with a draconian law prescribing death hanging round my neck. And I'm being asked to feel pity for a man who is free to do everything else but face the anger of young men whose future he has mortgaged?"[19]

Ken reminded Kobani that he, Birabi, and the others had placed the false radio announcement in Ken's name and, with that one selfish and illegal act, ruined Ken's plans for a peaceful election process. "It was so unnecessary!" he railed. "All those in Ogoni who wanted to vote were to be allowed to do so!"

Kobani protested that he had taken no part in the scheme, but Ken would hear none of it. Pleading ill health, Ken abruptly ended the visit.

|||||||

With Ken's arrest following hard on the resignations of Kobani and Leton, MOSOP was in disarray. Although Ledum Mitee had represented MOSOP at the UN conference in Vienna, little else had been accomplished. The steering committee had stopped meeting, and this concerned Ken greatly. In early July, he asked MOSOP secretary Dr. Ben Naanen, a young historian from the University of Port Harcourt, to call a steering committee meeting.

On July 6, 1993, the steering committee met to elect new interim leaders. Ken's name was put forward as president. Owens, alarmed for his brother's safety, said, "I would prefer he continues to be the spokesman of the Ogoni people. You should have another president. Why don't you put Ledum's name as the president?"

At the time, Ledum Mitee was still in Europe. His elder brother, Fegalo, said it was obvious Ken Saro-Wiwa should be president. There followed a spirited debate between the two, each arguing why his own brother should not become the next leader of MOSOP.

After the meeting, Owens drove to Owerri with Diana to see Ken. "They nominated you for president," Owens told him. "I didn't like it. I thought that you being the spokesman of the Ogoni people was more important than the presidency of MOSOP."

Ken urged Owens to get to the point. Who had been elected? Owens reported that six new vice-presidents were chosen (one from each Ogoni kingdom); Ledum Mitee was elected in absentia to the new post of deputy president; Father Kabari, a Catholic priest, assumed the role of treasurer; Ben Naanen remained secretary; and, in absentia, Ken Saro-Wiwa had been elected the new president of MOSOP.[20]

|||||||

Ken's health continued to deteriorate. The antibiotics and analgesics the police doctor prescribed showed no effect. (By some estimates, at that time 70 per cent of prescription medication in Nigeria was adulterated or diluted, or was simply talcum.)

During Ken's incarceration, Owens was in constant contact with a physician named Dr. Idoko,[21] a pleasant young man who was currently doing his Youth Corps service with the police. Striking up a friendship with the amiable Idoko, Owens asked if it would be possible for Ken to be moved to a hospital. Dr. Idoko offered to make the formal recommendation. The commissioner of police was less obliging. It took several days of lobbying before the police accepted the doctor's direction that Ken be hospitalized.

On July 9, Ken was transferred to the new police clinic in Owerri, about a hundred metres from the building in which he had been detained.[22] He was, in fact, the first patient ever admitted to this clinic. Although the new medical facility was sloppily constructed and poorly outfitted, Owens was greatly relieved that his brother would at least have a clean place to sleep.

Ken's Lagos lawyer, Olu Onagoruwa, came to Owerri to petition once more for Ken's release. This time, a judge ordered Ken to be produced in court on July 13. On the designated day, a festive contingent of eight hundred Ogonis gathered outside the Owerri courthouse in expectation of Ken's release and triumphant homecoming.

Inside the courtroom, the hearing was set to begin. Everyone was in place except for one person: Ken.

The judge demanded that the police produce him at once. The police, however, could not comply. Ken, they said, was no longer in Owerri. He was now an inmate at Port Harcourt Prison. At five that morning, mobile police had wrenched him from his hospital bed and taken him in shackles back to Port Harcourt to appear in another court. Without legal representation, Ken was hauled before a judge and charged with six counts of unlawful assembly and conspiring to publish a seditious pamphlet.[23] The judge refused bail and ordered Ken committed to prison, adjourning the case until September 21.[24]

Owens made the two-hour drive to Port Harcourt. He arrived at the grey fortress of Port Harcourt Prison about four in the afternoon and demanded to see Ken.

"I'm a physician and Ken is my brother," he said. "I have the drugs he was taking in Owerri. I need to give them to him." After checking Owens's

credentials, prison officials searched him and made him sign a stack of documents before they opened the massive gates. The guards led Owens to the
infirmary on the right side of the prison yard. He found his brother seated
beside Dube and Nwinee on a cot with a rotting mattress.

The infirmary was damp and leaking, with no ceiling to cover the
underside of the corrugated tin roof. Owens noticed that the other inmates
had moved their own narrow beds closer together to clear more space for
Ken. In honour of his presence, they had even attempted to tidy the filthy
ward. Ken was obviously revered at Port Harcourt Prison – easy to understand, given that a third of the 1,200 inmates were Ogoni.

Once again, Ken had to suffer the indignity of the toilet facilities, or lack
thereof. In the entire prison, there was but one flush toilet and that was for
the sole use of the prison warden. Even in the infirmary, inmates and staff
were required to urinate into a dirty container whose contents were poured
through a hole in the back wall of the prison. Defecation was accomplished
with latrine buckets passed around – a prospect that particularly distressed
Ken and his delicate sensibilities.

Owens gave Ken his medication and asked what else he could do. Ken
had a long list of errands. Among other things, he asked Owens to get him
a transistor radio to replace the one the Owerri police had confiscated. He
also wanted copies of all the latest newspapers and asked Owens to find
copies of William Boyd's articles on him from the London *Times* as well as
Tony Daniels's piece on his arrest in the *Times Literary Supplement*.[25]

In Port Harcourt, Owens contacted Ken's personal physician, Dr. Bobo
Ibiama, a cardiologist, and Ibiama came to see Ken in prison. After listening
to Ken's heart and taking his blood pressure, Ibiama announced to the
prison guards, "I will need to do some more tests. But from what I see, this
man should be in a hospital bed receiving twenty-four-hour care."

As in Owerri, convincing the authorities to release Ken to a hospital
was not simple. When Bayo Balogun, the commissioner of police, was contacted, he said the transfer would require an order from a high court judge.
The following morning, Owens obtained the high court order, but when he
presented it to Balogun, the commissioner threw the document on the
floor in contempt.

Two days later, Ken demanded an audience with the prison warden.
There, before most of the warden's senior staff, Ken virtually called the man

an assassin. "I cannot understand why you have refused to send me to hospital in spite of a consultant's order," he said. "The prison authorities will have a lot of accounting to do to my family and to the Ogoni people if anything should happen to me because you have been dithering over who is to sign what paper, or because you do not know what to do about simple matters of life and death."[26] Going further, he warned them, "If I die here, my blood will be on your head!"

Ken's theatrics worked. Six hours later he was informed that he would be admitted to the University of Port Harcourt Teaching Hospital. As the prison had no transportation – neither bus, nor car, nor ambulance – Ken's own office provided a car to take him the few blocks to the hospital. Many of the nurses on duty were the same women who had worked with Owens when he was an intern; Ken was given a prime spot in a secluded section of the hospital, tucked under crisp white sheets and comfortably sedated.

|||||||

Fifteen minutes after Ken left Port Harcourt Prison, a judicial order arrived demanding that he, Dube, and Nwinee be transferred to separate prisons elsewhere in the country. At midnight, armed security forces marched into Port Harcourt Teaching Hospital and awoke Ken from a drugged sleep. "Get up," they told him. "We are taking you." Ken groggily protested, but the police were insistent.

As they were hauling Ken from his bed, the ward matron appeared with several colleagues. "Sorry," the matron said. "You can't take away the patient without the specific instructions of the chief medical director."

"I have my orders, madam."

"And I have my orders, too!"[27]

A shouting match ensued. Forming a wall with their bodies around Ken's bed, the matrons and nurses told the police that Ken was staying put. The police demanded to speak to the head of the hospital, but the hospital director took the side of his nurses. The police had no choice but to withdraw.

The next morning, a letter arrived informing Ken he had been granted "bail" and was set free. The judge in Owerri – the one who presided over the trial that Ken had missed – found that Ken had been illegally arrested and ruled that he be released at once and paid compensation.

In truth, many people and international organizations had played a part in securing Ken's release: Amnesty International, which adopted Ken, Dube, and Nwinee as Prisoners of Conscience; PEN International's Committee for Writers in Prison; UNPO; Greenpeace; the UN Working Group for Indigenous People; the Association of Nigerian Authors; the BBC; William Boyd; and, from London, Ken Wiwa, Jr.[28]

When Owens came into the hospital room, Ken told his brother all that had transpired in the night.

"Ah, so you can go now," said Owens, greatly relieved. "That is wonderful."

"No," said Ken. "I am not moving out of this bed until all the confusion in this country has died down. Now hand me my transistor radio."

CHAPTER SIX

It was common for many Ogoni fishermen to migrate to Cameroon early each year and return in midsummer with income earned during the voyage and any unsold catch. On July 10, 1993, while doing clinical rounds, Owens had received disturbing news about some of these migrants. A story was circulating that a group of Ogonis making their way home along the Andoni River had been massacred. Nearly all the victims were said to be fishermen from the Gokana waterside village of Bodo.

Owens learned that a battered and frightened young woman carrying infant twins had stumbled, weeping, into Bodo. At the house of the village chief, she reported that she had been returning home in one of the fishing boats with her husband when they were set on by uniformed men wielding automatic weapons[1] who slaughtered everyone except her and her two children. She had no idea who the men were, where they were from, or what language they spoke.

The village chief reported the mass murder to authorities in Kpor, and the local police sent two officers to investigate. The police returned and said that when they reached the place the woman described, they were accosted and then pursued by "bandits" in speedboats. They said their attackers seemed to come from the villages on the Andoni shore but the officers couldn't be sure. They did not find evidence of a mass killing.

Bayo Balogun, the Rivers State commissioner of police in Port Harcourt, claimed that no one had died. "I have not seen any corpse," Balogun said in a press interview. "If somebody has died, I would see the corpse. There is no problem in the area."[2]

Yet 136 people were missing.

If something had happened to them, Commissioner Balogun suggested, it was probably the result of an "inter-communal crisis" between the Ogonis and the Andonis – a possibility that frightened many in Ogoniland. The Ogonis and the Andonis had not always been the best of neighbours. Back in 1922, a bloody clan war had broken out between them over fishing settlements along common creeks. To stop the fighting, the colonial government assigned tribal boundaries, but neither side respected the new borders. Another conflict broke out in 1937 and continued for fourteen years. In 1972, the Ogonis and Andonis went to war again over the ownership of some islands.[3]

Despite their fractious history, the Ogonis and the Andonis relied on one another economically. The Andonis, as well as several other delta tribes, frequented the Ogoni waterside village of Kaa to buy fruits and vegetables from Ogoni farmers and to pick up manufactured goods such as soap and clothing from the Igbo and Yoruba merchants. The Andonis themselves kept many prosperous stalls in the Kaa marketplace, so it was hard for the Ogonis to imagine what could have driven their neighbours to commit such an atrocity.

|||||||

Some nights, after visiting Ken in hospital, rather than make the long drive back to Bori, Owens would stay with a friend in Port Harcourt. As commander of the Port Harcourt Naval Base it was not unusual for Owens's friend to receive oil company executives at his home. One morning before Owens left, executives from Chevron came to see the commander. Chevron wanted navy personnel to accompany its boats into a particularly contentious area of the delta. The commander let them use armed sailors from the base.

Owens thought it ominous that Chevron would be enlisting the aid of the Nigerian navy, but what he learned next was even more discomfiting. Three weeks after the loss of the Bodo fishermen, one of Owens's patients, an Ogoni policeman, came to Owens's Bori clinic (recently re-named the Inadum Medical Centre) to tell him that he would not be able to continue receiving treatment for tuberculosis. Owens had been giving the man daily injections of streptomycin. He was in only the second month of a year-long TB regimen; it was important that he finish the full course of antibiotics. When Owens asked why he was stopping, the policeman said he was being transferred.

"Well, can't you talk to your boss? I could give you a medical paper saying that you can't leave."

"No," said the policeman. "All Ogonis in the police in Bori and other places in Ogoni have been transferred out." He showed Owens the transfer papers.

It was true. The state police had ordered all ethnic Ogoni members on the force to leave Ogoniland.[4] No justification was offered for this action.

Owens took a copy of the transfer papers to Ken in hospital. "You see?" said Ken. "The transfer of Ogoni security people from Ogoni – this is how genocide begins." In longhand, he wrote a press release and gave it to Owens to take to 24 Aggrey Road to be typed and faxed to UNPO and the media.

On August 5, 1993, five days after the transfer of all ethnic Ogoni policemen, Owens left the naval commander's house in Port Harcourt to return to his clinic. Although it was barely dawn, he found the medical centre in Bori filled with people screaming and dripping blood.

"What is this?" he exclaimed.

"Soldiers come shoot us!"

"Soldiers? Where?"

"Kaa!"

The wounded told Owens that the waterside market village had fallen under attack and been destroyed.[5] The assailants had hurled grenades and fired on the sleeping villagers with automatic rifles and mortars. The villagers said their attackers were wearing Nigerian military uniforms. One elderly woman blamed the Andonis, but others said no, it was soldiers who shot them. "The army and the Andonis." "Soldiers with the police." "The army and the Andoni people." Some victims swore they witnessed boats filled with heavily armed soldiers crossing to Kaa from the Andoni side of the river. These were "Shell boats," the villagers said, watercraft that had been rented in the past by Shell staff.

The clinic was in utter confusion; nobody was certain about what had happened. Bloodied villagers kept streaming in; it was madness trying to triage so many patients with grave injuries. When a man was brought in with a gunshot wound to the head, Owens ordered that he be taken straight into the operating theatre but there was little that could be done. As Owens struggled to get an IV started, the injured man looked up and said in Khana, "Doctor? Doctor?"

"Yes."

"These people must not go. No matter what happens, the Ogonis must not let Shell get away with this thing."

Owens held the man's hand and looked him in the eyes. "As long as I live, these people are not going to get away with this."

The patient nodded and died.

The parade of casualties did not stop. Owens's ambulance returned from trip after trip to Kaa filled with the raped, the wounded, and the dying. After three straight days of staunching haemorrhages and stitching wounds, Owens took time to visit Ken and to relate these horrors. While there, Owens ran into Karl Maier, an American journalist working as a correspondent for *The Independent* newspaper in London. When Owens described the flood of patients, Maier wanted to know where the attack had taken place.

"A village called Kaa on the banks of the Andoni River."

"Let's go there right now," said Maier.

They went to see Kaa for themselves. Although Owens had lived through the nightmare of the Biafran War, the devastation of Kaa was beyond anything he had ever witnessed. Dozens of homes, shops, and schoolrooms lay in ruins. In the rubble of homes, Owens and the journalist could see the remnants of meals that were being prepared at the time of attack, along with bloody shreds of clothing and bits of family photographs. The few walls left standing amid the shattered cement block and twisted tin roofing were scrawled with slogans such as "NO OGONI KINGDOM!",[6] "ADA-GEORGE IS IN CHARGE," "SHELL IS IN CHARGE" and "WE ARE MARCHING TO BORI." The marketplace was demolished, blackened. Charred carcasses of goats gathered flies next to the hacked-off stumps of plantain trees. Orphaned children poked through the rubble in search of dead parents.

As with the mysterious loss of the Bodo fishermen, the Nigerian government blamed the incident on a squabble between the Ogonis and the Andonis. But the sheer scale of the damage to Kaa made an ethnic clash difficult to credit. The Andonis were a small tribe of impoverished fishermen.

The evidence clearly indicated Kaa had fallen in a military-style assault using sophisticated weaponry.

In the attack, thirty-three Kaa residents were killed and eight thousand left homeless.[7] Most townspeople had been able to flee into the forest, where they remained in hiding. After a couple of days, many worked up the courage to go to the nearby villages of Eeke and Gwara to seek food and shelter.

With an ambulance following, Owens and Karl Maier drove to Eeke to visit the primary school, a makeshift refugee centre for more than five thousand displaced residents of Kaa. The people were in a pathetic state, hungry and frightened. Many recognized Dr. Wiwa and pleaded for help. Owens assured them that MOSOP would investigate.

An old man in the crowd raised his finger toward Dr. Wiwa. "We are ready to be the sacrificial lamb for the Ogoni people so that Ogoni people will get salvation. But we must not allow this thing to happen to any other community in Ogoni! If it happens again, the struggle may be destroyed."

When asked about his brother, Owens replied that Ken was still in hospital but was feeling better. "I am going to tell him what I have seen here," he assured them all.

The refugees were heartened to see that Owens had brought a white journalist with him, for this meant that people in the outside world would hear about what had happened. Maier interviewed many of the refugees and took photos while Owens attended to the injured, placing the most severe cases in his ambulance.

The Inadum Medical Centre was operating beyond capacity, so overcrowded that some patients were forced to lie on the floor. But the people of Kaa needed more than medical attention; they needed food and shelter as well. Owens turned to the Daughters of Charity, a group of Irish Catholic missionary nuns stationed in Eleme Kingdom. He had developed a working relationship with the sisters during his days at Taabaa General, helping them treat cases of tuberculosis and leprosy. Without hesitation, the nuns threw themselves into the relief effort, feeding, clothing, and finding housing for the thousands of Kaa survivors.

Faced with this tragedy, MOSOP was impotent. It had not been prepared for anything on this scale. Groups of Ogoni youths took matters into their own hands, searching for elderly and infirm Kaa residents hiding in the forest, carrying them to the next village for help. Other volunteers went to neighbouring villages and asked if displaced Kaa residents could be taken in.

Although the Nigerian media covered the incident at Kaa extensively, there was little response from the government. From his hospital bed, Ken wrote a letter to Governor Ada-George accusing him of neglecting the safety of the Ogoni people. Ken also organized a fact-finding mission to Kaa, to be conducted by himself, Bennett Birabi, and other Ogoni leaders, as well as officials from the police and the army. Ken convinced his doctors to release him from the hospital for the afternoon, promising to return that evening.

Owens accompanied Ken's entourage to Kaa. Birabi took photographs and a MOSOP cameraman shot videotape of the wreckage. Bullet casings littered the ground and empty dynamite cartons of European provenance bearing the name "Oxford" lay scattered about. Hearing that Ken Saro-Wiwa would be coming, some citizens of Kaa returned to the village to give him a full report of the tragedy. Their tales and weeping distressed him greatly, and he went back to his hospital bed in a highly agitated state.

Ken began sending word throughout Ogoniland, urging people to refrain from seeking revenge on the Andonis. "This is what the government and Shell want you to do," Ken warned. "They want you to retaliate. They want to convert our struggle to a violent struggle, and we cannot compete with them."

Despite Ken's message of non-violence, some Ogoni youths from Kaa went across the water to an Andoni village and wreaked as much damage as they could with clubs and slingshots. Some Andoni people were badly beaten and a great deal of pottery was broken.

The reprisal for the reprisal came quickly. The village of Eeke – where most of the displaced Kaa residents had taken refuge – was destroyed in an attack almost identical to the one at Kaa. This time, hundreds of people were killed and the number of displaced people in Ogoniland doubled. Owens and the Daughters of Charity converted the primary school in Luawii, a small village five kilometres from Eeke, into a refugee camp, but there was clearly a need for a more systematic way of handling the growing problem of internally displaced people.

At the MOSOP steering committee meeting following the Eeke raid, Owens proposed the formation of an ad hoc unit of MOSOP to be called the Ogoni Relief and Rehabilitation Committee. Owens was elected chair, with members of FOWA, COTRA, the Council of Ogoni Professionals, and NYCOP filling out the board. Owens also convinced the local transportation union to join the Ogoni Relief and Rehabilitation Committee, recognizing that its members' trucks would be useful in the relief effort.

Three days after the attack on Eeke, three more Ogoni villages – Gwara, Kpean, and Ken-Nwigbara – were destroyed in pre-dawn military-style raids. Inadum Medical Centre was thrown into pandemonium. Owens's ambulance was running non-stop, bringing in casualties faster than the small team of Ogoni doctors and nurses could care for them. The transport union ferried the most seriously wounded to Port Harcourt and evacuated refugees to villages that had not yet been attacked.

Before each of the nighttime assaults, many Ogoni had heard the sound of a helicopter. The sound would approach from the Andoni side of the water and buzz over a section of Ogoniland in the darkness. Shortly after the noise faded away, the attack would begin. When the carnage was over, the helicopter would return, as if to survey the damage. The ominous sound of a helicopter in the darkness became a harbinger of impending terror.

The attacks continued. In the ensuing weeks, 20 more villages were demolished, leaving 750 Ogonis dead and 30,000 homeless.[8] As committee chair of the Ogoni Relief and Rehabilitation Committee, Owens wrote to the Nigerian Red Cross and the World Health Organization for aid. The only response came from a Red Cross chapter in Port Harcourt, which sent merely a shipment of soap and buckets.

The Ogoni churches raised money, while the Daughters of Charity procured badly needed medicine and medical supplies. The women of FOWA spread out across Ogoniland to collect donations, food, and clothing for the displaced. Two rooms of the clinic as well as Owens's garage at home were soon stuffed with donated items. Every morning his ambulance left for the refugee camps, taking out relief supplies and returning with the wounded and the critically ill.

The lack of food and the unsanitary, overcrowded conditions in the refugee camps exacted a heavy toll. Dysentery, typhoid, and malnutrition became widespread. The Daughters of Charity set up a triage system to feed the most severely malnourished of the camps' children, who were developing kwashiorkor, the familiar pot-bellied look of starvation.

The Daughters of Charity also received funding for the relief efforts from Shell Nigeria. The nuns used Shell's donation to buy blankets, towels, knives, and small farming utensils that the people desperately needed to begin rebuilding their homes and livelihoods. When offered these items, the refugees of Kpean stated emphatically that they would accept nothing donated by Shell. The overwhelming belief among the Ogonis was that the army had committed these outrages on Shell's behalf. Even Ken Saro-Wiwa himself could not convince the Ogonis to swallow their anger and accept Shell's aid.

Soon after his release from hospital, Ken travelled to Lagos for a media blitz designed to bring world attention to the horrors being perpetrated in Ogoniland. Speaking on behalf of MOSOP, he blamed the military government for conducting the raids and fomenting this strangely unprovoked ethnic warfare. According to Ken, the government's goal was to stop the Ogonis from demanding their rights.

Although the Nigerian government publicly claimed that the outbreaks of violence in Ogoniland were the result of ethnic clashes, evidence emerged that suggested the government had played an active role in inciting the antagonism. An army corporal described to Human Rights Watch/Africa his participation in the raid on Kpean, detailing how he had been sent one night to a location in Andoni territory as part of a mission to maintain peace between the Ogonis and Andonis.

"When we arrived at the assembly point," he said, "they suddenly changed the orders. They said we were going across to attack the communities who had been making all the trouble." He said a force of approximately 150 soldiers crossed the river in speedboats to enter Ogoni territory, then marched inland to the village of Kpean. There, in the pre-dawn, they opened fire. "I heard people shouting, crying," he said. "I fired off about one clip, but after the first shots, I heard screaming from civilians, so I aimed my rifle upwards and didn't hit anyone." The corporal testified that, after the shooting, his troops moved into the village to burn and loot the homes.[9]

|||||||

The next time the Wiwa brothers got together, Owens said, "Dede, the people are very angry. Kobani is starting to spread the word that you are responsible for all of this."

"Me?" said Ken.

"Yes. He and some of the others are saying that you are the one who sent people to attack Andoni and that this is the result of MOSOP. They're saying, 'If there was no MOSOP, this place would not have been attacked.' I'm in the villages and I'm hearing these things, so I'm very concerned."

Edward Kobani and others were claiming that it was suspicious that Ken had not been in Ogoniland during any of the attacks. "You never see him," said Kobani. "He creates all the trouble, and then runs away when the attacks happen." (The fact that Ken had been hospitalized during the period in question didn't seem to matter.)

Ken made plans to visit the destroyed village of Kpean. Owens, however, counselled his brother not to go, saying that he was afraid the people would react badly. Kpean was one of the larger communities to have been attacked. Several churches and schools had been obliterated, along with a nearly completed health centre. More significantly, many Ogoni elites came from Kpean and had lost their homes in the attack. Knowing that these elites were blaming him, Ken decided to bring three companions on his tour: Owens, Professor Claude Ake, and a supportive Port Harcourt banker who was originally from Kpean and had lost property during the raids.

Ken's trip to Kpean was unannounced. Within ten minutes of his car pulling into the ruined village, nearly a thousand people surrounded the vehicle. Contrary to Owens's fears, the citizens of Kpean were not angered at the sight of Ken Saro-Wiwa. They were overjoyed.

The outpouring of affection stunned Owens. At the sight of Ken, these wretched people standing in the rubble of their homes seemed to forget their troubles and rejoice that the Great Ogoni Man was with them. Women danced and sang around Ken's car, and Owens recognized people sporting bandages over wounds he himself had stitched up, dancing in the streets because his brother had come.

||||||

At the time Ogoniland was under siege, Nigeria's federal government was itself in turmoil. The transition program to bring democratic rule back to Nigeria had been aborted. Chief Moshood Abiola had been declared the winner of the recent presidential election, but the elections had been annulled.[10] On August 22, 1993, the military dictator, General Ibrahim Babangida, yielded to heavy domestic and international pressure by relinquishing power to an interim government headed by his own appointee, Chief Ernest Shonekan.[11]

Four days into the Shonekan administration, Ken Saro-Wiwa and Ledum Mitee flew to Abuja aboard an air force jet sent by Nigeria's defence minister, General Sani Abacha. Ken was looking forward to this meeting. He had known Abacha since their days as neighbours on Nzimiro Street in Port Harcourt. In the late 1970s, the diminutive, introverted Abacha had been a guest at Ken's house,[12] and Ken Junior had been close friends with Abacha's son Ibrahim.[13]

Over a genial lunch, the three men discussed specific Ogoni demands as well as the call by opposition groups for a "sovereign national conference." Abacha feared such a conference would lead to the breakup of the nation along ethnic lines, but Ken and Ledum countered that without such an assembly, secessionist tensions would inevitably grow. Ken also complained about the repeated seizure of his passport. Abacha apologized, explaining that it was probably due to a few overzealous security officers. He assured Ken it would not happen again.[14]

When Ken returned from Abuja, he told Owens he had the odd sense that Abacha, not Chief Shonekan, was controlling the new government. Nevertheless, the new head of state seemed to be a good chap and it had been a productive trip. Owens bristled.

"Do you know that this man was the director of Shell Nigeria?"

"Shonekan?" asked Ken with surprise. "No, I didn't know that."

"If he resigned from Shell after becoming head of state, I didn't see it in the papers. And what did you talk about with Abacha?"

Ken replied that he had requested the Nigerian army be brought into Ogoniland to protect the people from further attacks.

"Why would you do that?" exclaimed Owens. "People are saying that they saw the Nigerian army come and shoot them. When you went and saw these places, you said yourself that this was a military-style attack – and the military people who followed us said that this was done with military weapons! One army guy said that even though he had been in Liberia, he had never seen this sort of destruction."

"Well, if the military is fighting for our side and they are also attacking from the Andoni side, then both of them will be fighting themselves."

Ken's logic escaped his pragmatic younger brother.

"Look," said Owens, "the mobile police who have been brought in to Ogoni since this has all started are not protecting the people. They are abusing the people and raping the women. You think the army will be better?"

"But the army has to be better than the police," said Ken. "It does not seem possible that anyone could be worse. In any case, something is better than nothing."

Essentially, the Ogoni people were now completely without a law-enforcement body. When the attack on Kaa had taken place, the villagers had implored the mobile police to protect them, but the police had refused. According to the villagers, a police officer in charge told them: "I thought you people said you didn't want the Nigerian police. You wanted Ogoni nation. Go ask the United Nations to come and fight for you."

The resulting sense of powerlessness and fear offered the perfect breeding ground for vigilantism. After the first attacks, groups of young males began promoting themselves as defenders against further Andoni raids. At Kaa and the other destroyed villages, these young people were invaluable in helping to evacuate the injured, the elderly, and the very young and in helping large groups of people make their way from one refugee camp to the next. The villagers gave the vigilante groups money and food and helped them to acquire weapons. Since guns were not affordable even if they had been available, the young men stockpiled spears, slingshots, and other primitive arms. The most popular weapon was the machete.

A machete, of course, is no match for a gun. To help even the odds in battle, the vigilante youths put their faith in magic. Ogoni medicine men created invincibility potions for the vigilantes that they inserted into the youths' bodies through deep slits in the skin. To activate the magic in their flesh, the vigilantes were instructed to paint their bodies with chalk, and to chant, dance, and drink copious amounts of liquor. Returning from these ceremonies, the self-mutilated youths felt intoxicated and exhilarated with their newfound supernatural abilities.[15]

The machete-armed youths, feeling magically invincible and revelling in their unaccustomed authority, quickly turned into a societal menace. Their solicitations of "contributions" for money and food soon turned into outright extortion, and their benevolent protection of the Ogoni villages degenerated into thuggery. Just as the government confused UNPO with the UN, it also confused these vigilante youths with NYCOP. Although NYCOP made it clear that it did not condone any of the actions of the vigilantes, the government seemed eager to use the unlawful actions of the vigilantes as an excuse to crack down on MOSOP.

After more mysterious "ethnic clashes," the Nigerian army moved troops into Bori, commandeering the primary school as their headquarters. When Ken heard about this, he told his brother to be civil to the soldiers. "If we are friendly with them," said Ken, "it will be a way to disarm them in case they want to do anything."

"I'm always civil to everybody," joked Owens, but Ken needn't have bothered cautioning him. Owens counted as friends several of the soldiers stationed at Bori whom he had known since his Youth Corps days at the Command and Staff College in Jaji. He had also met several other officers through his youngest brother, Letam, now stationed with army intelligence in Lagos. Per Ken's wishes, Owens began treating the soldiers for malaria and other medical problems, which angered many citizens of Bori. They felt that Dr. Wiwa was aiding and abetting the enemy.

||||||

In September 1993, a document known as the Ogoni-Andoni Peace Accord was signed. The accord came about in response to Ken's request that the Rivers State government convene a panel to look into the cause of the armed incursions into Ogoniland.

For the first session of this panel, the government asked Ken to bring with him a team of Ogoni representatives. To decide who would sit with him on the panel, Ken hastily arranged a meeting – strangely enough, not with MOSOP, but with KAGOTE, the club of Ogoni elders. The club selected Edward Kobani, Albert Badey, and three others to represent the Ogoni people with Ken. When MOSOP's steering committee learned about the government panel – and that their Ogoni representatives had already been chosen – they were understandably upset.

At the next steering committee meeting, Owens confronted Ken, saying, "You are going to a meeting with these people? What is there to discuss?"

"What is there to discuss?" Ken responded. "Ogoni is being destroyed. Anybody can see the destruction. We will go there to argue the point."

MOSOP's steering committee, however, was insistent that the point not be argued without them. It was decided that MOSOP's acting secretary general

Dr. Appolos Bulo, NYCOP's Goodluck Diigbo, and Owens Wiwa should also attend the government panel – even if they had to sit in the gallery.

On the morning of the meeting, Owens went to 24 Aggrey Road to meet up with Ken, Diigbo, and Appolos. To his surprise, Ken had already left for the meeting and the two other MOSOP members had decided not to show up at all. The day was not getting off to a great start.

When Owens arrived at Government House, he found Ken in the reception area of the governor's office, chewing on the stem of his favourite pipe. As they walked into the meeting room together, Owens stopped short and shot his brother a concerned glance. Not only had they not expected Priscilla Vikue to be at the meeting, but Major Paul Okuntimo of the Rivers State Internal Security Task Force was there as well, sitting beside two representatives from Shell. One of the Shell reps was M. Promise Omuku, a fellow Ken knew well. Ken approached the Shell delegation and greeted them warmly.

"You are out of the hospital," remarked Omuku. "How are you?"

"I'm still alive," Ken wryly replied.

The meeting opened with a short speech by a government representative who set the rules for the discussion and asked for each side to start its presentation. Kobani began with a short speech; this was followed by a few words from Ken, and then a statement from two men representing the Andonis: Chief Dr. Silas Eneyok and Chief A.M. Edeh-Ogwulle.[16] Eneyok was harshly critical of Ken. When one of the Shell reps rose to give his presentation, Ken interrupted to enquire why Shell was there at all. The representative responded that although his company had not been involved in the recent conflict, many people thought that it was somehow culpable. It therefore wanted to defend itself fairly.

After Shell's presentation and a brief discussion, it was announced that a document of agreement that had been prepared earlier was ready for signing. Dumbfounded, Ken listened to the wording of the agreement. One clause particularly worried him: "That it is in the interest of the Ogoni and Andoni Communities and all the Governments in Nigeria to ensure an immediate resumption of all full economic and social activities within Ogoni and Andoni areas."[17]

Now it was clear why Shell was at the meeting.

Ken scribbled a quick note to Albert Badey, explaining that a reporter from the *Wall Street Journal* was waiting for him at the MOSOP office. He then slipped out of the room, motioning for Owens to follow. As Owens was on his way out, he noticed Bennett Birabi (who had come late to the meeting) shaking the hand of one of the Andoni representatives, saying, "You've been doing very well. Everything went according to plan." Evidently pleased with how the meeting had been going, Birabi then went to shake the hands of Shell's PR reps.

In the reception area, Ken told Owens to let the Ogoni delegation know that he was not going to sign the document in its current form, at least not until it had been presented to the Ogoni General Assembly for later discussion. Ken then left and Owens returned to the meeting to relay this information to Kobani, Badey, and the other Ogoni chiefs. The chiefs as well as the other panel members, however, insisted that the Ogoni-Andoni Peace Accord be signed immediately and that Ken's name appear on it.

"But this is not possible," said Owens. "Can we not go and discuss these things with the people at home before the signing? All this is being done in too much of a rush."

Owens was overruled. The Ogoni-Andoni Peace Accord was signed immediately. Ken was not among the signatories.

The following morning, the newspapers contained letters to the editor describing the process of achieving peace in the Niger delta in spite of opposition on the part of Ken Saro-Wiwa.

"We would like to inform the general public that as a member of the Ogoni delegation, Mr. Saro-Wiwa, attended all sessions of the conference and fully participated in the deliberations," read a letter in the *Daily Champion* signed by Leton, Kobani, and four other Ogoni elites. "At no point in the deliberations did Mr. Saro-Wiwa object to any part of the accord that was reached and is now signed. . . . It is deeply regretted therefore that Ken Saro-Wiwa has so dishonourably sought to deny a document he actively contributed to. . . . Does not this act per se (which is vintage Saro-Wiwa) reveal his sinister motives? Back home in Ogoni, Mr. Saro-Wiwa's propaganda machine is busy telling the people that Mr. Ken Saro-Wiwa has single-handedly brought peace-keeping troops to halt hostilities. By refusing to sign a document the main thrust of which is a cessation of hostilities, he is inciting the people to resist the peace process. The Ogoni people should now be familiar with Mr Saro-Wiwa's 'double speak' and 'disappearing act' antics. He is never present or available during any crisis. He never commits himself to anything so that he can later exploit the situation if need arises. In that way, he is always the super hero while his victims are mere villains."[18]

Although Owens found the press accounts upsetting, Ken seemed almost to expect such treatment. After all, as he said, one could hardly hope to threaten the income of wealthy men without reprisal.

|||||||

During their joint birthday celebration on October 10, 1993, Ken said to Owens, "Shell would like to see me dead."

"What are you saying?" asked Owens.

"I'm saying they want me dead."

"But surely Shell wouldn't try to kill you."

"No," said Ken, "but I'm sure they'd love to see me hanged."

|||||||

Even though a peace accord had been signed between the Ogonis and the Andonis, Ogoniland remained in turmoil throughout the autumn of 1993, with sporadic shootings and abductions of Ogoni fishermen by armed bandits from Andoni territory. The Ogoni became so afraid of getting into their boats that fishing nearly ceased, and the livelihoods and food supply of a great many of the people evaporated. As the vigilante youths continued their campaign of violence and extortion in the smaller communities, the army also started demanding rations from the townspeople of Bori. In an effort to cow the populace of Ogoniland's largest town, army soldiers began indiscriminately raping women and young girls. The Inadum Medical Centre was soon overrun with women complaining of injuries, venereal diseases, and unwanted pregnancies.

At the same time, Diana's unplanned pregnancy was preoccupying Owens. Diana had planned to introduce Owens to her parents in June, but Ken's arrest and the subsequent wave of violence made it impossible for Owens to get away. By late October, Diana's parents still had no idea that their daughter was even acquainted with Dr. Owens Wiwa, much less pregnant.

Yet even with the crisis in Ogoniland, Owens had achieved two of his three New Year's resolutions for 1993. He had increased his real estate holdings and had opened a second clinic near the petrochemical plants in Eleme Kingdom. The only outstanding item on his list was to get married before year-end.

The major obstacle was Diana's father, Professor Cosy Nuaba Barikor, a devoutly religious Seventh-day Adventist church deacon. How could Owens possibly tell this man not only that he loved his daughter, but had gotten her pregnant?

According to Ogoni custom, the father of a prospective groom or an elder family member must approach the father of the intended bride to ask permission for the couple to wed. Owens explained his sticky situation to his brother, and Ken volunteered to talk to Professor Barikor on the couple's behalf.

When the professor was summoned by his housekeeper to greet a visitor at the door, he was astonished to find himself face to face with the famous Ken Saro-Wiwa. Hitherto, he had seen Ken only on television, in the newspapers, or on the podium at anti-Shell rallies.

"Oh great man!" he exclaimed. "This is quite an honour."

Diana's mother, Dora, demurely greeted Ken and ushered their guest into the living room.

"What brings you here?" asked the professor.

"I have come about your daughter."

"My daughter? You mean Diana?"

"Yes," said Ken. "I am very impressed with her. She's one of the brightest young girls of Ogoni. I have noticed that she has been very active in the movement and has been coming to my office often. When I was detained, she was one of the first to visit me in jail. She's a wonderful woman, your daughter, and you should be very proud. And so I was very pleased when my junior brother Monday came to inform me that Diana was his choice for a wife."

Professor Barikor's eyes widened.

"Therefore I am here today to request Diana's hand in marriage for my brother."

"Well, yes, I see. . . . Well," said the professor, giving a half-hearted, uncomfortable chuckle. As much as he admired Ken and the entire Wiwa clan, the thought of his daughter marrying into that particular family was troubling. Of primary concern was the matter of religion. Ken's father was Anglican and Ken was, by his own admission, merely an "occasional Christian."[19] The Barikors believed that the members of their sect should strive to marry within their own church. Nicknamed "Christian Jews" for their practice of keeping the Sabbath on Saturday (hence the "seventh day" in their name) and for their regimented, nearly kosher diet that forbids pork or shellfish, the Adventists were so strict that their women were even forbidden to wear earrings. Professor Barikor found the idea that his daughter had been engaging in an irreligious alliance disturbing indeed.

"I will need to talk with the girl and see what she thinks," said the professor. "Then I will let you know."

"That's fine," said Ken. As a gift, he had brought an expensive bottle of wine. He presented it to the professor and suggested that they open it and drink a toast to the happy couple. Casting a pained glance at her husband, Dora Barikor gingerly took the bottle from Ken while the professor looked a bit embarrassed.

"I'm sorry," he said. "In our religion, we don't take alcohol. But thank you."

When Diana's father later sat down with his daughter to determine her feelings for Ken's younger brother, he was livid when he learned she was nearly six months pregnant. Diana was tall and athletically built like her father, so her condition had not been readily noticeable. Of equal concern was her education. Diana had just started her final year as a political science major at university. With a husband and infant to take care of, how would she complete her degree?

Realizing the marriage – and the baby – were inevitable, Professor Barikor made the best of it. He received Diana's solemn promise that she would not abandon her studies; as for the religious differences, the possibility of converting Owens to the faith was nearly as pleasing as having a son-in-law already in the sect. The Wiwa family were informed of the Barikors' favourable decision and plans were made for a December wedding.

||||||||

Earlier that year, on June 12 (the same day as the contentious Nigerian presidential election) a Shell flowline located at Bunu, Tai, sprang a leak. Crude oil gushed from the rupture, devastating the surrounding farmlands and seeping into nearby streams, robbing the community of its fresh water and fish. For months Shell did nothing to stop or clean up the spill, claiming that it was not safe for its employees to enter Ogoniland. Interestingly, Shell had success-fully repaired a similar rupture without incident in nearby Nonwai, Tai, only a week before the Bunu rupture.[20]

In late September, after the Ogoni/Andoni Peace Accord had been signed, internal Shell documents show the company asked the Rivers State government for military protection to go to Ogoniland to "test" the accord. A small group, guarded by military personnel under the command of Major Paul Okuntimo, surveyed various Shell installations in Ogoniland on October 21. As in June, the Shell staff met no resistance from the Ogonis.

Two days later, Shell claimed that it received a telephone call from the Korokoro flow station in Tai Kingdom reporting a fire alert. Despite the area being a "no-go" for staff, the company dispatched two Shell fire trucks to the scene. Angered by the ongoing rupture at nearby Bunu, the people of Korokoro flew into a rage on seeing Shell employees in their town. A gang of knife-wielding youths apprehended the firefighters, trundled them into the village square, and threatened them with death. After four hours, the Shell employees were released. As for the fire trucks, these were confiscated by the community.[21]

Understandably, Shell was furious. The company let it be known that there had been no fire at Korokoro; the call had been a hoax.[22] (Indeed, there was no telephone service in Tai Kingdom. Where had the original dis-tress call come from?[23])

Three days after the dispute in Korokoro, Owens saw Major Paul Okuntimo and two Shell trucks filled with army troops pass along the road in front of his Bori clinic. A few hours later, Owens received word that Korokoro had been invaded.

Okuntimo had been invited by Shell to undertake an expedition couched as a "mission to rescue our fire trucks."[24] Villagers reported that at a demon-stration of Ogoni people outside the Shell flow station in Korokoro, Nigerian soldiers fired on the crowd.[25] In a letter to the Rivers State government, Shell claimed that the villagers had attacked the army, not the other way around: "We regret to inform you of the attack on the team comprising 24 armed per-sonnel and 2 drivers which went to Korokoro to dialogue with the chief of the community."[26] Another Shell report stated that on the way to the chief's house, the army was "attacked by a large number of villagers with guns, knives, and broken bottles. A soldier was hit by a bullet and treated at the nearest clinic, which happened to be Shell's. The soldiers showed restraint and retreated. No villagers were reported injured."[27]

Even though Shell insisted that no one had been injured at Korokoro, that day, Owens's ambulance arrived at his clinic bearing two Korokoro men who said they had been shot by Federal forces. One man, a seventy-year-old named Papa Ndah, had been hit in the stomach, the bullet perforating his intestines. The other victim, Monday Nte, was shot in the buttocks.[28] Owens took Papa Ndah directly to the operating theatre, where he and another surgeon repaired the abdominal wound. Later, as Owens was attending to the second injured man, weeping villagers came into the Inadum Medical Centre saying that someone had been shot and killed at Korokoro.

Owens got into his ambulance and drove to the besieged village, where he came on a group of women wailing outside one of the houses. Inside, a woman directed Owens to a back room, where the corpse of an Ogoni youth lay in a pool of blood.

"This is my son," said the woman. "They shot him and killed him."

"Who shot your son?"

"Okuntimo."

The other people in the house nodded and repeated, "Okuntimo. Okuntimo." They told Owens that the major had come into the town marketplace shouting, "Where is the chief? Where is the Shell vehicle?" Receiving no immediate response, Okuntimo began shooting into the air, and the other soldiers and policemen started firing as well. Panicked, the villagers screamed and fled. The old man, Papa Ndah, was at home in bed when he heard the commotion and asked what was going on. Before he could sit up, soldiers burst through his door and shot him.

The villagers told Owens that the dead youth – a seventeen-year-old schoolboy named Uebari Nnah Nziagbara[29] – had been killed by Okuntimo himself. The major, they said, had taken his pistol and shot the boy point-blank in the back of the head.

Owens looked down at the body. There was a large, gaping wound in the boy's forehead. Rolling the body over, he could see a small hole and powder burns on the back of Uebari's skull. From the look of the injuries, the villagers' account appeared to be accurate. Owens ordered the body taken to the police station in the nearby town of Kpite.

Immediately after the raid, the papers and TV carried interviews with Major Okuntimo as he sat in a Shell medical clinic, brandishing a bandaged wound, claiming he had "nearly been killed"[30] by the villagers of Korokoro. He failed to mention that the people of Korokoro had been unarmed and that the injury he sustained during the fracas was minor.[31] Nevertheless, he claimed to be enraged. Nigerian lawyer Oronto Douglas and British environmentalist Nick Ashton-Jones later heard him exclaim that he would never forgive Ken Saro-Wiwa for the way the Ogoni people had opposed Shell that day. "I ordered him to be taken to an unknown place and chained legs and hands and not to be given food," said Okuntimo. "Ken will never see the light."[32]

When he learned of the Korokoro melee, Ken Saro-Wiwa went to visit Okuntimo and his superior, Brigadier General Ashei, commander of the Second Amphibious Brigade. When Ken accused the major of accepting money from Shell to kill Ogonis,[33] Okuntimo sneered. There was no reason to deny the charge, for it was no secret that the armed invasion of Korokoro had been at the request of Shell Nigeria. Shell even acknowledged that it had provided the two transport vehicles that ferried the troops to Korokoro and that it had paid allowances to the soldiers of the Second Amphibious Brigade to carry out the raid.[34]

Later, in response to international criticism over financing a military assault on civilians, Shell issued the following statement: "The payment of field allowance to Nigerian military personnel happened only once, under duress at Korokoro in 1993. SPDC has made it clear that it will not happen again."[35]

|||||||

On November 17, 1993, General Sani Abacha staged a palace coup, forcing the resignation of head of state Ernest Shonekan. Many pro-democracy supporters welcomed the ouster because Shonekan was merely the political stooge of a deposed dictator. His Interim National Government had been little more than a toothless illusion. After the coup, General Abacha vowed he would rule for only a short time before installing the people's elected choice, Moshood Abiola, as president. To Ken and other Nigerian human rights activists, this was an encouraging turn of events.[36]

At the next steering committee meeting, MOSOP's official reaction to the coup was the most pressing item of business. Ken argued that a military government was preferable to the sham democracy Nigeria had been enduring. In a democratic Nigeria, he reasoned, the Ogonis would still be an insignificant minority without the protection guaranteed them by the current Nigerian constitution. A military government would be more likely than an elected one to listen and respond to Ogoni demands.

"Maybe we can use this military situation to negotiate some of the items in the Ogoni Bill of Rights, especially the issue of autonomy," he said. "We want an Ogoni state within Nigeria. All the other states were created by the military. Now perhaps it can create one for us." The committee agreed and decided to extend Abacha a cautious welcome.

Ken's predictions proved to be spectacularly misguided. Although Abacha appointed civilians to his cabinet, he dismissed most of them in short order.[37] The new dictator also wasted no time dissolving all political parties and democratic institutions, suspending the 1979 constitution, and cracking down on opposition. As his main decision-making instrument, Abacha created the military-dominated Provisional Ruling Council, which ruled by decree. He also passed a law declaring that the obstruction of economic activities – that is, impeding the flow of petrodollars into the federal coffers – was a crime punishable by death.[38]

Corrupt, cunning, and avaricious, Abacha could not wait to get his hands on the billions of dollars of oil revenue extracted from the Niger delta each year. Nothing would stand in his way. As soon as he seized power, he launched a full military crackdown on the activities of MOSOP. As Ken Saro-Wiwa became ever more outspoken on the issue of minority rights and was gaining international support for the Ogoni cause, Abacha became ever more determined to silence him.

|||||||

The most difficult part of planning Owens and Diana's wedding turned out to be finding a date that Ken would be available. He was forever flying back and forth between London, Lagos, and Port Harcourt, running his growing Saros International empire while raising worldwide awareness of the conflict between MOSOP and Shell.

Owens was just as busy, equipping and staffing his second clinic in Eleme, as well as overseeing resettlement of displaced Ogonis. By December 1993, he had three doctors working for him in his two clinics, but there were more than enough patients to keep him occupied around the clock. Whenever Ken was away (which was often), the other MOSOP sub-organizations such as FOWA, NYCOP, and COTRA would come to Inadum Medical Centre to ask Owens for advice, money, or adjudication of minor disagreements. Every Tuesday evening, Owens attended MOSOP steering committee meetings in Port Harcourt.

By November, Diana was into her third trimester and getting larger and more impatient by the day. Not pleased about playing second fiddle to Owens's clinics and MOSOP, she finally insisted their wedding take place soon. Ken and Owens looked over their schedules and determined that they were both free on December 12. And so the wedding date was set.

Traditionally, an Ogoni marriage takes place at the bride's family compound in her native village. Although the Barikors lived on the campus of the University of Port Harcourt, their family hailed from Bomu, Gokana. Professor Barikor's younger brother, who lived at the family compound, handled most of the wedding arrangements. Diana's mother, Dora, did the shopping for the wedding feast and procured the traditional bridal basin and wedding clothes. In Ogoniland, brides and grooms wear nearly identical outfits. Diana had decided she and Owens would be married in matching robes of white organza, but her own flowing robe would be beaded so that it would flash an iridescent sheen of peach, pink, and blue.

On the wedding morning, as Owens and Ken drove the short distance from Bori to Bomu, Owens marvelled at the number of people lining the roads. As they entered Gokana, a convoy of motorcycles converged in front of the car, tooting their horns to clear the people out of the way.

None of these people had come to see Owens and Diana get married. They had come to see Ken Saro-Wiwa. When Diana's uncle, the MOSOP chairman of Bomu, learned that his niece was marrying into the illustrious Wiwa family, he put on his MOSOP hat and decided that December 12 would also be the day his village would hold a gala welcoming ceremony for the Great Ogoni Man. The motorcycle convoy led Ken's car into the centre of Bomu, across the soccer fields of the primary school to a grand tent that had been erected. Owens left Ken on his own to receive the cheering crowds and went to Diana's family compound.

The Ogoni marriage ritual begins with the bride's procession from her new husband's home to her parents' compound. Because a twenty-mile walk from Bane to Bomu would have been impractical, Diana used her maternal grandparents' house in Bomu as her starting point. During the wedding procession, the bride carries a large tin basin on her head containing a trousseau of new native wraps purchased by the bridegroom, new shoes, a watch, a handbag, an umbrella, and a few other items to be used in her new life as a married woman. No matter how far the bride has to walk, her hands must never touch the basin, not even if it should fall. A toppling basin was the signal to the girl that her intended was not a suitable husband. Like most Ogoni brides-to-be, Diana spent many hours practising balancing the wide metal basin on her head. By her wedding day, she had perfected her talent to the point that she was even able to dance a bit to the drumming that accompanied her walk to the ceremony.

When the Ogoni bride arrives at her parents' compound, her brides-maids remove the basin so that the goods inside can be admired by all. Instead of exchanging vows, the Ogoni couple is cross-examined before the assembly by members of both sides of the family. Everyone takes a turn, asking the couple questions about how they met and fell in love, why they chose one another over other suitors, how they feel about getting married, and how they plan to uphold Ogoni family traditions.

When her father asked Diana, "Why do you love this man?" she looked at Owens in his beautiful white wedding garment and became tongue-tied. She wanted to say something sentimental and memorable, something akin to the florid prose found in Gothic romance novels, but the ceremony was performed in the Gokana language and she hadn't the vocabulary to express her feelings. Diana stammered a bit in the local tongue but could not find a way to formulate an answer without resorting to the familiarity of English.

To help his daughter out, Professor Barikor reworded the question. "Why is this man marrying you?" he asked in Gokana.

"Why am I marrying this man?" she nervously repeated – and the assembled wedding party burst into laughter.

It was a linguistic faux pas. In Ogoni, a woman doesn't marry a man; a man marries a woman.

Jessica Wiwa was questioned as well. Because traditional Ogoni practice requires that the bride move in with her husband's mother, both sides of the family need to know if the groom's mother supports the marriage. Living in the same house and cooking from the same pot would be difficult if the women did not get along. If bride or groom fails to answer their questions satisfactorily, or if the bride's mother-in-law is not in full support of the marriage, the bride must take her basin back onto her head and walk home again, perhaps to return another day when everyone feels more ready. Fortunately, Diana's basin stayed put.

Ken arrived from his own reception a half-hour into the question period. The chiefs of Bomu had presented him with a royal title, making him an honorary citizen of the village. With or without the title, Ken Saro-Wiwa's presence turned Owens and Diana's ceremony into the equivalent of a royal wedding.

After the question period, Diana's father slaughtered a goat, the traditional way of signifying that the bride's family was happy with the couple's responses. The goat was then roasted for the wedding feast while the family took photographs and toasted the happy couple. Normally, palm wine is the ceremonial wedding drink. Since the Barikors did not approve of alcohol, Ken saluted his junior brother and his new sister-in-law with a lukewarm cup of Kool-Aid.

|||||||

A few weeks after Abacha's rise to power, Ken was summoned to the presidential palace in Abuja to meet with the new dictator. Ken's initial impulse to support the Abacha regime was reinforced when he learned on this trip that Abacha was thinking of appointing Ken to the post of civilian administrator of Rivers State. Ken returned from the capital impressed and flattered; Abacha's offer never materialized. Instead, Abacha replaced Governor Ada-George with a crony, Lieutenant Colonel Dauda Musa Komo, on whom he conferred the new title of State Military Administrator.

Once again, at a MOSOP steering committee meeting Ken declared that having a military officer replace a civilian leader might be in the Ogonis' best interests. He pointed out that MOSOP's relationship with Ada-George had hardly been propitious. At Shell's bidding, the former governor had launched his state security apparatus against them, resulting in the deaths of hundreds of Ogoni people. Now that Ada-George was gone, MOSOP could only be friendly with the new military administration and hope for the best.

One of Lieutenant Colonel Komo's first actions after taking office was to summon Ken to Government House in Port Harcourt. Komo had been advised that he and Ken had much in common. Like Ken, Komo was intelligent and educated. Ken was especially impressed to learn that the new governor had studied in England and was an avid reader of the classics. In his report to MOSOP about this meeting, Ken announced that Komo was somebody that MOSOP could get close to, "at the very least to prevent further human rights abuses."

As with Abacha, Ken completely misjudged Komo. The first clue that Komo and his administration might not be trustworthy came the day after the wedding. At midnight on December 13, the Ogoni settlements in Port Harcourt's waterfront shantytowns known as the "Down Below" were stormed in a manner eerily similar to the military-style raids in Ogoniland. Dynamite, machine guns, rifles, and other sophisticated weaponry obliterated the meagre makeshift homes of the Ogonis.[39] The police did not respond.[40] In fact, when residents of the shantytown ran to summon them, the police barricaded the entrance to the station and told them to go away.

Sixty-three people are believed to have died in the attack.

||||||

Much of the talk during the Christmas season in 1993 revolved around preparations for the second annual Ogoni Day. January 4, 1994, was to be an even larger spectacle than the previous year. Documentary filmmakers Glenn Ellis and Kay Bishop were arriving from England to film the events for Britain's Channel 4. Representatives from UNPO and the Body Shop were also expected to take part in the celebrations.

As preparations proceeded, the Ogonis celebrated the season in traditional style: wearing new, brightly coloured outfits to visit the homes of family and friends for meals and conversation. Boxing Day found Diana eight months pregnant and in the kitchen cooking the holiday meal with Korkora, Owens's valet, when a surprise visitor paid a social call to the Wiwa home.

In full uniform, Captain Tunde Odina of the Nigerian army greeted the family cordially. Owens had known the captain for a few months; he had been giving Odina free treatments for malaria at the Inadum Medical Centre. In the spirit of Christmas, he welcomed Odina into his home.

Diana brought the captain a bottle of Maltex and a plate of chicken and jollof rice. The spicy West African rice dish was about the only recipe twenty-one-year-old Diana had mastered thus far, and the Maltex – a thick, malt-flavoured soft drink popular in Nigeria – was about the only beverage that complemented her early attempts at cooking. Taking a Maltex herself, Diana retired to the kitchen. The heat of the afternoon and the weight of her pregnancy were reason enough to leave the men to their boring political discussions.

Once the men were alone, Odina got to the point. "They're going to arrest you," he said.

Owens stared at the captain in stunned surprise. "Who?"

"Security people. They're going to arrest you."

"Why would anybody arrest me?"

"Take a holiday. Leave Ogoni. Go to Lagos or somewhere."

"I can't leave," said Owens. "And there is no reason why anybody would arrest me. I haven't offended anybody."

"I'm telling you, they're going to arrest you. You need to leave Ogoni now."

"But it's the end of the year," said Owens. "I have my patients. My colleague, the other doctor that I employ at the clinic, has gone on holiday so I'm alone. And this is my honeymoon period! I'm not even taking my honeymoon because I have work to do. I'm not going anywhere."

"Well, you've been warned," said Odina.

A few hours later, Major Akintola of the Army Intelligence Unit at the Brigade in Bori Camp also paid the Wiwas a visit. Unlike Odina, who was a mere acquaintance, Akintola was a colleague of Owens's younger brother and a friend of the family. He arrived not in uniform but in the mufti of traditional African dress; even so, his visit was official business.

"The governor of the state, Lieutenant Colonel Komo, wants to see you and Barika Idamkue," said Akintola.

As Owens called once again for his chauffeur, Diana appeared from the kitchen. "Where are you going?"

"The governor has sent for me. He wants to see Barika Idamkue, too."

Diana eyed Akintola suspiciously. "Why would the governor ask to see Barika and Mon?"

"We have written appeals for relief materials to the Red Cross, to the government, to everybody," said Owens. "This guy has resources. Well! Maybe our cries have finally got to him."

"I don't know," she said.

"I'm a physician, Diana. I take care of people. Why else would he look for me?"

The governor was a Christian from the Zuru ethnic minority in the overwhelmingly Muslim state of Kebbi. Owens reasoned that if anyone knew what it was to be an oppressed minority, it would be Komo.

Owens and his driver followed Major Akintola's car back to Port Harcourt, stopping to pick up his nephew Barika Idamkue along the way. When they reached Government House, the Rivers State governor's mansion, Akintola escorted the two Ogonis directly into Komo's office. The governor was also dressed in traditional African mufti. He greeted his guests cordially, with much respect.

After according the governor his official salutations, and thanking him for the invitation, Owens said, "So you got my letter."

"Letter?" said the governor.

"For the Ogoni Relief and Rehabilitation Committee. I wrote asking for funds to get relief and to help settle the people in Ogoni."

"Oh yes," said Komo. "Well, we will think about that. But that is not why I asked you here."

"It's not?"

"No. We are concerned about this Ogoni Day," said Komo. "When are you planning this thing?"

"The fourth of January," said Owens. "Why?"

"We don't think you should have it."

Owens glanced at Barika, then turned to Colonel Komo. "But Ogoni Day is just a cultural event," he said. "It's something that has been happening over the years in Ogoni, only this time we are all doing it together under the umbrella of MOSOP."

"We're concerned about the security risks," said Komo.

"But it will be non-violent," Barika assured him. "This is just a coming together of the people affirming their identity to themselves, dancing and speech-making, just like we did last year."

"And it has already been in the news that we are expecting guests," Owens added.

"What guests?"

"Journalists, things like that. It's nothing to bother about."

"I don't want it to go on," said Komo.

"Well, we can't stop it. Preparations have been going on for quite some time," said Owens.

The governor tried to convince Owens to put a stop to the celebrations, but Owens kept insisting it was not possible. Owens then leaned back and said lightly, "Oh, and I was told by Captain Odina that I was to be arrested. Why do you want to arrest me?"

Komo grinned, looking over at Major Akintola. "It is not true," he said. "That is just one of the rumours we hear now and then. We keep telling people there's nothing like that." Komo, still smiling, informed Owens and Barika that that was all he required of them. The meeting was over.

It was nearly seven in the evening by the time Owens returned home. He and Diana had a light supper of her holiday jollof rice before the chauffeur drove him to the clinic for evening rounds. After checking in on the patients, chatting with the nurses, and making sure there was enough diesel in the generators to light the clinic through the night, Owens returned home. He and Diana were in bed by nine.

The sound of pounding on their bedroom door and Korkora's shouts – "Soldiers! Soldiers!" – woke them around 2 a.m.

Three heavily armed troopers had vaulted over the garden wall, frightening away the night watchman. As the elderly watchman scrambled to find a hiding place, the soldiers pounded on the front door: "Open up! Where is Dr. Wiwa?"

Owens threw on some trousers and shoes while Diana grabbed a native wrap to cover herself. Korkora held off the soldiers at the front door as best he could, but he was quickly overpowered. The soldiers came directly into the bedroom. One was Captain Tunde Odina.

"Where is Dr. Wiwa?" shouted Odina.

"I'm here," said Owens, pulling on a shirt. "What's the matter?"

"By the order of the Federal Republic, you are under arrest."

Odina's two compatriots aimed their AK-47 machine guns at the couple, but Odina kept his pistol in his holster. All three soldiers reeked of marijuana and alcohol.

"What arrest? Tunde, what is wrong with you?"

"I told you to leave Ogoni. I told you go to Lagos. And what do you do? You go to the governor. You tell him that I alert you that you might be arrested! That was stupid!"

The soldiers hustled Owens into the courtyard, where more soldiers were slapping and taunting Korkora while the night watchman kept still in his hiding place. Owens was stuffed into the back seat of an army sedan and sandwiched between the two soldiers armed with AK-47s. In a panic, Diana gathered up Owens's wallet and slippers and raced outside, but her pleas to let her give them to her husband were ignored. Odina jumped into the front passenger seat of the sedan, and the other soldiers piled into a waiting jeep.

"Where are you taking him?" Diana cried, but no one answered her. In frustration, she threw Owens's slippers at the back window of the sedan as it sped away into the night.

Next stop was the home of MOSOP deputy president Ledum Mitee in the Gokana town of K-Dere. There, one of the soldiers remained in the car with Owens while the others repeated the arrest scenario. The soldiers forced Ledum into the back seat next to Owens. With the jeep in its wake, the army sedan drove the twenty-six miles to Port Harcourt at high speed.

The drive ended at the old Nigerian Air Force base just outside the city limits. All the military personnel climbed out of the sedan. Owens and Ledum waited and watched in the darkened vehicle as Odina spoke in low tones to some of the soldiers at the base.

Ledum Mitee was just as surprised and upset as Owens by what was happening. Like Owens, this was Ledum's first arrest. "Where are they taking us?" he asked in a half-whisper.

"I don't know," said Owens. "Maybe they're going to put us on a plane."

"Why? You think they're taking us to Lagos?"

"Maybe."

But Owens and Ledum were not put on an airplane. Another car pulled up beside the sedan and the two detainees were told to get inside it. Although the darkness made it difficult for Owens and Ledum to tell exactly where they were being taken, they both knew Port Harcourt well enough to realize they were headed somewhere in the "government reserve area" (G.R.A.): the relatively posh section of town where the governor's mansion is located.

Around 4 a.m., they arrived at a private house. Like most upper-class homes in Western Africa, it was surrounded by high stone walls topped by shards of broken glass and coils of razor wire. Once inside the iron gates, Owens and Ledum were escorted in the pitch-blackness to a small room devoid of furniture. The two men could see nothing but a square hole in the wall through which mosquitoes swarmed. Although dawn was only two hours away, the night seemed to last forever.

With the coming of daylight, Owens and Ledum got the first good look at their surroundings. They were in the remains of what appeared to be one of the government-issued homes typically supplied to civil servants. The previous tenants, perhaps upset by Abacha's coup and their subsequent eviction, had looted the place. Not only had every stick of furniture been carted off, nearly everything else was missing, too – light bulbs, power outlets, plumbing fixtures. Bare electrical and telephone wires dangled from holes in the ceiling and walls. The windows – panes, sashes, and sills – had been ripped out, along with all the doors and mouldings. The kitchen cabinets and appliances were gone, yet through some small miracle the sink and toilet remained in the bathroom.

In the backyard Owens and Ledum found trees bursting with mangoes, oranges, and plantains – more than enough fruit to feed two detainees for a while, but there was still the matter of drinking water and other necessities to worry about.

In Nigeria, detainees are responsible for their own food and amenities. Having no money with them, Owens and Ledum could not bribe the three soldiers on rotating shifts to buy them food. Yet not all hope was lost. The guards were good-natured, grossly underpaid men with no apparent antipathy toward their charges. Since the Nigerian army did not see fit to feed its troops, the mere promise of a bribe – money, food, anything – was enough to enlist the soldiers' aid.

As it happened, Ledum Mitee's brother owned a hotel in Port Harcourt. Borrowing a scrap of paper and a pencil from one of the guards, Ledum wrote a note and instructed the soldier to take it to the hotel. "Give this to my brother there," said Ledum. "He will give you food."

The guard returned with several bottles of drinking water and a large, expertly prepared meal from the hotel kitchens. The food was shared all around. With that feast, Ledum and Owens bought enormous goodwill.

For the next few days, the soldiers returned to the hotel and were given good things to eat as well as toiletries, blankets, and pillows to pass on to Ledum and Owens. Ledum had told his brother exactly where they were being held, and each day relatives arrived outside the walls of the house to shout news and messages to the detainees.

On the fourth day of their strange imprisonment, Owens began looking at the wires sticking out of the walls and wondering if perhaps the telephone service was still connected. The next time one of the soldiers went back to the hotel, Ledum sent another note asking to borrow a working telephone set. The soldier returned with the phone, and Ledum connected it to the line. Amazingly, he got a dial tone.

The government-owned phone company had not bothered to cut off the service from the government-owned house. Although Owens and Ledum couldn't receive calls (they didn't know the phone number), they were able to make as many free local and long-distance calls as they pleased. And they dialled everyone they knew – relatives, friends, reporters, NGOs. The only people Owens could not reach were Ken and Diana. Neither of them could be found.

Owens hoped his wife had gone to stay with her parents. That, at least, would explain why she was not home in Bori. Because the Barikors did not have a telephone at their university campus house, employees from Ledum's brother's hotel had gone to tell Diana's family where Owens was being held. Still, there was no word of how she was doing or where she was.

In detention, Owens had a vivid dream that their baby had been born and that it was a boy. He had been looking forward to delivering his first child himself. Now he didn't know if he'd be able to. Never were he or Ledum Mitee given any indication why they were being held or how long they might be locked up. Fortunately, the baby was not due until the end of January, so there was still time.

Ogoni Day, January 4, fell on Owens and Ledum's ninth day of captivity. They were understandably anxious about missing the year's most important event in Ogoniland, but they learned that they had not missed much. Ogoni Day had been drastically scaled back. In a further attempt to break the people's spirit, the military authorities did not allow the celebrations to proceed, although they did permit a modest religious service in a field surrounded by soldiers.

What most concerned Owens was that he heard Ken Saro-Wiwa had not attended the service. What could have caused him to miss such an important event? And why had Owens not heard from Ken? Where was he?

|||||||

On the morning of January 2, Ken had awakened to find troops outside his home. Ken's chauffeur, alarmed, had tried to lock the gates but was quickly overcome. As Ken watched, the soldiers beat the chauffeur, then placed Ken and his servants under house arrest. For the next three days, no member of Ken's household was allowed to leave the property.

At five in the afternoon on January 4, the fearsome Major Paul Okuntimo visited Owens and Ledum in detention. Okuntimo was in full uniform and jovial spirits, giving Owens and Ledum the friendliest of greetings. In no mood to indulge in pleasantries, Owens asked, "Why are we being detained?"

"I was following orders," said Okuntimo. "And now you are free to go."

After nine days of detention, Okuntimo's nonchalant attitude was galling. "But why were we here?"

"For Ogoni Day," said Okuntimo. "The governor, he did not want it to go on. And so we were keeping you here."

There seemed nothing more to say. As Owens and Ledum turned to leave, Okuntimo added, "You know, you're not going to get anything you want from Shell. Unless . . ."

"Unless what?" said Ledum.

Okuntimo motioned for the two to come back into the room. "We need to see that a Shell man be off," he said. "Since MOSOP is so big and powerful, you should join forces with us. He is standing in your way, too."

"Why do you want the Shell man off?"

"He is undercutting my payments," said Okuntimo. He explained to Owens and Ledum that he had been paid by Shell officials to crush Ogoni protests against the company. He was angry with the company, he said, because Shell was no longer paying for the upkeep of Okuntimo's men. Okuntimo placed most of the blame on Steve Lawson-Jack, head of Shell's public and governmental affairs in the eastern region, whom he named as his main link to Shell. Okuntimo felt that if MOSOP would "remove" Lawson-Jack, Shell money would flow into military pockets as before – and possibly into Ogoni pockets as well.

"We don't do things like that," said Owens. "We talk about ideas and how to change corporate behaviour in terms of the environmental devastation of Ogoniland. We are not in favour of removing individuals. Ken has said quite plainly we will achieve our goals peacefully and non-violently."

"Ken Saro-Wiwa is not the saviour of the Ogoni people," said Okuntimo with sudden ire. "The Ogonis know that I am their redeemer!"

"What are you talking about?"

"Ken Saro-Wiwa has turned these Ogoni boys into terrorists. He makes the lives of people who don't see things from the same perspective as himself very difficult. And this man, this Ken Saro-Wiwa, he writes a letter, a complaint to my superiors accusing me of human rights abuses and this and that. I am not the one doing human rights abuses. He is the abuser!"

"Forget about Ken and listen to me," said Owens. "MOSOP does not operate that way. We do not go about working to remove individuals from their posts. Our goal is more honourable."

"You are released," said Okuntimo with disgust. He walked out the front door, taking the guards on patrol with him.

Just then, Owens heard his name being called. From the other side of the high wall in the backyard, a woman's voice was shouting, "Monday! Monday!" Stepping outside, he realized it was his mother-in-law, Dora Barikor.

"I am here, Mother," he called back. "Where is Diana? How is my wife?"

"They are fine."

"What do you mean 'they'?" Owens asked. "I only have one wife."

"They are fine," she repeated. "You have a child."

"I have what?"

"A baby!" she said. "You have a son."

In his excitement, Owens started to tell his mother-in-law about his dream. Then it occurred to him that shouting over a wall was no longer necessary. He told Mrs. Barikor to come around to the front of the house. There, Owens learned that he had been a father for about three hours; Diana and the baby were resting comfortably at a private clinic in Port Harcourt. On the free phone line, Owens called for a car to pick him up.

When he and Mrs. Barikor walked into the hospital room, Owens froze. Diana was in the bed in front of him, the baby in an adjoining room. His mother and several MOSOP members were also there. Whom to greet first? The look on his face made Diana smile. Finally, he went to Diana's bedside, kissed her, and said, "Thank you very much."

Of all the things her husband might say, that was the last thing she expected. "Okay," she replied and burst into laughter.

New visitors kept arriving. Ken's girlfriend Hauwa Madugu showed up along with several MOSOP members Diana had never seen before. Owens's mother started to sing in Khana and perform the traditional dance to welcome the baby into the world. All the guests stayed late, and Owens spent the night in the clinic with his wife and child. The next morning, when he and Diana finally had a moment to themselves, she told him how their child had been born.

The night of Owens's arrest, the Wiwas' chauffeur was off duty. Diana had never learned to drive, so she had no way of going after her husband. "Korkora!" she called to the valet. "Go call Barika to come get me. We have to follow them!"

Korkora ran down the street to rouse Owens's nephew Barika Idamkue. In the meantime, Diana went inside to dress properly and gather up some things her husband would need in jail. Barika got to the Wiwas' house as fast as he could, bringing Owens's half-brother, Love Wiwa, who worked as a driver. Diana tossed Love a set of Owens's car keys and said, "Come on, quick! They went that direction!"

Barika, however, told Love not to go anywhere.

"What are you talking about?" said Diana.

"Look," said Barika, "if you follow them, they will just shoot you."

"But if we don't follow them, we won't know where they took him."

Barika stood fast and would not allow Love to get in the car. In tears, Diana told him that if he wouldn't drive her, she'd drive herself, and with that she opened the driver's door and tried to get behind the wheel. Barika grabbed the keys from Love.

"We will wait until it is light," he said. "Then we will go look for him. But not in the dark." Diana knew he was right. Driving around in the dead of night would be pointless.

At six in the morning, Love drove Diana and Barika to the Brigade in Bori Camp but there was no sign of Owens. They went to the air force base in Port Harcourt and to every other military installation they could think of, but no luck. Next they started checking all the hospitals in Port Harcourt to see if any bodies had been brought in.

The only thing left was to start informing the family that Owens was missing. Diana went first to Ken's house but he was not at home. She left a message with his household staff and then had Love drive her out to the university area north of town to see her parents.

Diana stayed at Ken Saro-Wiwa's house in Port Harcourt for the next few days while she went around the city looking for Owens. It didn't take long for her to realize the futility of her search. After asking around, she found out the name of the chief military officer in the city and his address in the G.R.A.

Diana instructed Owens's chauffeur to take her to the officer's house, but when they arrived, soldiers stopped them at the gates. Incensed, Diana reached over and started blowing the car horn. Mortified, the chauffeur pleaded with her to stop lest the army shoot them both. She continued to blow the horn incessantly.

For two days, Diana had Owens's driver take her back to the same house where she relentlessly sounded the horn. Astonishingly, she was ignored. On returning to Ken's after yet another fruitless day of honking, Diana received word that Owens was being held right next door to where she had been carrying on her honking campaign. Because of her advanced pregnancy, Diana's relatives made sure she was the last person to be told of her husband's location. Instead of being furious, she was almost amused to think that she had managed to get so close to Owens without realizing it. And at least he was all right.

The next morning she had their chauffeur take her back to Bori to pick up toiletries, shoes, and underclothing for Owens. On the way back to Port Harcourt, Diana's car was stopped by a soldier at a military checkpoint. Two soldiers farther down the road were also checking cars along the busy highway, but weren't asking any other cars to stop. For some reason, only the Wiwas' maroon Mazda was singled out.

"Show me your licence," the soldier asked the driver. The chauffeur handed over his driving permit. The soldier peered into the back seat.

"You, madam. Let me see your I.D."

"What kind of I.D. do you want?" asked Diana. "I don't have an I.D."

"I need to see your papers," he insisted.

"I'm a housewife. I've never heard of a housewife having papers."

The soldier demanded to see the vehicle's registration. Korkora took the documents out of the glove compartment and handed them to him.

"This say the car belongs to Doctor."

"Yes," said Diana.

"Are you the troublemaker's wife?"

"I don't know what you mean."

"Dr. Wiwa."

"Yes, I am Doctor's wife."

The soldier tossed the papers back through the car window. "Get out," he ordered. "All of you."

At gunpoint, Diana, Korkora, and the chauffeur were directed to stand at the side of the road. The roadblock was located beside a large, stagnant pool near one of the area's oil refineries and there was very little shoulder.

"Show me your identification," the soldier said.

Exasperated, Diana said, "I don't know what you mean by identification. I don't have any."

As the soldier grew more insistent, Diana became defensive and belligerent, which only enflamed the soldier. As the argument escalated, Korkora stepped between Diana and the soldier and held up a hand. "You should try to be a little gentle with the lady," he said. "Can't you see she is expecting a child?"

The soldier laughed. "That is not my business. I am not responsible for her pregnancy. She needs to show me her papers!"

"Look," she said, "there's nothing I can show you."

"If you don't show me some identification, then I will make you serve a punishment."

"But there's nothing I can do."

Pointing his machine gun at her, he said, "Then you must go into the pond."

The soldier could have picked no harsher penalty. Diana was acutely hydrophobic. Just looking into a body of water terrified her. The pool's stench was disgusting; heavily polluted with waste from the refinery, the water was covered with an iridescent slick of oil punctuated with globs of slime and floating garbage.

"Look, I can't go into that. Please, it's too . . ."

The soldier struck Diana in the back with the butt of his weapon. "Into the pond," he ordered.

She took a step and stopped. He struck her again. Weeping, she took a few tentative steps until she sank in the muck to her knees.

The two other soldiers approached. One said, "What's going on?"

As Korkora tried to explain, Diana gave a cry and grabbed her abdomen. A shocking cramp had overcome her. The newly arrived soldiers looked at one another; one said to Diana's tormentor, "Leave the lady alone. Let them go."

The first soldier shrugged and motioned for Korkora and the chauffeur to get back in the car. Korkora rushed to fish Diana out of the pond and help her into the back seat.

As they drove away, Diana asked, "Why did they let us go?"

"Because you're in labour," said Korkora.

"What do you mean?"

"You're going to have the baby."

"You don't know what you're talking about. I just got a cramp."

The chauffeur drove Korkora and Diana to her parents' house in Port Harcourt. Diana wasn't able to return to Ken's house because the army had put the entire place under house arrest. There could not have been a worse time for both Ken and Owens to be out of commission. Tomorrow was Ogoni Day.

Diana planned on attending the Ogoni Day ceremonies on her husband's behalf. Early in the morning, at around five, as she was getting ready for the drive back to Bori, she felt sharp pains in her side but assumed they would go away. When her mother discovered that Diana was having these recurring cramps, she asked her at which clinic she planned on having the baby. "We need to go there right away."

"I don't have time to go to the hospital right now," said Diana. "There is nothing wrong with me. I have to get to Bori."

When Diana picked up her bags and walked to the car, Mrs. Barikor got in the front seat.

"What are you doing?" Diana asked.

"I'm going with you."

"You're going to Bori?"

"We're going to the hospital," Mrs. Barikor said with maternal finality. Diana grudgingly assented, thinking she would pop into the clinic to appease her mother and then get on with her day.

At the clinic, when the doctor asked Diana how she was feeling, she admitted to a couple of cramps. He asked if she had felt anything wet. Yes, she said. The day before, standing in the fetid pond, she had noticed a sudden dampness between her legs.

"Your water broke," said the doctor.

Diana looked at him, puzzled. "Well, I have to go to this event for a while, but I'll come back."

Her mother howled with laughter. "You must be a crazy woman! You cannot go anywhere. You're having a baby."

By the time Owens arrived, gaunt and mosquito-stung from his nine-day detention, the crowd in the hospital room had christened the child born on Ogoni Day "Mosop." But as Owens gazed down at his first-born son, he did not feel political. He was not even resentful of his mistreatment at the hands of Komo, Odina, and Okuntimo. He felt nothing but joy and forgiveness.

"In spite of all the bad that has happened," he said, "God is now smiling on us, and we have been delivered of a healthy baby boy." He named his saro Baribefii, which in Khana means "God is victorious."

Baribefii Saro-Wiwa.

CHAPTERSEVEN

During the horrific autumn of 1993, the rates of disease and death in Ogoniland skyrocketed. This was hardly surprising, considering all that had just occurred. The problem was only exacerbated by the closure of the few government hospitals and health centres in the area, as well as the extreme overcrowding in many villages from internally displaced people. Despite such evidence, however, a growing number of Ogonis came to believe that the real cause their suffering came from the realm of the supernatural.

Tales circulated in Ogoniland of a mysterious wandering shaman who could identify witches. The vigilantes, who at this time were still acting as ostensible defenders of their villages, began to invite this shaman to come and cast out the forces of evil. In the village square, the shaman would perform sacrifices, recite incantations, arrange and rearrange his fetishes, then point at one house or another and declare, "That man is a witch!" or "That lady is a witch!" The unfortunate soul, exposed as the cause of the misfortune befalling the community, would be forced to confess and suffer a punishment. At times, such witches were beaten, and on at least one occasion the police were called in to investigate a witch-related homicide. The vigilantes called this practice "clearing the witch from Ogoni."

The vigilantes, of course, exploited these witch-clearings and profited from the resulting climate of fear. Witch hunts were also a source of vigilante revenue: a "witch" oftentimes could escape punishment after making a sizeable cash donation. As for village rivals, they found that the clearing of the witch could be an effective tool in settling old scores.

When Owens told his brother about the scheme, Ken refused to believe it: "You cannot be describing Ogoni to me."

"Well, that's the place it has become," said Owens. "That's what has happened to Ogoni. You have no idea, Dede, because you are always travelling abroad or in Lagos. You only come home to do your rallies and when they see you, the people are orderly and peaceful."

"Even so, this 'clearing the witch' sounds too strange."

"Yes, but it's true."

When a group of COTRA chiefs complained to Ken about the same thing, Ken decided to see what could be done to stop it. During a tour of the six kingdoms, he made it clear in his speeches that MOSOP would report to the police any lawless behaviour, and he asked people to identify the ringleaders. Ken's next move was to send the names of these identified vigilante leaders to the police and the army. It took three months for the army to act, and then only after severe pressure was exerted on them to do so.

The authorities were hesitant to crack down on vigilante lawlessness for good reason. By this time, the army and the police were collaborating with the vigilantes on their extortion schemes and other crimes. This was understandable, for the Abacha regime had recently stopped paying the delta's police and soldiers, and they needed the income.

Although crime in Ogoniland was statistically lower than in most parts of Nigeria, the international focus on MOSOP had led the media to cover the vigilante youth lawlessness in exhaustive detail. NYCOP, reeling from this negative publicity, felt it must do something to restore its reputation. In response to the extortions, witch hunts, and other crimes, NYCOP decided that all its members would be required to carry I.D. cards. This measure had little effect. The Ogoni elites, the government, and Shell continued to equate the illegal actions of any Ogoni teenager with MOSOP's youth wing. Increasingly, MOSOP was depicted and viewed as a violent organization that had to be quashed.

|||||||

On January 11, 1994, the Rivers State government set up a seven-member commission of inquiry to investigate the causes of December's violent Ogoni-Okrika clashes in the waterside shantytowns of Port Harcourt. Ken Saro-Wiwa delivered a presentation meant to show that the attacks had been masterminded by the government and by Shell. "The Okrika people who are busy laying claim to mud," he said, "would not be doing so if they were all the multimillionaires which the oil resources of their land entitle them to." He charged that a government besotted with Shell petrodollars was the party at fault, and that former Governor Ada-George and eleven others in government and Shell had played active roles in the events leading up to the disturbances – accusations that outraged the commission.[1]

Though MOSOP was increasingly under attack, Ken was feeling more confident in his ability to stand up to the Nigerian government. For one thing, his lawyer, Olu Onagoruwa, had been appointed the federal attorney general and minister of justice. Another close friend, Alex Ibru, was the new federal minister of internal affairs. Ibru had been publisher of the *Guardian*, Nigeria's most influential newspaper, and was a long-time supporter of the Ogoni movement. Because of Ken's personal relationship with him (and a tacit understanding that Ken would give Ibru's newspaper exclusive interviews and scoops), MOSOP had been getting a lot of positive coverage in the *Guardian*. Ken imagined that Ibru would continue this support on his first official visit to Ogoniland.

The Nigerian feds launched their own investigation into the crisis in the country's oil-bearing areas. On January 19, 1994, a federal ministerial committee began a tour of Ogoniland. For Ken, this was a momentous event, because it was the first time federal government officials had come to the area since the creation of MOSOP. The committee consisted of Alex Ibru, Minister of Petroleum Resources Don Etiebet, and Minister of Tourism and Commerce Melford Okilo under the escort of State Military Administrator Dauda Musa Komo.

The tour was Komo's opportunity to show the Abacha regime that he had the Ogonis under control. But things did not go well for the military governor. In Bori, a crowd of 150,000 met the committee waving placards with slogans like "The land is our life," "Rehabilitate Ogoni Refugees," "Pollution Kills Ogoni," and "Shell Stop Genocide in Ogoni."[2] As the crowd jeered, Komo, dressed in camouflage and sporting a beret with a tall feather cockade, stood with microphone in hand and stammered, "We want to be sure, and we want to, we appealed the last time and I'm appealing now, that most of the demands, including yours, can be made by the government, but that can only be made if there is a climate of peace, and fear and suspicions are removed from our minds."[3] He urged the Ogoni people to correct the world's impression that their struggle was synonymous with violence. The insult to their struggle only fuelled the crowd's angry response.

In contrast, when Ken Saro-Wiwa took up the microphone, the crowd ignited and the feds got a taste of Ken's messianic charisma. To the ministerial committee, he began, "On behalf of the entire Ogoniland, its chiefs, people, and spirits, along with the Movement of the Survival of Ogoni People, MOSOP, I say welcome to our Shell-shocked land."

Ken lambasted Shell's argument that because the federal government owned 60 per cent of their joint venture, the government therefore was responsible for the pollution, not Shell. "That is hogwash!" he said to the delight of the crowd.[4]

The thunderous ovation Ken received must have galled Komo. In front of these important committee members, the Ogoni people had lionized Ken and treated Komo with contempt.

On January 24, the three major oil companies in Port Harcourt urged Komo to take urgent measures in Ogoniland, estimating they had lost more than $200 million in 1993 because of "unfavourable conditions in tier areas of operation." This was all the reason Komo needed to begin his full assault on Ken Saro-Wiwa and the Ogonis. No sooner had the ministerial committee left Port Harcourt than Komo in his rage and humiliation created the Rivers State Internal Security Task Force, which he placed under the control of Major Paul Okuntimo.

Okuntimo's assignment was simple: box in the Ogonis and subject them to the authority of the Rivers State Internal Security Task Force.[5]

|||||||

After his recent arrest, Owens began taking precautions to ensure his family's safety and made plans to move Diana and the baby to Port Harcourt. The move would have practical value as well: Diana had kept her promise to her father and had gone back to university a few weeks after Befii was born. To allow his wife to be nearer to her school and parents, Owens rented a place in a new subdivision in the north end of Port Harcourt. Their house in this private walled complex was one of a dozen neat white bungalows tucked along winding landscaped lanes. All the other residents, ironically, were Shell employees.

Owens had begun splitting his time between his clinics in Bori and Eleme. In addition to his MOSOP-related meetings and activities, he continued his work with the Daughters of Charity, travelling twice a week to Kaa, Eeke, and Kpean to provide TB screening, disease tracking, and free medical exams.

At the Bori clinic one afternoon, Owens received a note from a doctor representing the Italian oil company Saipem, saying that he and his colleagues would like a meeting with the renowned Dr. Wiwa. The following week, the Saipem doctor returned to the Inadum Medical Centre in the company of two Italians. Taking a quick look around the clinic, the doctor asked, "Can we go somewhere else to talk?"

"Of course," said Owens. "Let's go to my house."

In Owens's home, the Italians got right down to business. They told Owens that Saipem wanted to construct an oil and gas pipeline through Ogoniland to the aluminium plant at Akwa Ibom State.

"What does this have to do with me?" asked Owens.

"We've been talking to some people," said one of the Italian oil men, "and they said we should talk to you. When we build the pipeline, we will need a hospital or a clinic to treat our employees. We'd like to discuss whether you and your clinic would be ready to do that."

The Saipem doctor said that his company was prepared to drill Owens a well to provide the clinic with running water, and give him another electrical generator, plus all the equipment necessary for Owens to have a medical laboratory on site. The pipeline was projected to take four years to complete and would likely employ between two hundred and three hundred people. Saipem would pay the Inadum Medical Centre a lump sum each month as a retainer and cover the cost of any drugs or medical supplies needed for Saipem's employees.

The offer was enticing. Owens thought about it for a moment. "Well, I am a physician and I treat people, but on the issue of pipelines and the rest, it would be better for you to talk to MOSOP." He gave the men directions to the office of MOSOP's deputy president, Ledum Mitee, thanked them for their time, and went back to his clinic.

When Ken later learned of Saipem's plans, he called on MOSOP to form a committee to deliberate the matter. At a subsequent steering committee meeting, the committee recommended that the Italians should not build a pipeline without first performing an extensive environmental impact assessment. After the committee made its presentation, Ledum Mitee made it known that he had been retained by Saipem as a solicitor.

"But you are the legal officer of MOSOP!" exclaimed Owens.

"Well, in my agreement with Saipem, I will not represent them in court against the Ogonis," said Ledum. "I will only give them legal advice on the Ogoni issue."

"Ledum, you cannot represent two opposing teams."

"How much are they paying you?" asked Ken.

"There is a retainer."

Owens glowered at Ledum. "When they came to me and said they'd be paying a retainership, the money they were promising me was mouth-watering. They were going to pay me something I do not even get as gross for three years! What did they offer you?"

After a bit of evasion, Ledum finally gave the figure: a half-million naira per year (approximately US $23,000).

IIIIIII

When a fact-finding delegation from the U.S. Army arrived in the spring of 1994 (one of many international delegations to investigate the crisis in the Niger delta), Owens escorted them to the destroyed communities of Kpean, Kaa, Gwara, and Biara. During the tour, he noticed the Americans picking up bullets and mortar shells and taking photographs. Back at the clinic after the tour, the Americans asked if they could visit the oil installations in Gokana on their way back to Port Harcourt. Before they had a chance to leave, however, a bus pulled up in front of the clinic and out spilled a group of frantic Ogonis bearing two wounded men. One was bleeding from a bullet wound to the thigh, the other from a serious wound to the chest. As Owens took them inside, the Americans snapped photo after photo.

The Ogonis said they had been among a group of traders travelling home by bus from Aba. At one of the four military checkpoints between Port Harcourt and Ogoniland, soldiers were extorting money from every motorist who passed. For no apparent reason, the soldiers opened fire on the bus. As Owens extracted a bullet from the man's leg, the American army major became furious: "Who would shoot unarmed people? I want to see who did this. Where can I find the people responsible?"

Owens asked a MOSOP activist who was with the group to take the Americans to the Bori police station. There, they were directed to see Major Paul Okuntimo at the Bori Camp army base in Port Harcourt. Having no knowledge of Okuntimo's reputation, the Americans went to Bori Camp to excoriate the head of the Rivers State Internal Security Task Force.

It was a decision they soon rued. When the Americans criticized Okuntimo, he pulled out a pistol, pointed it at the American major's head, and threatened to kill him. He then threw all three American soldiers into the guardhouse and held them there for several hours.

Another American who felt Okuntimo's wrath was *Wall Street Journal* reporter Geraldine Brooks. On April 9, 1994, the Australian-born journalist had just interviewed an Ogoni woman who had suffered extensive burns after mistakenly setting her lantern down at night on a pool of oil seeping from a broken pipeline. When Brooks approached an army officer to ask for the military's account of some of the violent incidents in Ogoniland, she was handed over to the sss and interrogated for two days before being deported back to the United States for "security reasons."[6]

||||||||

On Easter Sunday, April 3, 1994, Nigerian troops attacked the northern borders of Ogoniland. A security report claimed that ten thousand armed Ogoni youths were planning to besiege the Afam Power Station in Tai Kingdom. After receiving this report, Lieutenant Colonel Komo ordered his Internal Security Task Force to destroy all the Ogoni villages in the area.[7]

Owens was attending to his normal rounds that morning, when, suddenly, the Inadum Medical Centre was overrun with hysterical Ogoni people carrying in family and friends with gunshot and machete wounds. There were so many injured people that the clinic's stock of medical supplies became rapidly depleted. Before long, the only thing Owens could do was ask one of his doctors on staff to try to keep things under control while he drove to Port Harcourt for more sutures and intravenous drips.

Instructing his ambulance to follow behind his personal car, Owens left for the city. On the way to Port Harcourt, both sides of the road were lined with injured people streaming southward. As Owens's car and ambulance slowly passed, people waved and called out for help.

"Where is Bori?"

"Where is Inadum?"

When Owens pulled over, they told him that Nigerian soldiers were destroying their towns and killing any Ogonis they could find. Incensed, Owens ordered his driver to take him to see an army captain he knew in Port Harcourt. Finding the captain at home, Owens demanded, "Why are Nigerian soldiers shooting at these villagers in Tai?"

"You shouldn't worry," said the captain. "Everything is being taken care of."

"What do you mean? People are coming to my clinic. They are being shot. They are being killed."

"It is being taken care of," he repeated.

"Please," Owens said, "let's go together and see the place being attacked. Come with me."

The captain refused. Reasoning that if the army wouldn't do anything, the media might, Owens went to pay a call on Tayo Lukula, a reporter for the *Guardian* newspaper. When Owens told him of the attack on Tai, still in progress, Tayo said he couldn't write an article based on hearsay. He'd have to do his own investigation.

"You can come to my clinic. You can meet the people and talk to them."

"No," said Tayo. "I want to see the place that is being attacked."

"Fine. Let's go right now."

At eleven in the morning, Owens, the journalist, and the Inadum ambulance reached the village of Nonwa, Tai, where they stopped at the house of Noble Obani-Nwibari, the MOSOP vice-president representing Tai Kingdom. On hearing that they were heading for the battle front, Noble offered to escort them himself. After passing three small villages, they came on a sign that read "Operation Starts Here." Directly beyond was the village of Oloko I, where houses were in flames and the bodies of slain villagers and goats littered the ground.

"Oh my God," said Owens softly.

Driving past the ravaged village to its nearby twin, Oloko II, he witnessed an identical scene. Seeing the Inadum ambulance, bloodied survivors filtered out of their hiding places to plead for help. Owens's chauffeur and his ambulance driver gathered as many victims as fit in the ambulance and rushed them back to the clinic.

With the ambulance and two drivers gone, Tayo said, "We need to find the soldiers and ask why they are killing villagers."

"Then we must go on farther to see where the army checkpoint is," said Owens.

Noble pointed down the road: "It is near the Afam Power Station."

With Owens at the wheel of his maroon Mazda, the three ventured on. The checkpoint they encountered was merely a long wooden pole stretched across two oil drums. Four soldiers sat on a makeshift bench.

"Stop!" one of the soldiers called out. "Stop the car or I shoot you right now!"

Owens came to a stop. When he, Tayo, and Noble stepped out, they were ordered to put their hands up. As more soldiers arrived, the checkpoint guards searched the Mazda for weapons or hidden passengers. Finding nothing, they frisked the three Ogonis.

"Take off your shoes."

The detachment's lieutenant approached Owens and said, "Who are you?"

"I am Dr. Wiwa. People who were shot and cut have been coming into my clinic. I wanted to come and see if there were more people injured so that we can take them to my clinic."

"In your car?"

"My ambulance was following me. They have already taken some people but they will be coming back for more."

"Turn around and face the forest!" shouted the lieutenant.

Hands raised, Owens, Tayo, and Noble turned their backs to the checkpoint. Staring into the trees by the roadside, Owens heard the squawk of a walkie-talkie and then the lieutenant saying something unintelligible. The next sound was the cocking of rifles. Then the metallic screech of truck brakes.

"No! You can't do this!" a man's voice rang out. "You can't shoot them while we are here!"

Looking back over their shoulders, Owens, Tayo, and Noble saw a four-man firing squad aiming at their backs. Behind the squad was a Mercedes-Benz 911 military transport vehicle disgorging a dozen police officers.

"Stop!" yelled one of the policemen. "This cannot happen in front of us."

Owens gaped at the men who had disrupted the impending execution. He recognized the one who had just shouted as an assistant superintendent of police, and driving the police vehicle was the father of one of Owens's patients, a young girl suffering from sickle-cell anaemia. Owens had been giving the policeman's daughter free treatment at the Inadum Medical Centre in an effort to create goodwill between the Ogonis and the police. Apparently by doing so, he had created his own lion of Androcles.

"If you want to do this, you have to wait until we are out of here," said the assistant superintendent. He motioned for his fellow officers to gather around, assuming a stance that left no doubt they weren't leaving. A vicious argument broke out between the army and the police, with both sides waving guns at one another. It was so surreal that all Owens, Tayo, and Noble could do was stand with their arms raised and gape at the commotion.

Whipping out his walkie-talkie once more, the frustrated lieutenant walked toward his jeep to radio his superiors. Abruptly, he returned with his new orders: the three captives were not to be shot on site, they were to be arrested. The police, however, saw this somewhat transparent tactical manoeuvre for what it really was. Obviously, the army was simply going to take Owens and company a little farther down the road and shoot them elsewhere.

As the army loaded Owens and the others into their jeep, the police scrambled to get back in the Mercedes and succeeded in getting on the narrow road ahead of the army.

"Overtake them!" shouted the army lieutenant, but there was no room for the jeep to get around the truck. Whenever the road widened, the police vehicle swerved to block the way. After five kilometres of this game, the lieutenant waved his fist out the window and screamed in rage for the police truck to stop. When the police failed to pay him any attention, he pulled out his pistol and fired into the air. The Mercedes came to an abrupt halt, and the army lieutenant jumped out and ran toward it. Pulling open the truck door, the lieutenant dragged the driver out and began to beat him.

"Stay there!" he commanded the driver, who lay crumpled and gasping on the ground. "If you get in our way again, I will kill you!"

The lieutenant jumped back in the jeep and motioned for his driver to go around the Mercedes. As they drove away, Owens heard the policemen shout at the army lieutenant: "We know you are with them! We know! If anything happens to Dr. Wiwa, we know you did it!"

|||||||

Owens, Tayo, and Noble were transported to the Bori Camp army barracks in Port Harcourt.[8] The officers in charge recognized Tayo Lukula from his articles in the *Guardian* and released him. Owens and Noble, however, were taken inside and presented to Major Paul Okuntimo.

"We have been looking for you," said Okuntimo. "Now that we have you, we're going to put you away for a very long time."

Owens looked at the major as if he were insane. "This must be some sort of joke."

Okuntimo nodded to a soldier, who ordered Owens and Noble to step into the next room. When Owens hesitated, the soldier sent him flying through the door with a violent shove. In the next room, the two captives were commanded to strip naked.

"I am not going to strip," said Owens.

The soldier slapped Owens soundly across the face for his defiance.

Stunned, Owens said, "I know you, don't I? Your name is Kennedy, right?"

"So?"

Owens was dumbfounded by the man's attitude. He had once been a guest at Kennedy's house. He had treated Kennedy for free at the Inadum Medical Centre. Now the man was slapping him in the face.

"We have to search you. Now strip."

"Go on and search me, but I am not removing my clothes."

Kennedy turned his attention to Noble Obani-Nwibari, who submitted to a full strip search. Following Noble's lead, Owens relented and allowed himself to be strip-searched before both men were cast into the wretched netherworld of the army brig.

The holding cell of the cantonment was the former military base armoury, a large fortified area now filled with hundreds of desperate souls. The armoury was hot and cramped and stank of shit, piss, and the musk of unwashed men. Caked blood was everywhere: on the ground, on the walls, on faces, legs, and torsos. Many of the inmates were Ogoni villagers taken during the raids on Oloko I and Oloko II, arrested while trying to flee the army's extermination of their community.

The other prisoners were hardened criminals. A few were vigilante leaders whom Ken had written about to security officials and one Owens had identified to the police. When he realized that he was being put in the same cell with vigilantes, he felt abject terror. His fears were put to rest once the Oloko villagers began to crowd around the great Dr. Wiwa. All the prisoners – including the vigilantes – showed Owens and Noble great respect, even clearing space for them in the holding cell.

Some soldiers at Bori Camp recognized Owens as well. Like the policeman with the ailing daughter, they too felt a debt of gratitude. In the circumstances, there wasn't much they could do for Owens beyond giving him advice. A young private came into the holding cell and called Owens aside. "Doctor, doctor," he said, "the thing you have to do now is zero your mind. That's what you have to do when you're in detention."

"What do you mean, zero my mind?"

"Forget about everything outside this detention. Just feel that this detention is your whole life. This room is your whole life. Forget about your bed at home. Forget about your clinic. Forget about your mother, your wife, and everything."

After Ken's last detention, Owens knew he might be a prisoner at Bori Camp for a long time. The only way to stay sane would indeed be to let go of memories of the outside world. He would take the private's advice and "zero his mind."

Three days later, the soldiers flung a new prisoner into the guardroom. Seeing that the man was terrified, Owens said to him gently, "Come in. Relax. Here, come have a seat."

The new prisoner timidly sat down and said that he was frightened because he was the only non-Ogoni in the place.

"Who are your people?" Owens asked. "The Igbo? Okrika? Andoni?"

"Ndoki," said the man. "You Ogonis are supposed to be fighting people from our place."

Owens and Noble Obani-Nwibari assured him that the Ogonis had no quarrel with the Ndokis or anyone else. The "ethnic clashes" that had been occurring were orchestrated by the government as a way to return Shell to Ogoniland. Hearing this, the Ndoki man broke down and confessed a terrible secret.

"I was sent to give a false report to the police and army that the Ogonis were attacking the Andonis," he said.

Owens and Noble looked at one another, and then Noble said, "Why would somebody tell you to do that?"

The Ndoki said that he was an oil company subcontractor and was told the only way he could get his contracts renewed would be to file the false report. The Ndoki broke into tears as he revealed that when he did as ordered, the soldiers went to the communities he had falsely accused and began burning the Ogoni homes and shooting the inhabitants.

"So why were you arrested?" asked Owens.

The Ndoki shook his head in disbelief. "The police came to my house and arrested me for giving a false report," he said.

Before they had a chance to learn much more of the Ndoki's story, Owens and Noble were transferred to Port Harcourt's Central Police Station. As bad as the detention centre was, the Central Police Station was worse. When they got there, the police once again asked Owens to strip and again he refused.

"No problem," said the policeman doing the booking. Pointing toward the guardroom, he added, "You yourself will want to come and strip when you get in there."

The policeman recorded Owens's name, address, profession, and date of birth in his logbook. He also wrote that the two Ogonis were being charged with murder and arson.

"Murder and arson!" exclaimed Owens. "What are you talking about?"

"Yes, murder and arson," said the policeman. He ordered his fellow officers to throw Owens and Noble into the guardroom.

The large cell was dark even in the daytime – so dark, in fact, that Owens could hear the groans of men inside but could not see any of them.

"I'm not going into that place," he said. "No matter what you do to me, I am not going in there when I have not done anything."

One of the policemen in the reception area was an Ogoni Owens knew well. He called Owens aside and said, "There's no other place they can keep you."

"There is no way I'm going in there!"

"Well – let me call my supervisor."

The Ogoni policeman spoke to his superior, who allowed Owens and Noble to sit on a bench outside the door to the guardroom until the end of the policeman's shift. At the start of the next shift, the officers on duty entered the guardroom and counted the prisoners and found they were short by two. When they figured out that it was Owens and Noble, they ordered them into the cell. Again, Owens flatly refused.

"I'm not going in," he said. "I'm going to stay right here. Don't worry, I'm not going to run away."

This became the pattern at every shift. When the police saw that Owens was adamant, they let him and Noble remain on the bench. In addition, Owens soon started receiving visitors, who bribed the officers into sparing Owens and Noble the horrors of the guardroom. Owens's sister Barine was among the first to come, bringing money and food. Diana and her father came to see him as well.

When Diana walked in the door, Owens was taken aback. She was an absolute vision. Her hair was elegantly styled and she wore a beautiful new dress. Catching sight of her husband on the bench outside the guardroom, she approached him with a beaming smile.

"What is this!" he said sharply. "Why are you coming looking this way? Look at the way I am and you are so dressed up!"

Seeing Owens's annoyance, Diana's radiant smile turned to tears and she fled, weeping, from the police station.

"What is wrong with this girl?" Owens asked his father-in-law. "What was she thinking coming to this place looking so beautiful? And why is she crying?"

"It's her birthday," said Professor Barikor.

In his predicament Owens had forgotten his wife's birthday. He felt terrible. Diana had been working hard during his time in detention. She had made sure that food was delivered each day and was working with the Daughters of Charity to lobby the government and international human rights organizations for her husband's release. In fact, Diana had been much more help than Ken had been.

For some reason, even though Ken was in Port Harcourt, he had neither written nor visited his brother. Owens wrote him a letter: "Dede, what is the problem? Why have you not come? Do you think I was responsible for what they say I did? You cannot think I would ever commit murder or arson!"

Owens had real reason for worry about what the Nigerian government was doing to blacken his name. To help build its case that Owens was a dangerous felon, Okuntimo came to the Central Police Station and filed a report claiming that Owens had been the cause of the military action against Oloko I and II. Astoundingly, Okuntimo claimed that Owens was the leader of a raiding party of Ogoni youths who were set to attack the Ndoki villages just across the Imo River. In his statement, he said the Ogoni warriors were singing war songs and chanting, "Owens Wiwa! Owens Wiwa!" as they marched off to slaughter the Ndokis. Owens couldn't imagine that Ken would have believed this nonsense. Still, why had he not come?

After several days, Ken did finally come to see his brother. Unbeknownst to Owens, Ken had been working all along to secure his freedom. With the Daughters of Charity, he had been putting pressure on the governor and the bishop to intercede and had alerted Amnesty International. Amnesty sent out an urgent release naming Owens and Noble prisoners of conscience and asking people around the world to write letters demanding their release.

Naturally, Owens and Noble were not allowed to remain on a bench in the hallway of the jailhouse forever; despite their bitter protestations, they were finally thrown into the dark horror of the guardhouse. The cell had been built to accommodate seven prisoners, but with the addition of Owens and Noble, more than twenty men were now squeezed into this dank, fortified holding tank. Some of the men were suffering from infected lacerations from floggings; others had developed pustular rashes from sitting in their own filth. Latrine matters were carried out with a communal bucket.

The prisoners called the guardroom "Angola." Like the frightful country it was named after, it had its own form of government. The role of president was claimed by the most fearsome prisoner, who formed his cabinet. There was a minister of defence, a minister of justice, even a minister of health. At dawn, the "president" would lead his miserable constituency in the Angola prison anthem and then the prisoner with the lowest status would empty the latrine bucket. After rapping on the bars, the inmate would be let out by a guard and escorted into the inner courtyard, where he would dump the urine and feces on the open ground, causing the entire building to be permeated with the foulest stench imaginable.

When the prisoners of Angola were fed, which wasn't often, they received a crust of bread and a cup of soup containing so few beans the legumes could be counted through the weak broth in a glance. As often as they were allowed, Diana and Barine came to the jail with home-cooked food. They learned quickly to make enough for all the men in the cell. Any had to be offered first to the president, who then shared with the others whatever he left behind.

From Angola, Owens could see a smaller holding cell across the inner courtyard where women prisoners were kept. At the time, there were only two women in custody; one had an infant. Every night, the woman with the child was let out of her cell and forced upstairs to sleep with a different policeman. In the morning, she'd return to detention looking haunted and soulless. Worse was the arrival of the wheelbarrow in the night. Nearly every evening, the police would haul away one or more of the detainees, those too ill or injured to walk.

"Where are you taking him?" asked Owens the first time he saw one of his cellmates put in the wheelbarrow.

"The Teaching Hospital," said a policeman as he wheeled the gravely injured man out the door.

At first, this gave Owens some comfort. The University of Port Harcourt Teaching Hospital was only a five-minute walk from the Central Police Station and he knew the poor prisoner would receive good care there. But shortly after the injured man was wheeled away, a gunshot could be heard. One day, Owens asked a policeman if he knew what became of all these men.

"They take them to the cemetery down the street and shoot them," said the officer.

"Just because somebody is ill?"

The policeman shrugged. "They are all armed robbery cases anyway."

Owens knew that most of these so-called robbers were innocent men who had been picked up because of their ethnic background and then savagely beaten into confessing a crime. By the time they were cast into Angola, many were suffering from serious internal injuries or wounds that quickly became septic. Within a short time, these men would become so weak they could no longer stand. And the police would come with the wheelbarrow in the night.

||||||

After Owens and Nobel endured a week at the Central Police Station, Professor Barikor arrived to say that they were being released on bail.

Curiously, the police had asked for no payment toward this so-called bail. As Owens signed the zero-naira bail bond, he asked the officer in charge, "How come you are accusing us of murder and arson and you are releasing us on bail without taking any money?"

The officer motioned for Owens to step into his office. Out of earshot of the rest of the station, he said, "All the people who were killed at Oloko were Ogonis and the houses that were burned down were all Ogonis. Okuntimo wrote that you were attacking the Ndokis, but nobody from Ndoki was injured or killed. So we cannot charge you for these crimes." A further reason for their release was the recent publication of an article in the Nigerian *Guardian* by Tayo Lukula describing his harrowing experience with Owens and Noble at Oloko.

When the final tallies were taken, it turned out that between April 3 and 15, 1994, more than eight hundred Ogoni men, women, and children had been massacred and six villages were completely flattened by Okuntimo's forces. As part of this operation, women and children were raped and villages looted before they were burned down.[9,10]

"Life has become cheap in Nigeria," said Owens to his father-in-law as they left the Central Police Station. "It makes me sick."

"We should report it," said Professor Barikor.

"Yes," said Owens. "But who would listen?"

|||||||

On April 21, Lieutenant Colonel Komo drafted a memo to Major Okuntimo entitled "Restoration of Law and Order in Ogoniland, Operation Order No 4/94." In it, Komo detailed an imminent, extensive military operation that was to involve members of the Nigerian army, air force, navy, and police. The stated objective of Order 4/94 was to "ensure that ordinary law abiding citizens of the area, non-indigenes resident [sic] carrying out business ventures or schooling within Ogoni land are not molested."

The real purpose of Order 4/94 was clear to Ken Saro-Wiwa. He was convinced the drafting of such a large force into the small Ogoni area was meant to intimidate and terrorize the Ogoni people so that they would allow Shell to recommence its operations without carrying out the environmental, health, and social impact studies the Ogoni people had been demanding.[11]

In reply to Order 4/94, Okuntimo sent Komo a restricted memorandum entitled "RSIS OPERATIONS: LAW AND ORDER IN OGONI, ETC," which outlined strategy. More importantly, it confirmed Ken Saro-Wiwa's repeated allegations that the Ogoni crisis had been contrived by the government for Shell's benefit – and that the ensuing military crackdown was expressly intended to enable Shell to resume oil operations in Ogoni.[12]

Okuntimo's communiqué listed in point form several "observations," including the following:

- Shell operations still impossible unless ruthless military operations are undertaken for smooth economic activities to commence.
- Division between the elitist Ogoni leadership exists.
- Either bloc leadership lacks adequate influence to defy NYCOP decisive resistance to oil production unless reparation of 400 million dollars paid with arrears of interest to MOSOP and Ken Saro-Wiwa.

Among the memorandum's recommendations were these:

- Wasting operations during MOSOP and other gatherings making constant military presence justifiable.
- Wasting targets cutting across communities and leadership cadres especially vocal individuals, various groups.
- Wasting operations coupled with psychological tactics of displacement/wasting as noted above.
- Deployment of 400 military personnel (officers and men).
- Initial disbursement of 50 million naira as advanced allowances to officers and men for logistics to commence operations with immediate effect as agreed.
- Surveillance on Ogoni leaders considered as security risks/MOSOP propellers.
- Ruthless operations and high level authority for the task force effectiveness.[13]

A spokesperson for the Shell Petroleum Development Company in London later categorically denied that SPDC had any involvement in, or knowledge of, Okuntimo's plan to "waste" Ogoni leaders. On the contrary,

Jessica Wiwa with sons (from left) Owens, Letam, and Jim. Bori, Nigeria, 1966. (*Courtesy Owens Wiwa*)

The Wiwas. Bane, Nigeria, 1983. *Back (L to R)*: Ken, Letam, Owens, and Jim. *Front (L to R)*: Barine, Jessica, Chief Jim Wiwa, Comfort. (*Courtesy Owens Wiwa*)

Ken Saro-Wiwa, spokesman of the Ogoni people. Ogoni, Nigeria, January 1993. (*Greenpeace/Lambon*)

Oil pipeline blowout near Sime village, Eleme Kingdom, November 18, 2001. At the time the photo was taken, villagers reported that the fire had been burning for seven months.

(*Morton Beiser*)

Oil pollution from leaking pipeline. Ogoniland, Ebubu, 1993.

(*Detlef Pypke/still pictures/Alpha presse*)

Inadum Medical Centre. Bane, Nigeria, September 2000. (*Morton Beiser*)

Ogoni Day, January 4, 1993. Thousands of peaceful protesters take control of a local oil station. (*Courtesy Owens Wiwa*)

Diana Wiwa, Anita Roddick (former president of The Body Shop), and Owens Wiwa address Shell employees outside Shell International offices in London, December 1, 1995. (*Greenpeace/Robert Morris*)

Doris Lessing, Diana Wiwa, and Owens Wiwa in Lessing's London home, December 1995. (*Courtesy Owens Wiwa*)

Greenpeace protesters stage a re-enactment of Ken Saro-Wiwa's execution. San Francisco, November 1995. (*Greenpeace/Melanie Kemper*)

Banner erected by protesters at the symbolic funeral of Ken Saro-Wiwa. Bane, Nigeria, April 24, 2000. (*Morton Beiser*)

Funeral procession at the symbolic funeral of Ken Saro-Wiwa. Bane, Nigeria, April 24, 2000. (*Morton Beiser*)

Ken Saro-Wiwa's personal effects: The ring and watch that helped identify his remains. (*Courtesy Owens Wiwa*)

Symbolic tomb of Ken Saro-Wiwa. Bane, Nigeria, September 2000.
(*Morton Beiser*)

insisted the spokesperson, Shell had urged the Nigerian government to reduce military operations in the region. In response to Shell's claim of innocence, representatives of Greenpeace sardonically suggested that Shell go even further and pressure Nigeria to halt any killings carried out in its name, even if it were ignorant of their motives.[14]

A year later, after receiving a leaked copy of Okuntimo's heinous restricted memo, Ken observed in a letter to American environmentalist Stephen Kretzmann, "This is it. They are going to arrest us all and execute us. All for Shell."

|||||||

Ken Saro-Wiwa received a jolt when he opened the newspapers on the morning of May 14, 1994. Both the *Sunray* and the *Nigerian Tide* carried notices entitled "Statement by the Chiefs and Leaders of Gokana: The Giokoo Accord" – a document that announced that the people of Gokana had disassociated themselves from MOSOP.

"It is hereby published for the information of the general public that the people of Gokana have resolved to put aspects of the events of the past two years in Ogoni behind them and to pursue vigorously the peace, unity and progress of the local government area. The chiefs, traditional rulers, men, women, elders and youths regret that the generality of the people were misled through false propaganda and misinformation," said the announcement. Among other things, it resolved "that the Movement for the Survival of the Ogoni people (MOSOP) was formed as a non-violent organization committed to the pursuit of justice and fairplay for the Ogoni people within the Federal Republic of Nigeria. Gokana people remain committed to this course but reject the diversion from the noble ideal."[15]

Among the signatories of the Giokoo Accord were Edward Kobani, Albert Badey, Samuel Orage and his brother, Theophilus Orage – four men who would forever change the course of Ogoni history a mere seven days hence.

CHAPTEREIGHT

In early May 1994, an unsigned pamphlet circulated in Ogoniland. It carried a list of fifteen purported facts about the Ogoni "vultures" and Shell. Titled "NEWS FLASH," it accused Edward Kobani, Albert Badey, and Bennett Birabi of accepting lucrative contracts from Shell for upcoming petroleum development projects. It also claimed that Senator Birabi and the Gokana elders were holding secret meetings with the government and Shell in an effort to eradicate Ogoni opposition to the recommencement of Shell's activities in the area. Worse, the pamphlet announced that Shell promised "400 million Naira in foreign currency to Ogoni Vultures to destroy Mosop and Ken Saro-Wiwa." It closed with an ominous call to Ogoni youth to "pay vultures in their own coins."

Understandably, the signatories of the Giokoo Accord were terrified to see themselves accused of being promised the equivalent of $4.5 million by Shell to destroy both MOSOP and Ken Saro-Wiwa.[1] Although the author of the pamphlet was unknown, the suspects were countless; most people of Gokana had no intention of disassociating themselves from MOSOP. The only Gokana group dissatisfied with MOSOP was the small coterie of elites who signed the accord.

On May 19, Owens and Diana went to visit her family's ancestral home in the Gokana village of Bomu and came on a massive protest rally. Such rallies had become common, but this was different. Instead of berating Shell, the people of Gokana were protesting against their own leaders. Although the anti-Giokoo Accord rally was entirely peaceful, Owens could hear the anger in the protesters' voices.

That same month, Sani Abacha produced a new target date for the restoration of democracy in Nigeria. Since Abacha's rise to power, there had been little progress toward his early promises to return the government to civilian rule. If anything, the country had grown more fascist as Abacha dismantled all elected institutions, terminated all national and state assemblies, closed independent publications, and banned all political activity. Nevertheless, in May 1994, the military dictator announced that Nigeria would hold a constitutional conference, with one-third of the delegates to be Abacha's own appointments. The other two-thirds were to be elected by public ballot, with the government retaining the right to reject any decisions made by the conference.[2]

Ken Saro-Wiwa was determined to be a delegate to this conference. Two months earlier, at the Constitutional Rights Projects workshop in Abuja, he had given Nigeria a preview of how he planned to push the Ogoni agenda at the upcoming conference. "The definition of Nigerian democracy is the government of Ogoni people by Hausa people for Yoruba people," he said. To the workshop participants, Ken put forth his vision of a new Nigeria made up of a federation of ethnic groups, each constituting a unit of administration equal to the others, each having an identical structure, and each living in accordance with their means. He further proposed the radical notion that Nigerian citizens should pay personal income tax, which would make people hold elected leaders accountable for the spending of their own hard-earned naira.

"The responsibility of every citizen and every ethnic group must be to contribute to the nation, not merely to demand a living from that nation," said Ken. "Therefore, all citizens must work and pay their taxes. Nigeria must back away from the system of administration followed by military dictators, which has confused the nation, turned virtually every citizen into a beggar, holding out a bowl for oil money."[3]

It was during this period that the Gokana Council of Chiefs and Elders scheduled a May 21 meeting at the palace of His Royal Highness, Chief James Bagia, the Gbenemene (the king of) Gokana.[4] The purpose of the

meeting was to celebrate the appointment of two Gokana residents to prestigious positions in the Rivers State government.[5] After the meeting was
announced, rumours began circulating around Gokana that someone was
planning to murder the vultures. Although unsure of where or when these
murders might take place, Dr. Garrick Leton, along with a handful of other
Gokana chiefs, alerted Lieutenant Colonel Komo to the rumours as well as
the May 21 meeting. Komo promised to "take care of the situation."[6]

Unwittingly, Leton gave the governor exactly what he had been looking
for – the date and location of an upcoming Ogoni gathering, a confirmation
of the widening schism in MOSOP, and an unsubstantiated threat of violence. For Komo, it was the perfect opportunity to direct Paul Okuntimo to
begin the "wasting operations."

||||||

Around noon on May 21, 1994, one of Owens's colleagues[7] came to the
Inadum Medical Centre. "Where is Dr. Wiwa?" he asked one of the nurses.

"In the operating theatre."

"Will he be long?"

"No, I think he's almost through."

Upstairs, the doctor knocked on the door of the operating room before
poking his head inside.

"Dr. Wiwa?"

Owens was putting the last few stitches into a patient upon whom he
had performed an appendectomy. He looked up from his work. "*Ododowa*,"
he said in greeting. "How can I help you?"

"I need a word," said the doctor.

Owens pulled the suture tight and asked his assistant to snip the thread.
"I'll be with you in just a minute."

"Please," said the doctor. "It's important."

Owens saw that his colleague was shaken. "Would you finish closing for
me?" he asked his assistant. Stripping off his gloves, he stepped into the
hallway.

"There's been a tragedy," said the other doctor.

"What do you mean?"

"Something awful has happened."

"What? Where?"

"In Giokoo, in Gokana," said the doctor. "I was at a meeting when I noticed a commotion outside. I went out and saw that one of the leaders, Albert Badey, was being roughed up."

"By who?"

"The crowd," he replied. "There was a large crowd making a lot of noise. I saw that Mr. Badey was very terrified. He was gasping for breath. And then the people saw him reach in his pocket and pull out something like a whistle."

"He's asthmatic," said Owens. "He carries an inhaler."

"Yes, I know," said the doctor. "But someone yelled out that Badey was whistling for the police, and that panicked everyone so they took the thing away from him and assaulted him even more."

"How bad is he hurt?"

"Very bad," said the doctor. "He – when I left, he was no longer moving."

As Owens expressed his concern, the doctor continued, "Also, as I was coming to Bori, I heard that Ken was detained by the police."

Owens pulled off his surgical gown, washed his hands, and ran down the stairs. If Ken had been arrested in Gokana, the police station in Bori would be the logical place he would have been taken. Owens arrived at the station to find Assistant Superintendent of Police Stephen Hasso standing outside speaking to a Gokana elder named Mr. Giadom. Parked in front of the station was the long black American car belonging to Mr. Giadom's son.

"What is this I'm hearing?" said Owens to the assistant superintendent, interrupting. "I hear you've arrested my brother again."

"No, I haven't arrested him. Ken was not arrested. He was escorted to Port Harcourt."

In anger, Mr. Giadom turned on Owens: "It's your brother who is the cause of all the problems in Ogoni! We were all fine until he came with this MOSOP thing and organized the people. And we are having all these problems now!"

"Sir, I don't think so," Owens responded. "If anything has happened to my brother, you are the sort of people who have caused him to be harassed the way he is being harassed."

Suddenly, the rear door of the American sedan opened and Giadom's son emerged, rushed toward Owens, and slapped him hard across the mouth. "You cannot talk to my father like that!"

Seeing Dr. Wiwa assaulted, the people on the streets of Bori came instantly to his defence. Attempting to calm them, Owens said, "There is no problem! There is no problem!"

Just then, Barinem Kiobel, a commissioner in the Rivers State government, approached Assistant Superintendent Hasso and begged the police to go with him to Giokoo to attend to a major disturbance in progress there. Hasso did not seem interested in the slightest.

Frustrated, Dr. Kiobel pulled Owens aside from the group still bickering over Giadom's slap and said, "Don't get involved in any squabbles. I've come to tell the police to go to Gokana. There is some problem going on there."

"Yes, I know. I've heard that Albert Badey has been roughed up. How is he?"

Kiobel stared at Owens, at a loss for words.

"I don't know," he finally said.

Owens returned to the clinic to tell his nurses that he was going to Port Harcourt. On the way, Owens's car was overtaken by several vehicles travelling at a speed that seemed excessive by even Nigeria's breakneck standards. The cars were racing toward Port Harcourt.

Owens arrived at his Port Harcourt bungalow at about three in the afternoon and found Diana playing with the baby. "Have you heard anything from Ken?" he asked.

"No, why?"

"There has been some trouble."

"What has happened?"

"Some commotion in Gokana. I need to go to Ken's house to see if he's all right."

Diana insisted on making some food for Owens but she had scarcely gone into the kitchen when one of her uncles arrived. He too was bringing news of trouble in Giokoo.

Owens urgently asked him for more information.

"There was a meeting. I think there was about seventy leaders there," the uncle said. "Doctors, lawyers, civil servants, businessmen and all that. They were having a meeting at the Giokoo palace."

"I didn't know about this meeting."

"We had no idea," said Diana's uncle. "MOSOP didn't know. But from what I heard, they had sent a notice that they were having a meeting to honour Gokana people who had been given appointments by the governor of Rivers State, Colonel Komo. So the meeting was to honour Dr. Kiobel and one other person whom I cannot remember."

"I just saw Dr. Kiobel at the police station in Bori," said Owens. "He was telling the police to come and stop a big disturbance. What went wrong?"

Diana's uncle didn't know, but said he thought some people were dead. He asked, "Have you talked to Ken?"

"No, but I'm going right now. Diana!" he called toward the kitchen. "I'm going to Ken's house."

On his arrival, Ken's house staff informed him that Ken had gone to his office. Owens hurried over to 24 Aggrey Road and found Ken sitting behind his desk, pecking away at a typewriter. "Dede, are you all right?"

Ken finished typing the paragraph he was working on, then leaned back in his chair, took his pipe from his mouth, and said, "I was deported from Ogoni." He continued, "I was to address a rally in Sogho, but when Ledum Mitee and I got there, army troops prevented me from speaking. Then with security agents following me, I turned around and headed for Bori, where there was to be a seminar for delegates and polling officers for the constitutional conference. You remember."

"Yes, of course."

"When I arrived, a lieutenant told me, 'If the seminar turns into a rally, I will disperse it.' Now, I ask you, Mon: how could I turn a seminar into a rally? Since my presence at Bori was not welcome, I decided to return to Port Harcourt. It was at that point that Ledum Mitee invited me to drinks at his home in K-Dere. I asked the security agent for permission to go there. When they gave me their consent, we set off in that direction. Ledum and I were in separate cars. When we drove past Giokoo, the security agents overtook my car and stopped it. Then an army car went past and blocked the road. Troops jumped out and cocked their rifles. The security agents told me they had decided against my going to K-Dere."

"Why?"

"They said it was in my own interest. I only had time to wave goodbye to Ledum. I ordered my chauffeur to turn back.[8] Security agents followed me the entire way to Port Harcourt."

"Has anyone told you the news?"

"I drove non-stop to the newspaper house, *Sunray*. I gave them an interview about the military refusing me to go into the places I want to go in Ogoni – and the fact that they deported me from Ogoni."

"Yes, Dede, I understand that. But has anyone told you what happened at Giokoo?" Owens then related to his brother all he had learned from Barinem Kiobel and the doctor in his clinic about the violence at Giokoo.

Ken nodded and then in a tone of total disbelief said, "I just received a phone call. It appears that they were killed."

"Who?"

"Kobani. Badey. The Orages."

Owens was shocked. "Are you certain?"

"Not entirely. I have been so very, very anxious. I have been calling the police and everybody for more information. No one will tell me anything." Ken pulled the sheet from the typewriter and handed it to his brother. "Here, take this. I have written a press release about how I was not allowed to go and campaign and was sent back from Ogoni. See that copies of this get to all the newspaper offices."

Owens skimmed the press release. "What's going to happen now?"

"You shouldn't worry," Ken said. "We didn't have anything to do with what happened there. We didn't even know they were meeting."

Owens nodded.

"Mon," said Ken. "We are in the most difficult phase of the struggle now. The easy phase has been won. We've done this campaign in a peaceful way, a non-violent way. So far, we've been able to get our message out. Everybody knows me as a man of peace. In any struggle, the period of negotiation is usually the most difficult. Our adversaries will now do everything to make sure that they weaken us by every means. This is the time that the big battle will start."

Ken turned back to his typewriter. Owens rose to leave, but froze when he heard Ken's final words: "I cannot imagine that the Ogoni people would kill their leaders."

Owens left 24 Aggrey Road to take Ken's press release to the newspaper offices in Port Harcourt. It was the last time he saw his brother.

|||||||

After distributing the release, Owens went home and found that his neighbours were holding another of their regular communal barbecues. It was 9 p.m. and the baby was asleep. Owens joined Diana in the gardens with Shell employees who lived in the same complex. After sampling the barbecued goat, Owens sat under the stars and chatted with two of the other residents – a lawyer and a Shell staffer – over brandy.

"Some MOSOP leaders were injured or killed today in Gokana," said Owens.

The Shell employee looked at him askance. "MOSOP leaders? Who?"

"Edward Kobani, Albert Badey, Chief Orage, and his brother."

"You must be joking. They aren't MOSOP leaders. I see those men all the time in my office at Shell."

"Well, they were in MOSOP and some of them were leaders."

"Those men you named, if anything has happened to them, you can rest assured that the police will come after Ken again – and for you, too."

"Why would they want to arrest me? I'm a doctor." Owens paused a moment, took a swig of brandy. "Well, you never know. Listen," Owens said to the lawyer, "if they come for me in the night, please come and bail me!" The three men laughed at the joke, but their laughter was hollow.

A few hours later, Owens and Diana were in bed, asleep, when Ken's chauffeur came banging at the front door. Owens got up, slipped on short pants and an oversized shirt, and let him in.

"They've come for Ken!" said the driver. "The army people came with a jeep and a truck!" He described how soldiers with machine guns had scaled the walls of Ken's property. "And they knocked at the door to the main house very, very loud, and went through many rooms until they reached Ken's room. They burst in, took Ken outside, threw him into one of the trucks, and drove away."

"You say soldiers?" asked Owens. "Was it soldiers or police?"

"Soldiers."

"Let's go back to Ken's house."

At Rumuibekwe Road, Owens heard the same story from Ken's house manager. Ken's whereabouts, however, were a mystery. Owens spent the rest of the pre-dawn hours going from police station to police station in Port Harcourt, asking for news of Ken's arrest. There was none.

Riding through the empty streets of Port Harcourt in the back seat of Ken's car, Owens pondered his next move. The situation was eerily reminiscent of what had happened to Ken the previous June. That time, the police had driven him all the way to Lagos. Perhaps they had taken him back there again.

"Drop me back at Ken's house," Owens said. Once there, he told the driver to ask Diana to pack a small case with some toiletries, a few clothes, and some money. "When she gives you those things, please come back here and take me to the airport."

Still unshaven, dressed in the same shorts and sandals he had worn all night, Owens arrived at the airport at six and boarded the first flight to Lagos.

As soon as he reached Lagos, Owens headed for the home of Hauwa Madugu, Ken's mistress in that city. Hauwa was surprised to find Owens at her door. An exceedingly tall, twenty-eight-year-old Jenju from northern Gongola State, she was pregnant with Ken's child and looked tired and uncomfortable. Owens broke the news as gently as possible, but there was really no way to soften the blow.

"Oh," she cried, "what are we going to do?"

"I don't know," said Owens. "I don't know."

Hauwa invited him in to rest, bathe, and change clothes. An hour later, listening to the radio, they heard the first official reports of the incident at Giokoo. Four Ogoni chiefs had been set upon by a mob, murdered, and burned beyond recognition in a car. According to reports, MOSOP president Ken Saro-Wiwa had been arrested in connection with the murders. There was also a list of other suspects wanted for the same crime. They listened as the announcer ran through the list.

In shock, Owens heard that his name was number one.

||||||||

Next morning, Lieutenant Colonel Komo called a press conference to declare that MOSOP was responsible for the Giokoo murders and that he had arrested Ken Saro-Wiwa, Ledum Mitee, and others. As cameras broadcast to a national audience, Komo displayed his evidence to the assembled journalists. "In the bag there for you to inspect later are remnants of pieces of bones from various parts of the body that they have been able to recover from the scene where the burning took place." (He did not, however, elucidate how a few unidentified bone fragments in a burlap sack could suggest that Ken Saro-Wiwa had masterminded a murder.)

"Ogoni is bleeding," Komo said, "and not by federal troops – 'genocidal' federal troops, as some of the papers carried some days back – but by ir-responsible and reckless thuggery of the MOSOP element which, as I said, must stop immediately. I therefore call on you to report accurately this event and to stop you being used as propaganda tools conveniently for some dic-tator like Ken Saro-Wiwa."[9]

The same day, the headline of the *Daily Times*, Nigeria's largest news-paper, exclaimed: "11 Ogoni leaders declared wanted." Again, the first wanted person named in the story was Owens. "The commissioner pleaded with members of the public with information as to their whereabouts or have hints that could lead to their arrest to contact the State Investigation and Intelligence Bureau (SIIB) Port Harcourt or the nearest police station."[10]

Every Nigerian television and radio station echoed the same report: that Owens was now a wanted felon. Fortunately, all the stories omitted his middle name, referring to him only as "Dr. Monday Wiwa."

Owens had no idea what to do. Besides the imminent danger he faced as a fugitive, there were the more prosaic problems of existence to be solved. Returning to either Port Harcourt or Ogoniland was impossible. Where would he now live? What would he do about Diana and Befii? What would he do for money? And what could he do to help Ken?

Making Ken his first priority, Owens headed to the Saros International offices and drafted a press release, which he clandestinely faxed to UNPO from a business centre in another neighbourhood. Next, he turned his attention to securing legal representation for his brother. Ken's former attor-ney, Olu Onagoruwa, now Abacha's federal minister of justice, was out of the question. There was another respected human rights lawyer in Lagos: Gani Fawehinmi. Owens paid a visit to his chambers.

Fawehinmi, a pro-democracy activist, had himself been arrested more than forty times, spending as long as six months at a stretch in detention. At Fawehinmi's sprawling Lagos law offices, Owens was ushered in to see the eminent lawyer. A thin-lipped man of middle age with thick glasses and wearing a well-tailored blue suit, Fawehinmi congratulated Owens for coming so quickly and said he had been following Ken's case in the news. "So what happened?" he asked.

"From what I know, at Giokoo Ken did not even come out of his car, and he was escorted back to Port Harcourt by the police," said Owens. "Ken did not even know for certain that these people were dead until much later. He had nothing to do with it. I'm worried about him. I fear he is being tortured."

"Where is he now?"

"Nobody knows. I got through to UNPO this morning – the Unrepresented Nations and People's Organization, in Holland – and they said they had been calling Port Harcourt and even called the governor to ask where Ken was and to let him know that they are tracking him. But nobody seems to know where he is."

After Owens filled out a few forms and paid a token retainer, Fawehinmi's associates drafted documents calling for the government either to produce Ken in court or to release him. They also prepared an affidavit for Owens to sign stating what he knew about Ken's innocence. These were promptly hand-delivered to the courts in Rivers State.

The day after meeting Fawehinmi, Owens paid a visit to the leadership of the Ogoni Welfare Association in Lagos. The Ogonis living in Lagos had been receiving a lot of conflicting information. They told Owens that on the day of the murders, Senator Bennett Birabi had telephoned prominent Ogonis in Lagos, London, and the United States to say that Ken was responsible for the murders of the chiefs. Birabi even called Ken's wife, Maria, in London to tell her that her husband was guilty. The news was doubly upsetting for Maria. One of the victims, Samuel Orage, was her beloved brother-in-law.[11]

Despite Birabi's claims, many of the Ogonis in Lagos knew Ken person-ally and were convinced of his innocence. Even so, most were too frightened to aid either Ken or Owens. After the incident at Giokoo, the government had begun arresting anyone associated with MOSOP. Most of the Ogonis in Lagos were federal civil servants who did not want to lose their freedom or their jobs. They were so frightened that Owens could not convince a single one of them to visit Gani Fawehinmi's law firm to help with Ken's defence.

The matter remained of where Owens would live. Although Hauwa had offered to let him sleep at her house, Owens thought it best if he stayed at Ken's Lagos address. The residence was relatively safe because technically it did not belong to Ken. Legally, the house was owned by the late Nomsy Dickson.

In the early 1980s, as his marriage to Maria was falling apart, Ken began a romantic relationship with a woman who owned a business at 63 Tejuosho Street in Lagos. As their relationship grew, Nomsy Dickson merged her business with Ken's and became his business partner in Lagos. Ken moved Saros International to her Tejuosho Street building, and moved himself into her house as a live-in lover. In time, they jointly built a home in the Lagos suburb of Ogba where Nomsy owned a piece of land. Sadly, Nomsy developed breast cancer soon after the house was completed and died in 1990. Since then, Ken had used the residence as a Lagos *pied-à-terre*, but mostly it remained empty except for the house staff and, occasionally, Nomsy's children by a previous marriage, and her children by Ken: Singto and Adele.

Owens moved into the house and each day went to Fawehinmi's office and Ken's office, keeping a low profile. At Ken's office, Owens received many letters and phone calls from local journalists, activists abroad, and diplomats wanting news. At first, Ken's staff directed everything for Owens to Gani Fawehinmi's office. When it became no longer prudent for Owens to be seen at Gani's, he started looking for a place to while away the daytime hours in safety, yet remain available to help Ken. He soon found an odd sanctuary.

The Saros International offices were in a nondescript white storefront with a plate glass door bearing Saros's large "S" patented logo, vaguely reminiscent of Superman's initial. The other businesses on this narrow commercial block consisted of a small snack bar fronted by a caged-in porch for dining, the St. Valentine's medical clinic, the Sir Bashy Boy Cut barber, and small clothing boutiques. It was in one of these boutiques across the street from Ken's office that Owens created his hideout – on a stool, behind a rack of dresses.

Sitting for hours each day in the boutique, Owens had ample time for contemplation. For the first time, he felt powerless. In Ogoniland, he had two clinics and a team of doctors and nurses. He had servants, two homes, and a highly respected position in Ogoni society. Now on the lam in Lagos, he found himself surviving on the goodwill of others. Although Ken's staff at Saros agreed to courier messages to the dress store hideout, they refused to take any direction from Owens. Normally whenever Ken was not around – which was often – his staff worked independently and reported to no one. Ken's employees were certainly not going to put themselves in harm's way for the benefit of their employer's younger brother. The domestic staff at Nomsy's house were even less deferential. Without money to purchase their respect, Owens was at their mercy. After Owens's clinic forwarded him badly needed funds, Nomsy's house staff started demanding money; soon, the "free" accommodations at Nomsy's became an untenable financial haemorrhage.

|||||||

Owens's goal during his fugitive life was to spread the word about Ken's innocence. Surprisingly, he found it hard to interest the media and civil society. At the time, Ken Saro-Wiwa's arrest was a distant third on the list of the country's news-making events. Most of the attention was on the annulment of the June 12 presidential elections and Abacha's upcoming constitutional conference. Many Nigerians were upset over Moshood Abiola being denied his election victory, and most recognized the need for a new Nigerian constitution. The debate degenerated into squabbling when it became apparent that the two issues were irreconcilable – to draft a new constitution would nullify Abiola's election. For this reason, most Nigerian civil liberties organizations had resolved to stand behind Abiola and boycott the constitutional conference. These organizations were unhappy when Ken announced his decision to seek election as a delegate and were thus unwilling to rally to Ken's defence after the murders in Giokoo. Day in and day out, Owens visited the National Democratic Coalition (NADECO), the Constitutional Rights Project, and the Campaign for Democracy, meeting repeatedly with the leadership and rank and file to explain Ken's reasons for wanting to attend the constitutional conference and to assure them of Ken's – as well as his own – innocence.

During these meetings, Owens received a constant stream of information about the rash of human rights abuses back home. Since Ken's arrest, Ogoniland had been under siege. No one was able to travel into or out of Ogoni without enduring a series of military checkpoints set up expressly to cull MOSOP members. Every Ogoni man and woman under forty was required to disrobe entirely at these roadblocks so that the soldiers could search their bodies for "NYCOP marks" – scars from the ceremonies in which vigilante youths had inserted magic potions under their skin to make them invincible. Anyone found bearing these scars was arrested.

This campaign was masterminded by Major Paul Okuntimo, who was not shy about revealing his motives, methods, or ultimate goals. "I am going to sanitize Ogoniland," he boasted to a British journalist. "MOSOP will be no more when I finish."[12]

At a town meeting in the Ogoni village of Barako, Okuntimo told the villagers, "You are the worst type of people. You killed the Andonis. Then the Andonis let us know. So we came and chased you people." Okuntimo was a fearsome sight in his olive drab uniform draped with leather magazines of bullets. As he spoke, the feathered cockade on his red beret bobbed and his eyes narrowed to slits in his round, dark face. "After the Andonis, you fought with the Okrikas and then with the Ndokis. So they invited us to chase you people. So we are the people who chased you from your houses and destroyed them."[13]

At a press conference broadcast by the Nigerian Television Authority, Okuntimo cheerfully recounted the progress of his "sanitation" campaign. "The first three days I operated in the night," he said. "Nobody knew where I was coming from. What I will just do is that I will just take some detachments of soldiers; they will just stay at four corners of the town. They have automatic rifle that sound death. If you hear the sound you will freeze! And then I will equally now choose about twenty and give them grenades – explosives – very hard one. So we shall surround the town at night. The machine gun with five hundred rounds will open up. When four or five like that open up and then we are throwing grenades and they are making 'eekpuwaa!' what do you think the people are going to do? And we have already put roadblock on the main road; we don't want anybody to start running. So the option we made was that we should drive all these boys, all these people into the bush with nothing except the pant and the wrapper they are using that night. We should get everybody into the bush."[14]

Okuntimo described this mayhem as "psychological warfare" intended to facilitate constructive dialogue. A report from Human Rights Watch/Africa gave a gruesome account of the brutalities committed against the Ogoni people during this so-called psychological war. Besides the hundreds of Ogoni men, women, and children gunned down by Nigerian troops, many more Ogonis were maimed and raped, and hundreds of thousands were driven into the forest. Soldiers destroyed houses, looted property, and extorted villagers who wished to avoid the wrath of Major Okuntimo. Six hundred Ogoni men were detained and tortured in specially established detention centres at Kpor and Bori. The remainder of the Ogoni population was driven out of government and private employment, leaving every sector of society crippled.[15]

||||||

It became clear to Owens that Ken was going to be tried in the court of public opinion long before he saw the inside of a courtroom. Early on, the newspapers had divided into factions concerning Ken Saro-Wiwa's supposed culpability in the Giokoo murders. The independent press was convinced that Ken had been framed. The Igbo-controlled newspapers used Ken's arrest as an opportunity to avenge the wrongs they perceived Ken having committed during the Biafran War, filling their columns with stark, brutal lies about Ken's past and his recent actions. The government-controlled newspapers trod a middle ground. Many of the reporters were long-time friends of Ken and knew it would be out of character for him to have been involved in murder. Nevertheless, they were obliged to write the news the way the government dictated.

Some days, Owens visited as many as seven newspaper and magazine offices. He had started off slowly after his arrival in Lagos, building relationships in a systematic fashion. On his first visit, he would ask to meet the editor and give him a brief speech about Ken's innocence. During his second visit, he would search out someone on the editorial board who had a connection to Ken and implore them to use their influence to see that the stories about Ken were reported without distortion. On third visits, Owens would look for a specific writer to form a relationship with. Many of these reporters were happy to see that Owens was safe; some had slept at his house or been treated in his clinic.

The first court hearing in the case against Ken Saro-Wiwa, Ledum Mitee, and the others arrested with them, was scheduled for June 28 in Port Harcourt.

||||||

Since the night of his arrest, Ken had been shackled to a wall in a dank, crowded cell at the military barracks in Afam. Ledum Mitee had not been arrested, but had turned himself in and somehow managed to get himself placed in the marginally better Bori Camp detention centre in Port Harcourt.

On the day of his court appearance, Ken was transferred to Bori Camp where he was relentlessly interrogated for three days.

When word leaked out that Ken was at Bori Camp, Diana Wiwa and Ken's Port Harcourt mistress, Elfreda Jumbo, delivered food to the jail, posing as his sister and wife. To win sympathy, Diana came carrying her baby. Ken wanted the women to deliver some messages he had written; he was not permitted correspondence, and smuggling a letter into or out of Bori Camp was nearly impossible. Since Diana and Elfreda were thoroughly searched before and after each visit, Diana had the idea of hiding Ken's letters in the one place the soldiers never looked – her baby's diapers. Soon Ken's mail began to flow to and from the jail, strapped to Befii's bottom.

In these smuggled letters, Ken utilized a system of code names he created for himself, his associates, and his opponents. He referred to himself as "Lion"; Owens was "Eagle"; Ledum Mitee was "PWCI"; Sani Abacha was "Bully." Anita Roddick's socially conscious cosmetic empire, The Body Shop, which was becoming a rather vocal MOSOP ally on the international scene, was given the code name "Soup."

During her next school holiday, Diana took the baby and Ken's letters to Lagos. It was the first time Owens had seen his wife and child since the murders at Giokoo, but his happiness was bittersweet and short-lived. Soon after Diana's arrival, Nomsy's children returned from university and made it clear that there was no longer room in their house for the fugitive and his family.

When Owens protested that they had nowhere to go, the only person to take pity was Nomsy's night watchman, a kindly, middle-aged man with a nagging cough, who slept during the daytime in a small room in the boys' quarters behind the house. Since his bed was unoccupied during the night, he offered it to the Wiwas, provided they vacate the room an hour before dawn and stay outside during the day. Even though it was the rainy season, the Wiwas had no choice but to accept.

Owens, Diana, and Befii lived in these depressing circumstances for nearly two months, and Owens could not imagine sinking any lower – yet lower his family did sink. The night watchman's chronic cough was getting no better. When he returned from a trip to a local clinic, the night watchman reported the grim news that he was suffering from tuberculosis. In other words, for the past two months, Owens and his family had been sleeping on the same sheets as a TB patient.

At the same time, one of Nomsy's relatives came to say that the Wiwas would have to move. This was not altogether unexpected. Every small noise in the night frightened the household into thinking the police had finally come for Owens. At last, Nomsy's relatives had had enough and the house was rented to new tenants.

Fortunately, Hauwa Madugu, who lived in the same Lagos suburb, renewed her offer to take the Wiwas in. Owens was initially reticent; he was uncomfortable that she was a federal government employee, a customs agent for the Abacha regime. Although Owens was certain that she was trustworthy, what might happen if one of her co-workers dropped by?

Owens and Diana had little choice but to lodge with Hauwa. Fortunately there were compensations to offset the risks. Hauwa lived in a pleasant three-bedroom bungalow in a secure estate nestled in the back streets of Ogba. Owens, Diana, and Befii were given their own room with one bed in which they would all have to sleep together. Apparently, any conjugal urges would have to continue to be suppressed for the unforeseeable future.

As Diana grimly regarded the bed, Owens whispered, "Think of it as very good birth control."

||||||

A few days later, at a convent school in Lagos, Owens paid a visit to Sister Majella McCarron, an Irish missionary nun from the African Justice and Peace Network. Owens knew her from her work with the Daughters of Charity and the Ogoni Relief and Rehabilitation Committee, and Ken had once described her as "a true friend who stood by the Ogoni people in their darkest hour."[16] At the convent, Sister Majella sat down with Owens and asked him what had happened at Giokoo. After he related what he knew, she asked, "Owens, is it possible that during one of your MOSOP meetings the issue of these chiefs came up and someone may have said something like 'Let's do away with them'?"

"No, sister," said Owens. "Nothing like that was ever said."

"Are you certain?"

"It did not happen. It was not possible."

Sister Majella was a realist, well-versed in the corruptible influences of Nigeria. But Owens's assertions that Ken did not foment violence of any sort mollified her and removed any traces of doubt she had started to entertain.

In August 1994, Hauwa went into labour. Ken's last child was born by caesarean section. From prison, Ken chose the baby boy's name – Barikwameleete, or "God has created me well" in Khana. Kwame for short. One of Nigeria's widely read gossip magazines carried a feature about baby Kwame. It related Hauwa's life, where she was born, how she met Ken, and the intimate details of their relationship. When the issue hit the newsstands, Owens was furious and communicated his anger in no uncertain terms to Hauwa, which upset her greatly. She denied any involvement with the publication of the article. To compound matters, Hauwa had made plans to hold an elaborate naming ceremony for Ken Saro-Wiwa's latest offspring at her home. When Owens saw the extensive list of guests being invited to the very place he was supposed to be hiding, he begged her to call it off. Hauwa insisted on going through with the party.

On the day of the event, Owens left the estate early in the morning and concealed himself behind the dress rack on Tejuosho Street. After the boutique closed, he wandered the streets of Ogba for hours until he felt it was safe to return to Hauwa's. Unfortunately, he underestimated how long the party would last and encountered a group of stragglers in her sitting room.

As he hurried into the bedroom, he said to Diana, "There are people out there who saw me."

"Don't worry," she said. "They're all Ogonis."

"No, I will not be safe here again. I have to start looking for somewhere else for us to stay."

|||||||

The next morning, the scream of a newborn woke Owens at dawn. Raising himself on one elbow, he saw Diana standing by the bedroom doorway in profile, the morning light faintly illuminating her round, placid face.

"Is Befii crying?"

"No," she said, "it's Kwame."

"It sounds like he's yelling for his breakfast."

"Soon Befii will be yelling, too. Hauwa says the cooker is out of gas."

Owens sat up, reached for his shirt, and fumbled through the pockets for a cigarette. As a doctor, he knew he shouldn't smoke, but at the moment, quitting this one pleasurable vice was out of the question. Since Hauwa would not let him smoke in the house, the prospect of a little excursion seemed welcome. "Let me go to the petrol station and I'll get gas for the cooker," he said.

"Mon, don't be long. Somehow it's okay when it's your own child, but when it's somebody else's baby . . ."

"Okay," he said, putting on pants. "Do you know where Hauwa keeps the empty gas cans?"

"I think she's already taken them out to her driver."

Owens grabbed his shirt and pulled it on over his head as he hurried out. He wore dark pants and his rumpled, yellow and brown *adire* shirt. He liked the oversized Yoruba garment's roomy pockets and the way its bold African print made him fit in with the masses. Ever since he had gone into hiding in Lagos, he had dressed "rough." But a shirt can change only so much; his aviator glasses, trim moustache, and dignified bearing betrayed his education and upbringing. And nothing could hide his marked resemblance to his famous older brother.

It was Sunday morning; a light rain was falling. Owens got to the car just as Hauwa was handing the empty gas cans to Sake, her chauffeur.

"Eagle gas," Hauwa said to Sake. "You know where to go?"

"Wait, I'm going with him," said Owens.

"Don't be long," she said. "We have to make the baby formula."

As the chauffeur drove through the gates of the estate, he said archly, "In the villages, the people don't cook formula. The mother feeds the child her natural milk."

"Yes, but my wife's natural milk is no longer enough," said Owens. "Our baby is seven months now." He glanced at the propane tanks on the back seat. "Eagle gas," he said with contempt. "That is formerly Shell gas!"

About half a kilometre down Ogba Road, Hauwa's car was stopped at a police checkpoint. Approaching the driver's side, an officer armed with an automatic rifle ordered Owens and Sake out. As the two men stood in the drizzle, the police searched the vehicle and asked to see the driver's licence and the car's insurance papers.

"Where you dey go?" asked one of the policeman in pidgin.

"To buy gas for the cooker," said Owens. "There is no gas in the house. There's a young boy who has just been born. We want to make his food, and we have to boil water. So that's why we're hurrying this morning."

Finding everything in order, the policeman peered at Owens and said, "Wetin dey for dat pocket?"

"Nothing. Just cigarettes."

"Gif mi siga."

Owens handed the officer the pack. The policeman took a cigarette and asked for a light. "Gif me de matches."

Owens absent-mindedly reached back into his shirt pocket and withdrew a small cardboard matchbox, which he offered. On opening the matchbox, the officer found a small piece of neatly folded paper inside.

"Eh!" the policeman cried. "Exhibit! Exhibit!" he shouted, holding the matchbox in the air and waving it.

"Na wetin?" asked the second officer.

"Drugs. Indian hemp. Cocaine!"

"No!" said Owens. "No, this is just a paper that I picked. It is nothing."

The second policeman took the matchbox and carefully unfolded the small piece of paper. It was a letter. Squinting at the spidery script on the torn sheet of ruled notebook paper, he read aloud: "Eagle, keep in constant touch with the following organizations – the BBC, UNPO, the VOA, Amnesty, Greenpeace, PEN International. . . . Keep them abreast of the human rights violations in Nigeria." The policeman looked up and said, "It is signed *Lion*."

"Eh?" said the first policeman. "The BBC? The Amnesty? Ah! You are the people spoiling the name of this country, O! Wetin you name?"

"Nwiado. Deebii Nwiado," said Owens, blurting out the first name that came to mind. "That letter was in my pocket. I . . . I just picked it up from an office . . ."

"What office?" demanded the policeman.

Owens had told the police his name was Deebii Nwiado. The real Deebii was Ken's personal assistant in Lagos. Unable to think of a convincing lie, Owens continued appropriating the man's identity. "63 Tedjuosho Street," he answered. "Saros Publishing."

"You people spoil the name of the country," repeated the policeman. "I am sure you are one of these people they are looking for."

"No! Nobody's looking for me," Owens insisted.

"Get in! Get in! Get in dis moto!"

Forcing Owens into the back seat of Hauwa's sedan, the policeman climbed in beside him and commanded the chauffeur to take them to the nearest police station, an obscure little outpost in the back streets of Ogba. When they arrived, the officer dragged Owens inside, presenting him before the superintendent of police.

"Here is one who destroy the name of the country!" the arresting officer proclaimed.

The imposing superintendent reached underneath the reception counter and pulled out a sheaf of papers. "You are under arrest," he said. "Make your statement."

"What statement?" asked Owens.

"Go ahead! Make statement. Just fill those things. Your name, where you live, how old, where you work, and write everything, where you went to school and everything."

"What about the driver with me?"

"He can go," said the superintendent.

Owens motioned for Sake to come near. He whispered to him to fetch Hauwa. "Have her tell my brother Letam what has happened. And remember! My name is Deebii Nwiado. You got it?" Sake nodded and ran back to the car.

"In here," said the superintendent, pointing to an empty cell. "You finish your statement here."

This is it, Owens thought as he began to write out his falsified police statement. *This is it.*

Hauwa Madugu soon arrived at the police station. As a customs officer, she knew how to handle Nigerian uniformed personnel. It also didn't hurt that she was a beautiful woman who spoke Hausa, the mother tongue of most of the police at the station.

"Deebii is a low-level clerk at my husband's printing business," she said in Hausa. "He couldn't possibly be mixed up in anything illegal. If he said he picked up that piece of paper off the ground, then that's what he must have done."

Still there was something about "Deebii Nwiado" that the police found suspicious. It was going to take more than a tall Jenju beauty's assurances to convince them otherwise. It would take "*dash*."

Owens was released eleven hours later after Hauwa came up with a bribe just large enough to buy Owens his freedom.

CHAPTER NINE

Back at Hauwa's house, Owens told his wife that she and the baby would have to leave Lagos; it was too dangerous for the family to remain together. Despite Diana's protestations, Owens sent her and Befii back to Port Harcourt the next day.

As for his own safety, Owens made a bold, counterintuitive decision – he would hide from the military in the midst of the military itself. Recently, junior brother Letam Wiwa had been transferred to Dodan Barracks, an army base in the vibrant, working-class Lagos neighbourhood of Obalende.[1] Dodan was the official residence of the military's elite forces as well as the army's commander-in-chief. Until 1991, it had also been the official residence of the president. When the presidential palace was transferred to Abuja,[2] Dodan remained home base of the presidential Brigade of Guards and Letam was an intelligence officer in that brigade. Because the authorities would never think of searching for Owens at Dodan, he decided to move in with his junior brother, and Letam readily agreed to assume the personal risk.

Flat D3 at Dodan Barracks was a small, brown cottage in the enclave of officers' housing. Although a high, well-guarded wall surrounded the former State House, the perimeter of the officers' quarters was ringed with a modest fence accessed through a poorly patrolled gate. With a bit of caution, Owens found, he was able to come and go with anonymity.

Besides being an ironically safe hiding spot, Dodan Barracks was ideally located. The neighbourhood was one of the safest in Lagos, with a large market and a number of good chop bars where one could always find an inexpensive bowl of pepper soup. Obalende was also just across the bridge from posh Victoria Island, the residential area of the cream of Lagos society and the location of most of Nigeria's foreign embassies.

Much to the chagrin of Ken's present lawyer, Gani Fawehinmi, Owens arranged an interview with Ken's previous lawyer, Olu Onagoruwa, now the attorney general and minister of justice. Only a year had passed since Onagoruwa had represented Ken during his "month and a day" incarceration. When Owens phoned the attorney general's house, Mrs. Onagoruwa answered the phone and became frightened when Owens said he wished to see the minister at home, privately.

"Owens! It is you! Oh, I don't know," she said. "I'll have to talk to Olu first. You should call back later."

Two days later, Owens finally spoke with Olu Onagoruwa directly. "Yes, do come by," he said, "but be very careful. I have police guarding my house."

At the appointed time, Owens arrived at the residence of the attorney general. A door in the iron gates swung open, and a policeman stuck his head out and asked Owens what he wanted.

"I have an appointment with the minister," he said. "You can confirm with the wife or with the minister himself."

Owens was prepared to give the officer a false name, but realized it would be foolhardy. He'd brought with him various documents – the press release he had drafted the morning after Ken had been arrested, a published article Ken had written from detention, and a stack of press clippings. These documents would expose his false identity. Fortunately, the policeman merely told Owens to wait while he verified that the minister was indeed expecting someone. He returned and escorted Owens to the front door.

Olu Onagoruwa welcomed Owens and invited him to sit down. "I would like to hear about what happened on May 21," said Onagoruwa, and Owens told him exactly what had transpired from the moment he first heard about the disturbance in Giokoo to his last conversation with Ken.

After listening to Owens's story, Onagoruwa said, "I have no doubt that Ken was not involved. Ken is a literary person. As a lawyer, I know that men of

Ken's profile, literary people, do not kill. They are against violence. Sometimes they may write about it, but they don't do it. That I know for sure."

"Ken is very anxious about what you are going to say or do about this event," said Owens.

Onagoruwa nodded. "Officially, I don't know anything about the case yet. There has been nothing on my desk. I'm going to find out what I can do."

"Please," said Owens, "I have heard that they have not allowed Ken to have access to family, to lawyers, to doctors. Ken has not been well. They tell me he has been held in leg irons and chained to a wall. He is being tortured."

"Are you sure about the torture?"

"I have information that they had tortured him, both on the first day and also the time they took him to Government House to interrogate him and where he was accosted by the wives of those who were killed. I understand they did terrible things to him."

"I will look into the issue of torture and whether he has been chained," Onagoruwa assured Owens, "but before I can do anything, you should get me some documentation."

"How can I do that? In case you haven't seen the newspapers, I am also wanted for murder."

Onagoruwa laughed and said, "You're here in the house of the attorney general. If you were a wanted man, I would arrest you immediately." Turning serious, he added, "But don't let people know you visited me. Don't tell anyone that we have spoken."

Owens did not keep his pledge. The following day, he met Gani Fawehinmi and related details of his visit. Fawehinmi said he didn't think Onagoruwa would do anything to help.

"Why?" asked Owens.

"This man is in a military regime," said Fawehinmi. "He was once a human rights activist. He has thrown away all his credibility."

Owens was aware that Fawehinmi and Onagoruwa had been close friends. After Abacha offered Onagoruwa the office of attorney general, the Nigerian press recounted the bitter schism that arose between the two lawyers. "In any case," said Owens, "he gave me his office number in Lagos and in Abuja. It's at least one inroad into the government."

"Owens, he will not help you," repeated Fawehinmi.

Undeterred, Owens turned his attention to courting the federal minister of internal affairs, Alex Ibru, former publisher of *The Guardian* and Ken's long-time friend. Like Onagoruwa, Ibru seemed genuinely sympathetic to Ken's plight. Owens concentrated on the issue of Ken's health and the fact that a military doctor had recommended that he be transferred to a hospital. Ibru said that if Owens could have someone fax him the doctor's recommendation, he might be able to do something.

Heartened, Owens sent a message to Port Harcourt and within a week received a copy of the papers Ibru needed. Ibru, however, did not think the doctor's recommendations were clear enough; he promised to put the matter before the next cabinet meeting.

Next, Owens tried to contact a few of his military friends from his Youth Corps days in Jaji who were now posted to the Ministry of Defence. Understandably, these officers were extremely reticent; one, however, did meet secretly with Owens only after Owens promised never to reveal his identity.

"I don't have any proof to give you," the officer said, "but from the discussions I hear, it appears that this thing is from a very, very high level and from people who have been trying to get rid of Ken."

"What people?"

"People who have big interests in the oil business." The officer looked away and then added, "All I can tell you is, I don't think Ken is going to get out of this."

||||||||

Once again, Owens was running low on funds. Though he was living rent-free at Dodan Barracks, he was using business centres to make daily telephone calls and send faxes to Port Harcourt and abroad. There was transportation and food to be purchased, not only for himself, but for Ken's staff and other Ogonis he conscripted to help, not to mention the daily doling out of bribes, *dash*, that was the only way of getting anything done in Lagos. Once again, Owens sent instructions to his clinic staff in Bori to do whatever they could to forward more cash to Lagos.

Ken was also going through a financial crisis. From the prison at Bori Camp, he sent instructions to his staff at 24 Aggrey Road to sell two of his real estate holdings so he could pay his lawyers and continue to feed himself and his fellow Ogoni inmates. Ken knew that one of the Abacha regime's goals was to bankrupt him before destroying him judicially. Although he hated playing into the enemy's hands by liquidating his assets, Ken had little choice.

Then, as if by magic, Ken's money problems were temporarily solved. In October, it was announced that he and MOSOP were one of four winners of the 1994 Right Livelihood Award, with the award committee singling out Ken for his courage in striving non-violently for the rights of his people. The award had been established in 1980 by Jakob von Uexkull, a Swedish-German writer who sold his valuable postage stamp collection to provide the original endowment. Believing that the Nobel Prizes ignored many of those whose knowledge and work were vital to the future of mankind, von Uexkull created his award to honour individuals and groups he felt offered answers to crucial world problems. Widely known as the "Alternative Nobel Prize," the Right Livelihood Awards are presented annually at a ceremony in the Swedish Parliament on the day before the Nobel presentations.[3]

The 2 million Swedish krone ($360,000 Canadian at the time) prize is divided equally among four annual recipients. Because the money is not intended for personal use but meant to strengthen or expand the honorees' work, one of the 1994 RLA recipients – "Pippi Longstocking" author Astrid Lindgren – was deemed not in need of financial support. Thus, Ken and MOSOP received one-third of the money.

The $120,000 windfall had immediate effects. It brought international attention to Ken's inhumane detention. It awakened people around the globe to the existence of MOSOP and to Shell's dismal record in the Niger delta. But it also roused rampant greed. Nearly everyone connected with MOSOP felt entitled to a cut.

The financial situation of MOSOP had always been murky. Up to this point, MOSOP did not have a bank account because the Nigerian government refused to recognize the organization as an NGO. There had never been much money to bank anyway. The proceeds from the One Naira Ogoni Survival Fund had been handed over to Ledum Mitee's brother, Batom, who deposited the sum in the bank where he worked. MOSOP's day-to-day financial responsibilities were funded by Ken directly from his personal accounts. Since the movement started in 1990, Ken had been footing the bill for MOSOP 100 per cent. For four years he had paid for all the overseas and domestic travel he and others had undertaken on behalf of MOSOP, and for office expenses such as phone, faxes, stationery, printing, and staff salaries.

He paid for MOSOP radio announcements and printing costs, and for the entertainment of journalists, foreign dignitaries, international activists, and federal politicians when they visited Ogoniland.

As for the Right Livelihood Award money, Ken thought it should be deposited into his account in the United Kingdom, where it would be safe. In a letter to Owens, he indicated that in accordance with the wishes of the Right Livelihood organization, the entire sum would be used solely for the purposes of MOSOP. He personally would not touch a cent.

At least two factions of MOSOP wanted authority over dispensation of the prize money. The first consisted of Ken and Ledum (president and vice-president of MOSOP), who were in detention at Bori Camp. The second was made up of active MOSOP members clandestinely meeting in and around Port Harcourt. Those imprisoned alongside Ken, those living underground, and those related to the victims at Giokoo also felt entitled. The pressure on Ken to begin doling out the wealth was enormous and depressing. Suddenly, no one would do anything for the cause without demanding payment upfront. Devastated, Ken wrote to Owens, "I did not know that all these people around me were just in it for the money."

Owens wrote back: "Why are you surprised?"

There was still hope for MOSOP. The women of FOWA and the students of NYCOP had continued to carry on their activities without complaint. Heartened, Ken delegated a committee (he was not a member) to look after the prize money and spend it according to MOSOP's need, with every penny to be accounted for.

Ken's first choice to travel to Sweden to accept the prize on MOSOP's behalf was his eldest son, Ken. Surprisingly, Junior wrote to his father from England declining the offer, saying that it was an inconvenient time because he had just taken a holiday to celebrate his engagement to his girlfriend. Ken wrote back and pleaded with Junior to go to Sweden and "sit in the back of the hall." Junior, offended, shot back an angry note asking his father not to tell him what to do with his own life.[4]

Ken's wife, Maria, also refused to represent her husband in Sweden but grudgingly agreed to send the twins, Zina and Noo, as long as they stayed in the background. In the end, MOSOP-USA president Vincent Idemyor, Nobel laureate Wole Soyinka, and exiled NADECO stalwart General Alani Akinrinade agreed to represent MOSOP at the Right Livelihood ceremonies, as did Barika Idamkue and Batom Mitee. Even Sister Majella McCarron went.

The plan to have the award money administered by a committee did not last long. Within weeks, Ken blocked the committee's access to the funds. A review of the statements showed that the committee had been anything but responsible with the prize. Instead of seeing that the money lasted MOSOP a long while, the committee went through a great chunk of cash in a spending spree. Ken wrote to Owens, explaining that he had decided to assume total responsibility for any future disbursements because the Swedes expected an accurate accounting. If it came to their attention that MOSOP had spent the money foolishly, the blame would no doubt come down squarely on Ken's head.

In prison, money remained a source of discord between Ken and the other Ogoni prisoners. In Ken's account to Owens, Ledum Mitee hounded Ken daily for cash. Ledum still had a law firm to run and a family to support. He gave Ken no peace about being denied his share of the Right Livelihood Award money. One day, having had enough of Ledum's nagging, Ken asked that his staff send him the accounting records of the RLA money and other personal expenditures that Ledum had made on behalf of MOSOP during their detention. The statement showed that a substantial part of the RLA fund disbursement went to Ledum Mitee, his wife, and his chambers. According to Ken, when he handed the statement to Ledum, the MOSOP vice-president glanced at it, crumpled it up, and threw it on the ground. He never mentioned money again.

||||||

Like Owens, many Ogoni activists wanted in connection with the Giokoo murders went underground. Those who made it to Lagos usually began their fugitive life by searching for Dr. Owens Wiwa.

One of the underground Ogonis who came calling was NYCOP deputy president John Kpuinen. After being contacted via Ken's staff at Saros, Owens met the man at a nondescript tavern not far from Dodan Barracks. Good-natured and garrulous, John was eager to fill Owens in on gossip from home.

"I was on the entourage with Ken and Ledum and all those when they left Bori, you know the day the chiefs were killed."

"Hey, keep your voice down," whispered Owens, glancing around the bar to see if any other patrons were listening.

John shrugged. Lowering his voice a notch, he added, "After leaving Bori, Ken wanted to go see his mother, but he decided not to. The soldiers were following him everywhere, right from the time he came to Ogoni. That was when Ledum told him to come to his village, that's Ledum's village, K-Dere. When they were going to K-Dere he was stopped by the soldiers. Ken decided to go back to Port Harcourt. It was right after that that this thing happened."

"Did you see what happened at Giokoo?" asked Owens.

"No. Ledum gave me a note to take to Bodo – that's one of the places where Ken was supposed to go and address the crowd – that I should go and tell the people that Ken was no longer coming. Bodo is very far from Giokoo and K-Dere. So I said okay, and I went to Bodo and explained that the rally was not going on. It was when I started to go back to my village, I noticed people running away from Giokoo. So I went toward Giokoo and the people said that there was a lot of problems, that Albert Badey and others were being beaten up and the soldiers and the police were there, that there was utter confusion."

Owens asked, "What happened to the chiefs?"

"I don't know," said John. "After hearing this, I stayed away, but I can tell you what I heard. Some people said they were having this meeting and then a fight broke out. Some people were playing football on the field by the Giokoo Palace, and there were market people, because there was a market around there. And all these people moved to the palace where they were having the meeting and began to beat the people there. Albert Badey fell down. He had one of those – what are those things called? The thing he used for his asthma?"

"An inhaler, a nebulizer."

"Yeah, he tried to used it, but the people must have thought he was trying to bring out a whistle to attract the police or whatever, and that made people more afraid and continue to beat him instead of helping him."

"What about the police?"

"I heard they just stood there and watched."

It should be noted at this point that others were less circumspect than John Kpuinen about the events that had occurred at Giokoo. One, Desmond

Lera Orage, the son of Giokoo victim Chief Samuel Orage, later presented a graphic and gruesome account of the murders to an assembly at Duke University in North Carolina. Although Desmond Orage admitted he did not personally witness the killings, he had no hesitation about recounting the local gossip and hearsay in detail.

After Badey was dead, Orage said, the crowd grabbed Samuel Orage, who cried, "Cut off my hands! Cut off my legs, but spare my life!" but the mob did not listen and they beat him to death. Desmond Orage said that Kobani hid in one of the rooms of the palace while the mob took Desmond's uncle Theo, chopped him up with knives, and ate him. Theo was supposedly decapitated and his head delivered to his wife and children. As for Kobani, Desmond Orage stated that the mob found Kobani, then smashed in his brains with a rake and drove a spear through his private parts. Before the four chiefs' headless bodies were put into a Volkswagen and burned in the middle of the forest, the crowd gathered around and sang victory songs while Dr. Barinem Kiobel prayed over the dead.[5]

It is impossible to discern what – if any – truth there is in this account, although there is no doubt that Dr. Kiobel was at the police station in Bori while the attack was in progress and thus could not have been involved in such a heinous scene. Kiobel also lacked any motive whatsoever.

Back in the tavern, John Kpuinen delivered some particularly astounding news: "They have arrested Kiobel," he told Owens. "He is now in jail with Ken."

"But why would they do that? Dr. Kiobel is a politician. In fact, I heard he was one of those who the murdered chiefs were honouring that day at Giokoo! Why would he want to kill them? And anyway, Dr. Kiobel isn't even a member of MOSOP. This whole business lacks sense."

"Lieutenant Colonel Komo doesn't care about sense. He doesn't care that Ken and MOSOP didn't know anything about this meeting in Giokoo. He doesn't care that most of the MOSOP activists were waiting for Ken in Bodo, which is miles away from Giokoo."

John Kpuinen launched into a graphic description of the atrocities being committed in Ogoni under Komo and Okuntimo's command – the rapes, the arrests, the looting, the random shootings. It appeared that the military's priority was to find people who would testify against Ken or MOSOP, but they had not yet succeeded in coercing anyone into agreeing to commit perjury.

A few weeks after this visit with John Kpuinen, a Gokana man from Kpuinen's village came to Saros International looking for Owens. A member of Ken's staff escorted him across the street to Owens's hiding place in the clothing boutique.

The man said soberly, "I have bad news."

"What is the news?"

"Kpuinen has been arrested from the place he was staying."

"And where was that?" asked Owens. As with the other Ogonis in hiding in Lagos, Kpuinen and Owens had kept the locations of their hiding places a secret from one another.

"He was staying near Dodan Barracks."

"No!"

"Yes."

Like Owens, John Kpuinen had a relative who lived near Dodan, a policeman, who had reluctantly offered John a place to sleep, hoping he would keep a low profile. Unfortunately, remaining inconspicuous was not in Kpuinen's nature. According to Owens's visitor, Kpuinen never kept his mouth shut the entire time he was in Lagos. Whether playing checkers in a barroom or sitting with strangers in a restaurant, Kpuinen constantly talked about his exploits as deputy president of NYCOP and boasted of how MOSOP had driven Shell out of Ogoniland. Unfortunately for Kpuinen, one of the people he confided in turned out to be a close friend of a Lagos policeman who was related to one of the murdered chiefs. Though Kpuinen's name had never appeared on any wanted list, the policeman reported Kpuinen's whereabouts to the Federal Investigation Bureau. The FIB raided his hiding place, arrested him, and transported him back to Port Harcourt. He was now in detention alongside Ken, Ledum, and Dr. Kiobel.

Owens became terrified for his own safety. Dodan Barracks was obviously no longer safe and Owens would have to move immediately. Desperately scrounging for a new place to sleep, he made the acquaintance of one Dr. Osaro-Edee, an Ogoni physician living in Lagos. Osaro-Edee, a jocular, imposing man of middle age, was a long-time fan of Ken Saro-Wiwa's writing and had always considered Ken a sort of mentor, though they had never met. After hearing Owens's story, Osaro-Edee said, "Why are you roaming about and living in fear? I have a house. You come stay with me!"

Dr. Osaro-Edee lived in a Lagos fringe area called Ebutta Meta and worked at the nearby Railway Hospital. Although his family lived in a small hospital staff flat, Osaro-Edee generously gave Owens a private bedroom.

Owens desperately wished he could reunite with his family. In late November 1994, his wish was granted when Diana brought Befii back to Lagos during her school holidays. There was no question, however, of Owens keeping them with him at Dr. Osaro-Edee's. With Kpuinen's recent capture and his own terrifying brush with the law, he could feel the noose tightening. Though Diana longed to be with him, she understood that she and the baby would have to stay elsewhere.

IIIIIII

Any inconvenience during the holidays with her husband, however, would be preferable to the autumn she had just spent in Port Harcourt. Diana had returned home after Owens's arrest in July determined to finish her political science degree and to work for Ken's freedom. She also wanted some semblance of normal life for herself and her baby. Naturally, her parents were thrilled that their daughter was coming to live with them once again, and Dora Barikor was a willing babysitter. Diana longed to return to the beautiful home Owens had rented for them in the quiet estate on the north end of Port Harcourt, but it was too dangerous. Her only option, it seemed, was to remain with her parents and try to be invisible.

Diana's low profile ended with the arrival of some Canadians. In the autumn of 1994, Gerald Ohlsen and Flora MacDonald arrived for a fact-finding tour of the crisis-stricken area. Ohlsen was Canada's acting high commissioner to Nigeria; MacDonald, a former Canadian external affairs minister, was head of the Commonwealth Human Rights Initiative.[6]

Diana and several other Ogoni women arranged for the Canadians to be guests of honour at a community ceremony to thank the Canadian government for helping to rebuild the ravaged village of Kaa. Hopefully, this would give MacDonald and Ohlsen the opportunity to weave unescorted through the crowds and surreptitiously interview and photograph victims while viewing first-hand the lingering damage to the devastated community.

The women organizing the event knew that someone had to be delegated to meet MacDonald and Ohlsen at their hotel and escort their entourage through Ogoniland. Diana offered to do it.

On the appointed day, she and several nuns from the Daughters of Charity arrived in the lobby of the plush Hotel Presidential in Port Harcourt. Dressed in a smart black and red pantsuit, Diana introduced herself to sixty-seven-year-old Flora MacDonald. With her fair skin, red hair, and prominent chin, MacDonald seemed an exotic creature in a lobby full of Nigerians. As for the avuncular Ohlsen, Diana had met him the previous January when Owens had taken him on his first tour of Ogoniland. Ohlsen's thick white hair, glasses, and middle-aged spread were as Diana remembered, yet for some reason, Ohlsen was now acting as though he had never met her. Puzzled, she moved on to the tall black man standing next to Ohlsen and began shaking his hand. As she did so, Gerald Ohlsen casually passed behind her and gave her a covert pinch on the arm. "Don't say your name," he whispered. A wave of under-standing suddenly washed over her. The man standing before her was one of Okuntimo's security officers. Taking the hint, Diana smiled at the plainclothes agent, said hello, and suggested they start.

As far as the security forces knew, Flora MacDonald and Gerald Ohlsen were going to Kaa to attend one ceremony and making no stops along the way. Lieutenant Colonel Komo had no doubt been troubled by the prospect of foreign dignitaries poking around an area that his troops had wiped out. Nevertheless, a well-chaperoned trip to Kaa was acceptable to him. Ironically, the village was now an example of stability and calm in the region because the little community had been more or less rebuilt with funds donated to the Catholic Diocese by the Canadian government.

Diana was aware that the Canadians would receive a skewed view if they saw only one rebuilt village. But Diana had a plan: she would show the Canadians Kaa Waterside as well. Unlike the village of Kaa, the separate and remote waterside area remained a haunting testament to the horrific mili-tary-style assaults that had plagued Ogoniland in the latter half of 1993. To get to Kaa Waterside, one had to drive south out of Bori, whereas Kaa Village lay along a different road heading east. Diana hoped she would be able to direct the military escort anywhere she pleased because Ogoniland was largely devoid of road signs. Once the entourage passed the turnoff for Kaa Village and headed for Kaa Waterside, she knew Okuntimo's men wouldn't have a clue where they were going.

After what seemed an excessively long drive, one of the security officers became impatient. "Have we not reached the place?"

"Not yet," Diana said coolly.

Before long, the Canadian entourage arrived in Kaa Waterside and were confronted with the decimated waterfront village. Homes were rubble; a school building lay collapsed; a church, roofless; a marketplace, destroyed. Flora MacDonald hopped out of the SUV and began snapping photos while Gerald Ohlsen poked around the ruins of the school. At the time of the attack on Kaa, a military attaché had characterized the occurrence as a mere tribal clash. Now, as Ohlsen viewed the evidence of mortar shelling, there was no question in his mind that the attack had been perpetrated by Nigerian forces.

It didn't take long for the security officers to realize they had been duped. They were in the wrong Kaa. Angrily they demanded that everyone get back in their vehicles. Diana apologized and said it had been an innocent mistake; along the way she had missed the turnoff.

In Kaa Village, hundreds were gathered to pay tribute to the Canadian delegation. At the ceremony in a church, MacDonald and Ohlsen were showered with speeches and songs. After the festivities, they went out to mingle and chat. Diana and the Sisters of Charity had arranged for several Ogonis who had been victimized by the army's recent offensive to be stationed at strategic points so that the Canadians could interview them on the sly. As Ogoni students diverted the attention of security officers, Diana whisked MacDonald into a nearby Sunday school room for a private talk with a young rape victim. Ten minutes later, MacDonald was out in the crowd again, the security agents none the wiser.

When it dawned on Okuntimo's men that Ohlsen and MacDonald were talking with amputees and taking pictures of their wounds, they announced that the Canadians had accepted the thanks of this community and it was time to go.

Gerald Ohlsen got wind of the security agents' intent to arrest Diana the moment the Canadian delegation was out of the picture. "Ride back to Port Harcourt with us," he whispered to her. "It's a diplomatic car. They can't arrest you from inside there. And outside the car, stick to us like glue."

Okuntimo's security forces shadowed the two Canadian SUVs on the way back to the Hotel Presidential. There, Diana dined with the Canadians as the security officers bided their time in the parking lot. Unable to remain at the hotel forever, Diana needed a means of escape. Again, the diplomatic immunity offered by the Canadians' SUV provided a solution. Ohlsen helped her into the front seat and told his chauffeur, "Keep driving till you lose them."

The diplomatic SUV took off like a shot with two state security vehicles in pursuit. At first, Ohlsen's driver stuck to main roads, weaving in and out of the congested evening traffic. Quick U-turns, unexpected detours, short-cuts through back alleys – he tried everything, but Okuntimo's boys would not give up. Diana directed the driver deep into the back streets of Port Harcourt, a poorly lit, confusing jumble of narrow, unmarked dirt lanes separating dilapidated storefronts, impromptu garbage dumps, and tumble-down housing. When he finally shook off their pursuers, the driver dropped Diana at the house of one of her uncles who lived in the bowels of the city, where she hid for over a week.

When no police or secret service agents came to nab her on her return to her parents' house, Diana assumed the matter had blown over. She could not have been more wrong. The following day, as she returned to university, a gossipy acquaintance stopped her and said that she was in the newspaper.

"What?" said Diana. "What am I doing in the newspaper?"

"They said you disguised yourself as a nun and deceived the entire secret service!" her friend gushed.

"What kind of news are you spreading now?"

"It's in today's *Monitor*."

Diana went to the student bookstore and bought the latest edition of the national tabloid. Her friend had not exaggerated. There in a two-page centre spread was the story of how Diana Wiwa, wife of the fugitive Dr. Monday Wiwa, had tricked the secret service into taking Canada's foreign minister and acting high commissioner into forbidden Ogoni territory and then, after a high-speed chase through Port Harcourt, escaped capture in the guise of a nun. The governor of Rivers State, Lieutenant Colonel Komo, had announced a one-million-naira reward for her capture – dead or alive.

Reading the story, Diana burst out laughing. The day she'd spent with the Canadians she'd worn a red pantsuit with black polka dots. This is what Okuntimo's men considered a nun's disguise? And a million-naira bounty? The government would never pay that much for her capture. They might have offered it for someone important like Ken Saro-Wiwa, but for a polka-dot nun? Ridiculous.

Yet Diana was deeply concerned. The secret service knew her name, her relationship to Owens and Ken, her activities in NYCOP and MOSOP, and that she was a student at the University of Port Harcourt. The Christmas holidays at school would start soon. Perhaps going with the baby back to Lagos for a couple of months might be prudent. It would be a different experience than her last trip there, however.

This time, she and Owens would both be fugitives.

CHAPTER TEN

Lynn Chukura was a white American teacher in her mid-forties. Born in Pennsylvania, she had moved to Nigeria in the seventies to teach American literature at the University of Lagos. Lynn was a friend of long standing to Sister Majella and in fact had been the person who introduced Majella to Ken Saro-Wiwa several years earlier. As a professor, Lynn lived in one of the staff apartments on the university campus. When she was approached by Majella about the possibility of taking in Ken Saro-Wiwa's sister-in-law and infant nephew, she agreed without hesitation.

"Oh, ab-so-lut-til-lee," she said in her odd, pizzicato way. After living in Nigeria for so long and spending an inordinate amount of time trying to make the locals understand her American accent, Lynn had fallen into the habit of speaking in a slow, excessively concise manner with each consonant over-pronounced. She was also one of the few women drivers in Lagos. Whenever she tootled around the city in her beat-up blue Toyota, the male drivers on the road would honk and taunt her in pidgin as "Madam I-Drive-Myself." She had loaned out her spare bedroom often to students in need of temporary lodging. As long as Diana took pains to use her maiden name, she and Befii would attract little attention at the intrepid Lynn Chukura's place.

Wife and child safely ensconced at the university, Owens began courting the diplomatic community, starting with Canada's acting high commissioner to Nigeria, Gerald Ohlsen. After his visit to Ogoniland, Ohlsen had been worried about Diana. When Owens came calling, Ohlsen ushered him into his office.

Owens was seeking the aid of the Canadian High Commission in securing Ken's release. Ohlsen, though sympathetic to the plight of Ken and the Ogoni people, harboured doubts about the true role of MOSOP and NYCOP in the murder of the four chiefs. He recalled that, during his 1993 visit, he had attended a NYCOP rally in Korokoro and been upset by the angry rhetoric of the leaders. Such inflammatory orations, he felt, could easily have inspired the people to violence. "The potential for violence was there," he told Owens.

"Of course the potential was there," said Owens. He chose his words carefully. "But NYCOP is not violent in nature and has never endorsed violence. There are over fifty thousand members of NYCOP. The fact that one or two might have been inclined to violence does not mean that the organization is violent. It is like any other organization – if one member beats his wife, you would not say that the organization promotes wife beating."

Owens laid out his case that the murders at Giokoo could not have been engineered by MOSOP or NYCOP; Ohlsen took notes and read the clippings and press releases Owens had brought. Ohlsen then turned to a more pleasant topic. As the holidays were nearing, he invited Owens and Diana to his home for dinner; he also said that he and his wife, Mavaia, would be honoured to throw Befii a first birthday party on January 4.

Owens happily accepted. After months on the run in Lagos, skulking about, sleeping where he could, he ached for a normal family life. At the Ohlsens' home, safe from the prying eyes of police, bounty hunters, and paranoid citizens, Owens, Diana, and Befii would have a chance to relax and express their affection for one another, if only for an hour.

At Lynn Chukura's house, whenever Owens braved a visit, he was expected to behave as if he and Diana were in-laws. Lynn had been telling people that Diana was a single mother and that Owens was the brother of Befii's father. Sometimes a week or more would go by between visits, and although Diana and Owens longed to embrace, they dared not. Owens found it excruciating to carry Befii around the house and play with him as if he were a nephew.

On Christmas Day, 1994, Owens and Diana could stand the lack of physical intimacy no longer. After paying a holiday visit to the Chukura household, then pretending to leave, Owens snuck into Diana's room and spent the night. Despite their attempts to be discreet, a young co-ed lodging with Lynn caught Owens slipping out at dawn. He did not stick around to find out if his cover had been blown.

Owens had good cause to suspect that the girl might realize the Wiwas' true identity. After the cancellation of the presidential elections, student activism on the campus of the University of Lagos was on the rise, and student groups had begun voicing their opinions about the Ogoni issue and the arrest of Ken Saro-Wiwa.

As Owens walked across campus that Boxing Day morning, he ran into a childhood friend of Ken's, now on the faculty of the university. Owens had tried to contact him after Ken's arrest to see if he would write an op-ed piece but the professor never responded. Now, when he saw Owens in the flesh, Owens thought the man was going to faint.

"Did anyone see you come here?"

"No, sir," said Owens.

"Are you sure nobody saw?"

"There is no one around. I don't know who would see me."

"There are spies everywhere!" said the professor. "They know that I am Ken's friend, and I'm sure they are looking to see if any of you people are coming. If they see you here now, that will be the end of my life."

"Sir, I am seeing you here only by accident. The only reason I contacted you earlier is that Ken said I should approach you to write a piece. We would still like you to do it. You don't have to use your real name."

"No! Everybody knows my English! I am even sure Abacha knows the way I write."

Although it seemed to Owens unlikely that the dictator would be familiar with anyone's style of prose, he said, "I understand. It's just that Ken really needs this to be done now."

Seeing Owens's disappointment, the professor dug deeper. "Ken is my friend, but if I write this thing for Ken, I might get in trouble. But I'm always available. This is my phone number. The next time, phone my office and if there is any problem I will send somebody."

"I'm sorry to have disturbed you," said Owens dejectedly, walking off without even taking the number.

||||||

Owens next contacted a friend who had become a brigadier in a sensitive position at Defence Headquarters in Lagos. When he phoned, the brigadier suggested he come to Defence Headquarters. "I am aware that you cannot be involved in a crime," he assured Owens. "It would be no problem if you come to my office. I don't think many people will recognize your face."

At their meeting, Owens relayed what he'd heard about the government's desire to silence Ken Saro-Wiwa. The brigadier nodded. "You are right, Monday," he said. "I have to tell you I think this is very serious. It's not likely that Ken is going to get out alive."

The brigadier's bluntness sent Owens into panic. He couldn't imagine things getting worse, yet bad news kept coming. A smuggled letter from his mother informed him that, following Ken's arrest, Okuntimo's forces had come to Bane and shot villagers at random. For nearly three weeks, the entire population of the village had lived in the forest "like animals." Once tensions died down, the women of FOWA resumed their regular community meetings even though public gatherings had been banned in Ogoniland by Lieutenant Colonel Komo. The women defied this order out of sheer necessity. With all the Ogoni menfolk in hiding and MOSOP indefinitely suspended, the women had to keep society functioning. Now meeting ostensibly as a religious group called the Zion Women, FOWA held prayer vigils for Ken and the other detained Ogonis. The Zion Women also started a chain of fasting, with one village praying and fasting for one week, the next village the following week, and so on. In her letter, Jessica wrote that in every village, one could hear women praying through the night. "I think with these sort of prayers, there is no way God will not listen and let the truth be out so that my sons will be back. I am not feeling too well with all this, but you must stay in hiding and everything will be okay."

Owens was devastated to learn that his ninety-year-old father and seventy-six-year-old mother had been forced to live out in the bush. His spirits were further dampened by the Sisters of Charity, who wrote describing the abuses at a detention centre in Kpor; every morning detainees were beaten with a *koboko*, a bullwhip made of cane and rope. Some were hung upside down before being thrashed. Okuntimo's soldiers bludgeoned prisoners in the chest with rifle butts. Soldiers were arresting children as young as twelve; little girls and old women were being raped; looting was out of control.

An Ogoni man wrote to Owens describing a note he found on the ground shortly after Okuntimo's soldiers raided their village. It was a looter's shopping list. "On this operation you are going," wrote one soldier to another, "this is your chance and ask you take the things that you see, please remember to pick sewing machines and some things for my wife."

From his half-brother Harry Wiwa came an urgent plea for Owens to flee Nigeria. According to Harry, some people in Ogoniland were saying they wished Owens had been arrested with Ken, insisting that Ken was not the real power behind MOSOP – it was Owens who masterminded the murders and was responsible for the ills that had descended on their land. Even before the Giokoo murders, an Ogoni elder had told Owens that some people were saying that one way to destroy Ken would be to eliminate his younger brother, the doctor. It was enough to inspire paranoia.

In growing panic, Owens began spending each night in a different location. His hideouts included the home of his former medical school roommate, Chuks Mordi, a doctor with a clinic in Lagos. Owens's former girlfriend, Ethel, took him in on certain nights, as did journalist friend Sam Olukoya at his small apartment near the State Abattoir. Owens also found a haven at the home of Monday Abua, an Ogoni customs officer who lived in Badagry, two hours outside Lagos near the border with the Republic of Benin. And his room at the flat of good-hearted Dr. Osaro-Edee was always available. Some days, Owens hung around Osaro-Edee's hospital or Chuks Mordi's clinic to talk about medicine and lend a hand. It felt good to be useful, to be a doctor again.

At the Canadian High Commission, Gerald Ohlsen acted as a mentor and protector to the entire Wiwa family. He coached Owens on his presentations, instructing him in the protocol and language of diplomacy. Gerald and Mavaia Ohlsen also threw Befii that first birthday party on Ogoni Day, January 4, at their colonial two-storey house in Ikoyi. Befii spent the day in the shady backyard gardens, riding happily on the Ohlsens' pet giant tortoise.

Despite the ban on public gatherings, despite the imprisonment of MOSOP's leadership, and despite the threat to their lives and liberty, the

Ogonis turned out to celebrate the third annual Ogoni Day. To mark the occasion, Ken Saro-Wiwa smuggled out a poem, which was read aloud to the thousands gathered in all six Ogoni kingdoms:

> *My brothers and sisters, my beloved children –*
> *Dance. Dance.*
> *Dance this 4th January, 1995 as we inaugurate the United Nations decade*
> *of the world's indigenous people.*
> *Dance your anger and your joys.*
> *Dance the military guns to silence.*
> *Dance their dumb laws to the dump.*
> *Dance oppression and injustice to death.*
> *Dance the end of Shell's ecological war of 30 years.*
> *Dance my people, for we have seen tomorrow –*
> *And there is an Ogoni star in the sky.*

|||||||

In Lagos, nearly all the embassies, consulates, and high commissions are located on one looping street on Victoria Island. Owens had visited nearly every embassy to drop off copies of MOSOP press releases and news articles about the dire situation in Ogoniland. In January 1995, he began seeking audience with the ambassadors themselves.

MOSOP secretary Ben Naanen, who had been spending much time in The Hague, introduced Owens to his contacts at the Dutch embassy, but the Dutch political officers in Nigeria were hardly sympathetic to the Ogoni cause. To them, MOSOP was a violent organization and Shell should be allowed to recommence its oil operations in Ogoni. Because of the Netherlands' disappointing reception, Owens abandoned any hope of aid from Royal/Dutch Shell's other partner – the British – and avoided the U.K. embassy.

Elsewhere on embassy row, the Swedes and the Danes were compassionate and highly concerned. The Belgians, Norwegians, Italians, and even the Russians were helpful and sympathetic. But next to the Canadians, Owens could find no better friend than the Irish. Through Sister Majella, he developed a warm relationship with Irish ambassador Brendan McMahon,[1] a lover of literature and a fan of Ken's writing.

As for the Americans, during Owens's first formal meeting with a U.S. political officer, he spelled out his reasons for requesting diplomatic assistance. "Our colleagues who are in detention may be killed," he said. "Those at home who are not in detention, more than half of Ogoni, are not sleeping at their homes. They are sleeping in the bush because of the activities of Okuntimo. The first time the army came to Ogoni, it was Shell that asked them to come. The army was not there before 1993. And ever since they've been coming it has been at the request of Shell."

The American political officer said, "Then why don't you go see Shell and talk to them?"

Nearly every diplomat had suggested the same thing. Did these people not understand that Owens was a fugitive wanted for murder? That he suspected Shell had been involved not only in the murders at Giokoo, but in the government's rush to pin the crime on MOSOP? To pin the crime on Owens himself?

Owens had heard that a Shell lawyer was going about Ogoniland questioning people and, in some cases, taking them to the late Edward Kobani's house where they were being prepared as witnesses for the prosecution. Why was Shell involved in preparing the case against Ken? The idea that Owens could waltz into Shell's headquarters and speak with the managing director seemed absurd.

Several organizations in the United Kingdom had been pressuring their embassy in Nigeria to help Ken Saro-Wiwa. Ken sent Owens a note saying that the British were interested in exploring the matter and asking his brother to visit the First Secretary Political at the British High Commission. After phoning for an appointment, Owens was met in the lobby by the secretary himself. For some reason, this well-mannered Scot held their discussion in the lobby, in full view of the other visitors to the embassy. Although Owens was uncomfortable, he did not complain. The secretary asked questions and took notes. At the end of the interview, he told Owens to say hello to Ken and let him know that they were watching.

Soon, Owens was visiting the British embassy frequently. But he was never allowed to speak with the high commissioner, and all meetings were held in the lobby. Not once did a British diplomat invite him to step into an office. It was hard not to feel insulted.

One day, as one of the British first secretaries came into the waiting room, Owens lost his temper. "Sir, why is it that in all the other embassies that I go to, they ask me into their offices where we can discuss matters in private? I am concerned about my security. Here, you keep me sitting in the lobby, and I am thinking that maybe somebody will recognize me and alert the authorities. Why don't you ask me into your office? I'm not coming to ask for political asylum!"

"I'm sorry, but this is our policy. It's the way we treat all our visitors."

"No," said Owens. "I see the white businessmen and the Nigerians who go in there. So it is not true that you keep everybody here."

His outburst must have been effective. The next time he came to the British embassy, a young political officer he had not met before ushered him into a private office. The young chap was fresh out of Oxford. "This is my first foreign posting," he beamed.

Owens began his well-rehearsed speech about the history of Ogoni and Shell, how the military came to Ogoniland at the invitation of Shell, the human rights abuses. "Why don't you guys do something? Shell is your company. Why don't you talk to Shell and ask them to stop using the army on us?"

The young man said frankly, "Owens, Shell is Britain's biggest investment in Nigeria. In fact, it's 90 per cent of the British investment in Nigeria. I'm in this office because of Shell. In a way, Shell is paying my salary. We don't tell Shell what to do. Shell tells us."

Through his conversations with the British, Owens learned that Shell Nigeria's managing director, Brian Anderson, regularly met with Sani Abacha. If Anderson could be convinced to put in a good word for Ken, Abacha might relent and let him receive proper treatment.

Owens approached Gerald Ohlsen for help in arranging a meeting with Brian Anderson, but the Canadian acting high commissioner got nowhere with Shell. Owens next sought help from the Swedish ambassador, an aged man considered the doyen of the Lagos diplomatic community, but not even he could be of help. Anderson was unreachable.

|||||||

On November 21, 1994, the Federal Military Government of Nigeria had appointed a three-man special tribunal to try Ken Saro-Wiwa, Ledum

Mitee, Barinem Kiobel, and John Kpuinen. The chairman was to be Justice Ibrahim Nadhi Auta, a federal high court judge from Lagos. The other two members were a judge from Cross Rivers State and an army lieutenant colonel serving under Abacha. The three members were hand-picked to ensure a conviction. Their judgement was to be final. Abacha gave the tribunal the power to impose the death penalty.

In January 1995, a Gokana man named Beribor Bera joined the ranks of the accused at Bori Camp. According to the newspapers, Bera had made a full confession and planned to testify that Ken Saro-Wiwa had ordered the murder of the chiefs at Giokoo. In reality, Bera was an Ogoni rustic who had been arrested at random and tortured into signing a confession. Bera's tormentors extracted his teeth one by one until he agreed to become a witness for the prosecution. After coercing his confession, Okuntimo's men converted him from a prosecution witness into one of the defendants.

No formal charges were laid against any of the five in detention until eight months after the incident at Giokoo. On February 6, 1995, the tribunal charged Ken and the others with four counts of murder. According to the prosecution, Saro-Wiwa, Mitee, and Kiobel "counselled and procured" Kpuinen, Bera, and thirty others to commit the murders. The thirty men referred to in the charge as "still at large" were not at large at all, but had been languishing in detention at the State Investigation and Intelligence Bureau.

In London, Ken Wiwa, Jr., had finally come out of his existential funk and begun lobbying for his father's release. At well-publicized meetings around the world, Junior met with U.S. vice-president Al Gore, Ethel Kennedy, the Sierra Club, the World Bank, Canada's Environmental Defence Fund, TransAfrica, the political director of the U.S. Oil, Chemical and Atomic Workers Union, and the International Union for Conservation of Nature and Natural Resources' Commission on Environmental Strategy and Planning.[2] Things started looking bad for Shell when Lloyd's announced that it was selling its stock – held in Lloyd's TSB Environmental Investors Funds – because of Shell's poor environmental and social policies in Nigeria.[3]

In a move that many environmental activists labelled a shameless PR stunt, Shell timed the launch of one of its "green" initiatives to coincide with the start of Ken Saro-Wiwa's trial. With great media fanfare, the oil company announced it was embarking on an initiative called the Niger Delta Environmental Survey (NDES), a $2-million study to determine the causes of the delta's environmental degradation that would enable the "stake-holders" to find solutions to whatever problems might be uncovered.

No doubt the oil giant needed to shore up its image after the onslaught of bad press it received following the television documentary *The Drilling Fields*, not to mention the relentless anti-Shell media campaigns being waged by several environmental groups.

The NDES was created in the spirit of Shell's Better Britain Campaign, an ongoing public service project crafted by Ogilvy and Mather (now Ogilvy PR Worldwide), an initiative of similarly suspect motives. Some argued that the Better Britain Campaign was intended to deflect public awareness from the fact that Shell had been selling the organochlorine pesticides that were destroying the biodiversity in the British countryside.[4]

Certainly, Shell selected the members of the NDES steering committee carefully. As survey chairman, Shell appointed Gamaliel Onosode, head of Dunlop Nigeria, a company that uses oil by-products for the manufacture of tires. Also telling was the environmental survey's mandate: instead of concentrating on the ecological and social impact of Shell's activities, the survey sought to highlight environmental changes due to non-oil activities, such as farming and overpopulation. The NDES was to be performed without input from MOSOP or Ken Saro-Wiwa.[5]

Not that Ken was in any condition to contribute. At Bori Camp, the treatment of the Ogoni defendants was brutal. During the early days of his detention, Ken was repeatedly kicked and beaten. Spending prolonged periods shackled to a wall, he was deprived of food, medical care, exercise, access to counsel, and family visits. After bouts of heart arrhythmia and angina, he was finally allowed a visit by a military doctor. The doctor rec-ommended that he be transferred to hospital, but the military would not hear of it.

The tribunal met in Port Harcourt. Outside the courtroom, the defen-dants' family, friends, and legal council were regularly harassed and physically abused. Although Justice Auta ostensibly granted the accused unfettered access to counsel, members of the Ogonis' defence team were often pre-vented from meeting with their clients. On occasion, the lawyers were even barred from attending the trial. Soldiers physically assaulted members of the legal team, and Dr. Kiobel's wife was stripped naked and assaulted when she tried to deliver food to her husband in jail.

After the holidays, Diana braved a return to Port Harcourt to resume her studies at the university and to monitor Ken's trial. On February 21, 1995, as she stood outside the courtroom with Jessica Wiwa and Elfreda Jumbo, waiting for the tribunal session to begin, a soldier approached the three women and began thrashing them with a *koboko*. In terror, Diana and Elfreda fled to their car. Jessica retreated to her own car and ordered her driver to take her around to another entrance. No one was going to keep her from her son's trial. (Lieutenant Colonel Paul Okuntimo, who witnessed the event, said it was a "blatant lie" that the women were horsewhipped, claiming that Saro-Wiwa's mother was refused admittance to the trial because she was late.[6])

As Diana and Elfreda sat bruised and bleeding in their car, busloads of supporters amassed in front of the assembly house and began singing Ogoni solidarity songs. At Okuntimo's order, soldiers lobbed tear gas into the crowd and arrested more than 150 protesters.

In Lagos, Owens listened in stunned silence to Diana's voice on BBC radio as she described to reporter Janet Anderson the horsewhippings and the tear gas attack. Elfreda Jumbo described how soldiers had struck her with the *koboko*, breaking the skin on her hand and her arm and scarring her lower back.

Two days later, seventy-six-year-old Jessica Wiwa was horsewhipped again. Attempting to visit her son in prison, she was beaten, arrested, and transported in a military transport to a detention centre in Kpor. When Owens learned about the outrageous treatment of his mother, he became furious and planned to go back to Port Harcourt.

He did not make the trip. Astonishing as it may seem, Dr. Owens Wiwa, the fugitive, was suddenly invited to make his debut into Lagos high society.

|||||||

Every spring, the Irish embassy in Lagos held a lavish St. Patrick's Day party. Feeling that Owens needed to network among the diplomatic community in a more informal setting, the Irish ambassador, Brendan McMahon, invited him to the upcoming St. Patrick's bash.

On the night of the party, Owens nervously arrived at the ambassador's private residence in a borrowed African outfit. The party was in full swing with smartly dressed couples filling the house and spilling out into the back-yard, where chairs and a tent were set up. Musicians played in the hallway as foreign diplomats and Nigerian military brass mingled around tables of Irish and African food.

In attendance was Joelle Druml, wife of the Austrian ambassador. Druml was a journalist who had visited Ogoniland several times to report on the Ogoni crisis for the French newspaper *Le Monde*. Seeing Owens at the door, she whisked him over to introduce him to the French ambassador. Owens launched into his appeal: "I've been trying to get the diplomatic community to put pressure on Abacha to allow Ken to go to the hospital, and to put pressure on him to restore democracy to Nigeria so that the Ogoni struggle and the matter of Shell and the environment can be resolved."

The ambassador smiled and said that he was a great fan of Ken's writing. He especially liked the section in *On a Darkling Plain* where Ken and Maria escaped by night to Bonny Island in a canoe. "It really showed the courage of the man," said the ambassador. "So tell me, have you talked to the head of Shell yet?"

When Owens expressed his frustration at trying to contact Brian Anderson, the ambassador told Owens to make an appointment to come by the French embassy.

Just then, Brendan McMahon came over and spirited Owens away to meet U.S. Ambassador Walter Carrington.[7] Owens had met Carrington once before, casually, in the corridors of the U.S. embassy. At the party, the jowly African-American ambassador was clearly up to speed on the Saro-Wiwa situation. "You're doing well," said Carrington. "You have to continue." He told Owens that Gerald Ohlsen had been keeping the diplomatic community informed.

"I'm happy to hear that," said Owens. "I've also been trying to meet the head of Shell for quite some time. Many people have told me that he is close to Abacha and might be able to do something. But so far I've not had much luck getting through to him or to the British high commissioner."

The U.S. ambassador took Owens by the hand. "Come with me," he said. Leading Owens across the crowded room, Carrington walked straight up to the British high commissioner to Nigeria. "Lord Mansfield," he said, "this is Dr. Owens Wiwa. He has been trying to meet with you for a long time."

Lord Mansfield looked at the two men and responded coolly, "Well, how nice."

"I think it's important that you two have a nice long chat," said Carrington.

"Yes, of course. Why don't you come and see me in my office sometime?"

"When?" asked Owens.

"Well, anytime. Within the week?" said Lord Mansfield.

Owens smiled and shook his hand. "Wonderful. I will come. I hope that you remember, sir."

"I'll remember."

Owens left the party happier than he had felt in a long time.

|||||||

A few days later, the British high commissioner welcomed Owens into his office and asked his secretary to come in and take notes. Owens was not fooled by the ruse of having a secretary record the meeting. He had been to enough embassies to know that the room was bugged.

Owens reviewed the history of Ogoniland, pointing out that during the period of colonialism, the Ogonis built their own schools and hospitals, and that their people had survived better under British rule than they had fared since independence. Owens next introduced the issue of Shell, and described how MOSOP held a peaceful demonstration in January 1993 that was followed by a plague of terror. "From what I know, the first time there was any human rights abuse by the military, it was Shell that invited them there. I want to discuss these issues with the head of Shell Nigeria, Brian Anderson," Owens said.

Lord Mansfield listened attentively. "I think it would be better if you talked to Mr. Achebe, the general manager of Shell Nigeria."

"This is the man who is second to Brian Anderson?"

"Yes."

"But I want to see Brian Anderson himself."

"Well, that may not be possible. I think an appointment with Achebe would be more realistic."

"Sir," said Owens, "the slave traders who came from Europe did not go into the bush or into the villages to capture the slaves with their hands. They used middlemen, other Africans, to do that. There is no need for me to meet an African middleman. I want to see the head of Shell."

"I will need to get back to you on that," said Lord Mansfield, standing up. The recording secretary looked at Owens, giving him a clear signal that the meeting was over. Owens did not rise from his chair.

"Sir, I would really like to get a definitive response."

"You will hear from us," said the high commissioner firmly. "I will see what I can do."

|||||||

In March 1995, Nigeria's former head of state, Olusegun Obasanjo, went to Copenhagen to attend a UN international peace conference. Obasanjo had become Nigeria's military dictator in 1976, following the assassination of General Murtala Mohammed. After three years of mostly benevolent rule, he had become the only military head of state in Nigerian history to hand over power to an elected civilian government. Since his retirement, Obasanjo had become well known for his anti-corruption campaigns as well as his work for peace, good governance, and accountability. On his return from Copenhagen, Obasanjo was arrested by Abacha's forces.

Ten days earlier, Obasanjo's former deputy head of state, retired Major-General Shehu Musa Yar'Adua, had also been arrested, along with dozens of others accused of treason and related offences. Allegedly, they were involved in a plot to overthrow the government, but the government's only evidence rested on the statement of one man who said that he had discussed the possibility of a coup with the former dictator. Obasanjo, Yar'Adua, and forty-three others were tried by a special military tribunal similar to the one convened for Ken Saro-Wiwa and his compatriots. Obasanjo's tribunal ignored all the evidence presented in favour of the accused. It found every defendant guilty even though the only piece of evidence – the incriminating statement – had been retracted on grounds it had been made under duress. Convicted of treason, Obasanjo and Yar'Adua were sentenced to death.

According to Amnesty International, the real reason for the former leaders' imprisonment was their call for a return to civilian rule. During the previous "transition to civilian rule" program, aborted in 1993, Yar'Adua had been a presidential candidate with support in both north and south. However, Obasanjo's attempts to rally non-violent opposition to Abacha's government in both the north and east of Nigeria, as well as in Obasanjo's own western homeland, were seen as more of a political threat. This threat became magnified in December 1994 after Obasanjo attended the National Constitutional Conference as a delegate and a leading opponent of military rule.

After their convictions, Obasanjo and Yar'Adua were sent to prisons hundreds of kilometres from their homes and kept in harsh conditions to await execution. With the Special Military Tribunal verdict, there was no right of appeal to a higher or independent court.[8]

In Port Harcourt, Saro-Wiwa's military tribunal expanded when ten more Ogoni men were charged with the murders at Giokoo. On April 7, 1995, Pogbara Afa, Saturday Doobee, Monday Donwin, Felix Nwate, Nordu Eawo, Paul Levura, Joseph Kpante, Michael Vizor, Daniel Gbokoo and Albert Kabrara officially joined Ken, Ledum Mitee, Barinem Kiobel, Beribor Bera, and John Kpuinen on the dock.[9]

In charging these men, the tribunal had to satisfy itself that each "appeared to have committed an offence." The prosecution achieved this by dividing the pool of Ogoni defendants into three groups and producing for each set a "summary of evidence" against them. Eleven of the fifteen defendants were not even mentioned by name. The four mentioned – Saro-Wiwa, Mitee, Kiobel, and Bera – were accused of committing the murders in spite of the fact that there wasn't a shred of evidence against them. Although the eleven unnamed defendants had no case stated against them at all, Justice Auta not only ruled that they all had cases to answer, but approved the prosecution's application to hold a simultaneous trial.[10]

|||||||

Owens's preferred hideout was at Dr. Osaro-Edee's apartment near the railway hospital. In the evenings, Owens would sometimes accompany him to the hospital to do his rounds, assisting the doctor and savouring the sense of normality that medical work provided. At the clinic one evening, Owens received word that Dr. Osaro-Edee was looking for him. Finding the doctor in his office, Owens learned that someone had recently come with a delivery for him. Osaro-Edee handed him a luxurious cream-coloured envelope with gilt edging and Owens's name in fancy calligraphy. The stiff card inside was embossed with the insignia of the British Empire.

"The Right Honourable Lord Mansfield requests the honour of your presence at the celebration of the birth of Her Royal Highness Queen Elizabeth II. . . ."

The British High Commission had sought the aid of a Nigerian journalist who then traced Owens to Ken's office on Tejuosho Street. The staff at Saros International directed him to Osaro-Edee's clinic where he delivered this impressive invitation. The Queen's birthday party would be celebrated at the private residence of the British high commissioner.

What was Lord Mansfield up to? Had this something to do with Brian Anderson and Shell? Was it a friendly gesture? A trap? There was only one way to find out. The party was April 21, 1995 – three days away.

It was spring break and Diana had come back to Lagos to be with Owens. Like her husband, she was conflicted about attending the formal party. It was enticing to think of mingling with high society. But high society in Lagos included senior military officials from the Abacha regime and European oil company executives.

"Mon, I'm quite uncomfortable about this."

"I know," he said. "So am I."

On the night of the party, Letam Wiwa, in military uniform, drove Owens and Diana in his wheezing Volkswagen Golf to the sumptuous residence at number 3 Queen's Drive in Ikoyi. Letam had loaned Owens his smartest African traditional outfit of brightly coloured cotton with loose trousers and matching tunic top. From Letam's wife, Diana had borrowed high-heeled shoes and a dressy *bubu*, a smock-like African gown in a light moss hue trimmed with gold braiding. On her head she wore a head wrap in matching fabric.

Owens and Diana stepped out of Letam's Golf and approached the British soldiers at the gates. "May I help you?" asked one of the soldiers, giving the Wiwas a disapproving glance. As Diana fingered the front of her gown nervously, Owens handed the guard his invitation. With furrowed brow, the young British soldier read the card and then handed it to another soldier. He read the invitation, too, and looked the Wiwas over. He handed the invitation to a British security agent in a booth just inside the gate. The security agent studied the official envelope and its contents as well. Finally, he said, "Let them in."

In the foyer, guests were lined up to make their entrance. When it was their turn to step into the main reception hall, the Wiwas were met by Lord and Lady Mansfield. The high commissioner and his wife welcomed them to their home, and then Mansfield turned to his deputy high commissioner and whispered something. The deputy high commissioner shepherded the Wiwas into the thick of the party. From out of nowhere, servants in tuxedos materialized with trays of drinks and hors d'oeuvres. Owens and Diana each took a flute of champagne and anxiously scanned the other guests. Most were familiar faces from the diplomatic corps, but there were quite a few of Nigeria's top lawyers and businessmen on hand, as well as several senior army officers, one or two generals, and a federal minister.

The deputy high commissioner took Owens and Diana into a room where tables had been laid with food and flowers. One of Ken's lawyers, Olisa Agbakoba, was talking with a middle-aged Caucasian couple. The deputy approached this small clique and made his introductions.

"Dr. Owens Wiwa, Diana Wiwa, I'd like you to meet Mr. and Mrs. Brian Anderson."

Owens couldn't believe it. Here at last was the elusive managing director of Shell Nigeria. In person, this man whom every Ogoni feared was merely a tall, sandy-haired fifty-two-year-old with a plain face and a kind demeanour.

"Very pleased to meet you," said Anderson.

"Yes. Hello. How are you," said Owens.

After brief pleasantries, Mrs. Anderson espied someone across the room. Owens asked Diana if she, too, could excuse them for a little while.

With the wives departed, Owens and Olisa steered the conversation toward more serious matters. Anderson, of course, did not need a briefing on the Ogoni situation or the seriousness of Ken Saro-Wiwa's current state. "Right now, Ken is not feeling well," said Owens.

As Anderson was about to respond, a white American businessman came over and interrupted the conversation. He was the head of one of Shell's rival oil companies, and he was in a state of alcohol-fuelled exuberance.

"Brian, Brian, who do we have here?"

"I'd like you to meet Dr. Owens Wiwa and Olisa Agbakoba," said Anderson before introducing his rival.

"Have you tried the shrimp?" the oilman asked Owens.

"No thank you. I am more into suya."

The American picked up a plate of cocktail shrimp. "Here, have one of these," he said, spearing a shrimp with a toothpick and holding it up to Owens's mouth. Having both hands full, Owens let himself be fed by the oil executive. It was a surreal moment, but Owens shrugged it off. In Africa, Caucasians sometimes behave very strangely when they talk with the natives.

The American staggered off and Owens returned to his discussion with Brian Anderson, talking generally about the state of the world and Nigeria before turning to the topic of Ken once again. "The Ogonis respect Ken and Ledum very much," he said, quickly adding, "As I mentioned earlier, Ken is not feeling well. He is having trouble with his heart. I thought that maybe you could help to see that he is taken to hospital."

With that, Anderson reached into his dinner jacket and pulled out a business card. Handing it to Owens, he said, "Give me a call so we can discuss this some more." Then thinking better of it, he went through his pockets and procured another card. "The first card has only my office number. Here, take this one. This other one has my home address and phone number, too."

As Brian Anderson moved off to mingle with other guests, Olisa Agbakoba struck up a conversation with G.O.S. Bensen, a famous Nigerian lawyer, and five or six other influential Nigerians at the party. For the next hour, Owens and Olisa impressed on these men that Ken needed help.

"Don't worry," said one businessman. "We all know Ken is a man of peace. He could not kill anybody."

"We all know that," said another. "It's no problem. Don't worry. Nothing will happen to him."

Bored and uncomfortable, Diana rejoined her husband with her shoes and the hem of her gown ruined from walking across wet lawns. With no one to talk to, she had spent most of her time at the party in a far corner of the manicured back gardens on a tree swing. As for Owens, he was feeling better than he had in months. The party was a smashing success.

Just as the two of them began to say their goodbyes, Owens spotted Anderson and his wife. "Wait, Diana," he said. As they walked over, Owens whispered to her, "See if you can find some way to mention Ken's heart condition again."

As they said their farewell to the Andersons, Diana pricked up her courage. "My brother-in-law Ken has heart trouble," she said. "The doctors have recommended he be taken to hospital, but he is not being attended to in that way. Is there anything at all that you can do to help him?"

"I will see," said Anderson. "I'm not promising anything, but I will see."

As Owens led his wife away, Diana excitedly whispered, "Brian Anderson is such a nice man. He must not know what's going on in Ogoni. Somebody this nice cannot possibly be involved. Now that he understands what's going on, surely things will change."

CHAPTER**ELEVEN**

Brian Anderson came to Royal Dutch/Shell in 1968 as a trainee petroleum engineer. For the next twenty-six years, the British father of two worked around the world for Shell in such diverse places as Norway, Brunei, Sarawak, the Netherlands, Oman, and Australia. In 1994, Shell elevated Anderson to the key position of managing director of Shell Nigeria.[1]

Like Diana, Owens was impressed with Anderson's courteous, approachable manner. It was good that Anderson was relatively new to the Nigerian operations and had not been responsible for running the company during the 1993 Ogoni crisis. In the past, MOSOP's real beef had been with Anderson's predecessor, Philip Watts, Shell Nigeria's chairman and managing director from 1991 to 1994.[2] Perhaps Anderson would be someone the Ogonis could look to for help.

After leaving the Queen's birthday party, Owens wrote a long letter to Ken, reporting what had transpired. Early next morning, he put Diana on the first bus to Port Harcourt, entrusting her with the letter. Three days later, word reached Owens that Ken had been transferred out of his cell to a well-guarded hospital room. Overjoyed, Owens wrote again, saying he was certain Brian Anderson must have had something to do with this fortunate turn of events.

A week later, another letter arrived through the underground from Ken encouraging Owens to meet with Anderson as soon as possible. "Time is of the essence here," he wrote. He gave Owens explicit instructions on how to act and what to say. "The fact that the MD of Shell is willing or is being made to speak with you is a most important sign," Ken wrote. "It is possibly the best thing that has happened so far. That is if you handle it maturely with a sense of humour without ever losing your temper. If you get to see him, remember that you are dealing with one of the most powerful organizations on earth. The Dubai Royal Family, the British Monarchy both have substantial shares in it. The Br. Govt depends on Shell to survive – in that it provides employment and a lot of earnings. Diplomats from Britain will value it if at the end of their career, Shell can appoint them to directorships."[3]

Ken listed in point form the issues Owens was to cover. First, he should make it known that Ken was interested in dialogue. Second, he should impress on Anderson the common knowledge of Ken's innocence. And third, he should impress on Anderson that Shell must urge Abacha to call off this charade of a trial.

Concurrent with Ken's letter, however, Owens received disturbing news. The Nigerian minister for information had released a booklet entitled *Crisis in Ogoniland: How Saro-Wiwa turned MOSOP into a Gestapo*. On the cover was a picture of Ken and a swastika. The booklet claimed that Ken Saro-Wiwa "established the NYCOP (National Youth Council of Ogoni People) Vigilante, as a terrorist organisation well-trained and equipped in the art of terrorism." The Ogoni crisis, the booklet said, revolved around one man alone – Ken Saro-Wiwa. The Nigerian government, it claimed, was not involved in any detentions or extra-judicial executions.[4] If Anderson hadn't yet seen this libellous document, he would see it soon. Ken was right: time was of the essence.

Telephoning Anderson, Owens said, "I've been in contact with Ken, and I'd like for us to continue our discussion."

"Yes, of course. Come to my office and we'll talk," said Anderson.

Knowing that there would be policemen at the gates and inside the Shell office building in Lagos, Owens said, "There is no way I would feel secure there."

"Can you come to my home?"

Owens thought a moment. "Yes, I can." But just to be safe, Owens told himself, it would be wise to let a few people know where he was going just in case anything happened to him at the Andersons' home.

The two men set a date to meet.

In the late afternoon on the first Sunday in May, Letam Wiwa drove his brother to Anderson's house just off Awolowo Road, the main thoroughfare of Ikoyi. At the gates were two police officers. When questioned, Owens gave them his middle name only and said that Brian Anderson was expecting him. After checking with Anderson's staff, one of the officers escorted Owens to the front door. A Nigerian house servant dressed in white coat and trousers, typical of servants' uniforms during colonial times, ushered him inside.

Anderson greeted Owens in the foyer. As Owens stepped into the living room he was struck by the grandeur of the view from the many plate glass windows. The Andersons' gardens and lawns flowed to the edge of placid Lagos Lagoon, creating an elegant framework for the downtown skyscrapers of Victoria Island beyond. Owens sat on the settee and the servant approached respectfully to inquire if he would care for a drink or something to eat.

"Oh, I am not hungry, thank you," said Owens. "I'll just have some water."

"You can have a drink," said Anderson, smiling. "I won't poison you."

"No, water will be fine. I'm – I'm anxious to discuss what I've come to discuss. First of all, I'd like to thank you. After our discussion at the Queen's party, Ken was taken to the military hospital for the first time. So whatever you had to do with that, I'm grateful."

Anderson looked at Owens with a noncommittal expression. Owens gave him some background about Ogoni life and culture, describing what Ogoniland had been like when he was a child and what it had become, a land of extreme poverty where the people had not benefited from the vast wealth taken from their land.

Anderson agreed that, in the past, the distribution of oil profits had been unfair to the people of the Niger delta, but he defended his company's policies, saying it was not the place of private industry to assume responsibility for social programs like health, education, and infrastructure – these were clearly the government's domain. "Shell spends more on community projects in the Niger delta than it spends anywhere else in the world," said Anderson.

"Shell has renovated three school buildings, supplied teachers, and donated school furniture and science equipment. We've built roads. We've started immunization programs and we're committed to doing more."

To Owens's surprise, Anderson said that he himself had grown up in Nigeria. Much of his childhood had been spent in Jos, where his father worked at a tin mine. Anderson went to England to study at the University of London's Royal School of Mines, graduating with a master's degree in petroleum reservoir engineering. After working all over the world, then, his recent posting in Nigeria had been rather like a homecoming.

"Tell me a little about yourself," said Anderson. "Tell me about your family."

Owens described the polygynous household in which he had grown up. "My father has five wives, and from my mother, we are six children," he said. "Ken is the Saro, the oldest. My sister Comfort is a lawyer in Kaduna state. The next one, Barine, runs a catering school in Kano. My immediate brother, Jim, is in London, so we consider him lost to the family."

The servant entered to inform Anderson that he was wanted on the telephone. "Excuse me a moment," Anderson said as he picked up the receiver of the living-room extension. "Yes?" he said. He listened a moment and then lowered his voice. "Be patient. I'm at it. I'll get back to you."

Hanging up, he turned back to Owens. "Sorry. You were telling me about your brother in London. Why do you say he's 'lost' to the family?"

"Well – he's in England," said Owens, perturbed by the cryptic phone call. "He's not at home. So that means his contribution to Ogoni will be limited. He's a musician. He plays the guitar." Owens told Anderson a few things about Letam, his junior brother in the Nigerian army, and about his own two clinics in Ogoni. He described the military-style assaults on Ogoni communities in 1993–94, as well as Major Okuntimo's reign of terror. Anderson listened sympathetically as Owens described Ogoni citizens who had been raped, shot, or maimed at the hands of soldiers and police, not to mention the overwhelming caseload of Ogonis suffering from illnesses caused by four decades of oil-industry-generated pollution.

"Ken and Ledum are the only two that have the trust of the Ogonis," Owens continued. "They are innocent of their charges. It is important that they are free so that they can have discussions about this."

"I don't trust Ken," said Anderson.

"How can you say you do not trust a man you have never met? What has Ken said or written that makes you not trust him?"

"The claim of environmental devastation that Ken wrote about is one example," said Anderson. "From all reports, the Ogoni environment is the same as anywhere else in Nigeria."

"Well, maybe you're hearing from people who are just passing on the main road, not somebody who has gone in to look at the pipelines and the sites of the oil production and the oil spillages."

"No, the reports I've read are very detailed. They do not support any sort of claim of 'ecological devastation.' It's this sort of unfair, alarmist accusation that is precisely the reason why we feel we cannot trust someone like Ken Saro-Wiwa."

"If you will see that Ken is freed, he will be happy to engage you in discussion."

"First of all," said Anderson, "I don't have the sort of power you think I have. I cannot just phone up Sani Abacha and tell him what to do. Besides, Abacha would like to arrest me as well. The last time we met, he accused Shell of trying to destabilize his government by supporting the oil workers' strike."

"Anyway, I believe you *can* do something to help Ken," said Owens firmly. "Shell produces half the oil in Nigeria, and oil money is keeping the military in power."

Anderson thought about this. "I suppose it would be difficult – but not impossible. But you have to show goodwill."

"What is the goodwill?"[5]

"For one, you will have to write a press release on MOSOP letterhead saying that there is no environmental devastation in Ogoniland due to Shell's activities. Second, you will have to call off the anti-Shell campaigns you've got going in Europe and elsewhere. It's hurting Shell and it's hurting the Nigerian government. It would be for your own good. There's no telling how Abacha is going to react as long as this campaign continues."

Owens sat up straight. "You mean I should write that all the things we've been saying for years is a lie? I do not have the power to stop the campaigns. I can't do that."

"Well, Owens, you can."

Swallowing his anger, Owens replied, "I'm going to relay what you have just said to Ken. I'll get back to you."[*]

||||||

A week later, Owens received a letter from Ken: "I got your notes on the discussions with Brian Anderson. Surprising that you should spend three hours making no progress. That Shell can be thinking of token measures in Ogoni shows that they are still not aware of the depth of their problems. Even their discussions with the Nigeria High Commission in London on 16/3/95 shows that they are only interested in getting their money; they said nothing whatsoever about doing something about Ogoni, only to produce more films to counter the Ogoni campaign publicity. They still do not realize that the propaganda war was lost since 1993. Their PR dept is out of touch with reality. When the minutes of that meeting get to the Campaigning Group, Shell must blush."

In his letter, Ken mentioned Owens's next meeting with Anderson. "You must give him the idea we are very strong and very determined to have our way." Enclosed was a second letter Ken wished Owens to give Brian Anderson. Owens called the Anderson household again, and Anderson invited him back the following day.

On his second visit, Owens was met at the door by both Mr. and Mrs. Anderson. "Welcome back, Owens. I'm sure you remember my wife."

"Yes, hello," said Owens, shaking her hand. "How are you?"

This time, the Andersons served beer and peanuts in the living room and chatted about local events. When Owens tried to steer the conversation toward Ken, Anderson asked him to wait. After a few minutes, an African man in a business suit entered the room.

---

[*] This account is based on my interviews with Dr. Owens Wiwa, his letters from Ken Saro-Wiwa in prison, and the transcripts of Dr. Wiwa's extensive testimony before Greenpeace. When contacted, Shell spokesman James Herbert asserted that Wiwa's account of this matter is "misleading and contrary," and that Shell "rejects the claim that we offered to help ensure Ken Saro-Wiwa's release if he stopped the campaign." Similarly, Eric Nickson, another Shell spokesman, told the *New York Times* that Owens's version of these encounters is "misleading" and that Anderson never indicated he could be of assistance. On further questioning by the *Times*, Nickson conceded, "I wasn't present at the meetings, unfortunately."

"Oh, that looks like Ken," said the man, proffering Owens a handshake.

"Owens Wiwa, this is Nnaemeka Achebe, the director and general manager of Shell Nigeria," said Anderson.

"I was saying you look like your brother."

"I didn't know someone else was going to come," said Owens.

After Mrs. Anderson excused herself, the three men sat down to talk. "I have received a letter from Ken that he asked me to show to you," said Owens. As he withdrew the letter from the large front pocket of his *adire* shirt, he added, "Now, I must tell you I have already read this letter. Before I show it to you, I want you to recognize the situation under which Ken is writing this. He's in a hospital surrounded by soldiers. You should have this as background."

Owens handed Anderson the letter and nervously watched as the Shell executive silently read Ken's message.

"Dear Eagle," the letter began, "Every day we spend in detention enhances the Ogoni cause and puts Shell on the block. So, detention is not a bad thing for us as such. I always knew it was a part of the struggle. Even death itself will only make me a martyr and enhance the cause. I'm surprised that Shell should be talking of trust and goodwill. If they had listened to me early in 1993, all this would not have arisen. It would have cost them far less. Anyway, I'm looking for solutions, not trying to cast blame. Look at the following scenarios:

"(1) We remain in jail. The cause grows. Junior's recent visit to the US was a smasher. . . . Shell could be facing an expensive call for the boycott of its products. . . . Within Nigeria, the other oil-producing areas who have been working to see if MOSOP's non-violent stand will draw Shell out of its cocoon, decide that force is what Shell wants. There will be trouble on the oilfields, losing Shell a lot of its investment. Military force will not secure them peace on the oilfields. Everyone knows that now. As to the 'Major Environmental Survey', it will solve few problems if it does not have the support of authentic community leaders or if Shell tries to play politics with it. In the end all this would be extremely expensive for Shell particularly if as is being [illegible], we sue them in the U.S. Punitive damages could be imposed on them. Our sponsors are thinking in that direction. Is this what Shell wants?

"OR THIS?

"(ii) Shell uses its considerable clout to (i) CREATE an Ogoni State. We will have something to show to the Ogoni people & assuage their anger; (ii) Shell gets the Tribunal stopped or the Fed Attorney General to enter a 'nolle prosequi' using the fact of my ill-health and inability to go on with the trial. Shell's stock in Ogoni rises. The human rights and writers' lobbies which have stood solidly behind us abroad are disarmed. Shell gets a breathing space.

"(iii) Only the environmental lobby now remains to be satisfied. Once it is known that I am beginning to speak with Shell both in Nigeria and abroad, hostility against Shell will diminish somewhat. I'd do everything to douse the fires we have lit.

"(iv) Ogoni is only 32 miles by 12. With the confidence the people have in me, we can work to get whatever compensation is due settled over a period of time. But Shell would have to ensure that the Fed Govt pays rents and royalties to landlord communities. I can convince Ogoni people to lower their demands, although I'd have to be extremely careful before they lynch me. Even the govt. would welcome the better image this would create for them.

"If I were Shell, I would settle for this second scenario. If they do not have direct access to Abacha (but I don't believe that), they can work through the British High Commission. Abacha relies on the BHC. And they can get anything from him. With us [busy?] on the Ogoni affairs, Shell will find respite on the rest of the delta whereas while I remain in jail, I am a symbol of all the oil-producing areas. Time is of the essence in these matters. The campaign abroad is widening and is likely to move to Australia soon. In any case, the Sierra Club is determined to secure the best for Ogoni. You may show this letter to BA [Brian Anderson] and I pray he listens to us. And acts quickly. In the interest of Shell. Shell are the ones who need to show trust and goodwill. They have screwed Ogoni enough in the past. Regards, Lion"

As he read, the colour rose in Anderson's cheeks. By the time he finished, his face was flushed. Without a word, he handed the letter to Achebe, who also spent a couple of minutes absorbing the contents. After finishing it, Achebe handed it back to Owens. "Well, that's Ken. Concise and straight to the point."

Sensing Anderson's anger, Owens tried to mollify him. "Ken wrote what he wrote because this is generally the agreement of the Ogoni people. It's what is written in the Ogoni Bill of Rights, you understand. But Ken is willing to enter into negotiations on any of this. Please, please do whatever you can to see that Ken and the others are freed."

Achebe, also seeing how angry Anderson was, underlined the importance of continued dialogue. Anderson glowered and said nothing.

"Are you going to help and talk to Abacha?" Owens asked.

Anderson fixed Owens with his gaze. "If your brother wants to be a martyr," he said, "that's his business."

Anderson rose from his chair and informed Owens that he would no longer be available for further discussion. Henceforth, all further contact would have to be through Nnaemeka Achebe.

After Owens detailed the disastrous meeting with Anderson and Achebe, Ken wrote back, urging him to try to re-establish communication with Shell. Owens found it impossible. The next time he contacted Anderson's office, he was told that the managing director was not in. When he phoned Anderson's home, the house staff said that he was at lunch. When he called that evening, the house staff said Anderson had gone out for dinner. Owens phoned four or five times a day, but never again connected with the managing director. It was the same with Achebe – he had gone to Abuja, gone to Port Harcourt, gone to Germany . . .

Owens informed Ken of these rebuffed attempts, and Ken replied with a note from the hospital: "Give up. You shouldn't bother again."

||||||||

In the spring of 1995, Royal Dutch/Shell was being attacked from all sides. Hot on the heels of the European Parliament's declaring the effects of oil extraction in the Niger delta an "environmental nightmare"[6] came an unrelated Shell PR crisis known as the *Brent Spar* incident.

In February 1995, Greenpeace received word that the British government had granted Shell permission to sink a contaminated oil storage platform into the North Atlantic west of Ireland and Scotland. Decommissioned in 1991, the 4,000-tonne *Brent Spar* had been a loading buoy and crude-oil storage tank for fifteen years. Shell was hastening to sink the *Brent Spar* because North Sea environment ministers were scheduling a meeting in Denmark to discuss measures to eliminate the discharge of hazardous substances from all sources into the North Sea and the marine environment.

To prevent the sinking of the *Brent Spar*, Greenpeace invaded the rig. Shell managed to evict more than two dozen of the activists, but Greenpeace re-occupied the rig and continued its protest. Because of Greenpeace's action, consumers worldwide boycotted Shell products, and a Shell station in Germany was firebombed (an action condemned by Greenpeace). Because of the boycotts, the violence, and the damning publicity, Shell dropped its plans to sink the *Brent Spar*, and none of its oil structures have been dumped at sea since.[7,8]

To the Ogonis, *Brent Spar* was a beacon of hope. If Shell could act responsibly in the northern waters, it might do the right thing in Ogoniland. In any event, it was becoming more difficult by the day for Shell to portray Ken Saro-Wiwa as the leader of a violent organization. Amnesty International had adopted Saro-Wiwa as a prisoner of conscience, proclaiming him a staunch advocate of non-violence. Then the Right Livelihood Award was awarded to him for being "a star in our human cosmos." And on April 17, 1995, he was given another prestigious award – the Goldman Environmental Prize.

The Goldman Prize, the world's largest cash award for environmental activism, annually honours one "grassroots environmental hero" from each of the six inhabited continents. San Francisco philanthropists Richard and Rhoda Goldman created the prize to recognize the efforts of ordinary people to protect the environment. As the 1995 winner for Africa, Ken received US$75,000 and was lionized in *Time* magazine as "a hero who dares to raise a voice against the partnership between the government and a foreign oil company."[9]

||||||

Owens spent the late spring alternating between Sam Olukoya's small apartment near the State Abattoir and Monday Abua's home in the border town of Badagry. After months in hiding, Owens was getting bolder. He even went so far as to visit the Saros International offices instead of hiding at the women's boutique across the street. Part of this recklessness came from the fake I.D. he had obtained courtesy of an editor at *Tempo*, a popular Lagos newspaper. Owens's false papers identified him as "Tony Uche," a *Tempo* sub-editor.

While he was suffering from a bout of malaria, Owens hailed a taxi one afternoon and asked the driver to take him to a clinic where one of his friends was an attending physician. Nearing the clinic, the taxi was stopped by police at one of the many checkpoints in Lagos. After being ordered out of the car, Owens was subjected to a full body search. Normally he had been vigilant about not carrying anything incriminating, especially after the police had discovered the letter from Ken concealed in a matchbox. Armed with his fake I.D., however, Owens had recently let his guard down. On this day, the police discovered in his pocket a small address book filled with high-profile addresses and phone numbers.

"You have all these numbers," said the policeman, thumbing through Owens's little book. "Look! This one is to London. Amnesty. And this one to the U.S. What is this? Ah! You must be a four-one-nine person!"

Section 419 of the Nigerian criminal code – "obtaining goods by false pretences" – is the pidgin term referring to the classic Nigerian con game currently spreading like wildfire via international e-mail spam.

"I am not four-one-nine," said Owens.

"You are four-one-nine! What are you? What is your name?"

"Tony Uche. I'm a journalist for *Tempo*."

"You work for the *Tempo*? I never seen your name in this paper. You are four-one-nine, you!"

"But I'm an editor. You wouldn't see my . . ."

Before Owens could protest further, the policeman commandeered the taxi to drive Owens to the nearest police station. Within minutes, he found himself standing once more before a desk sergeant, who ordered him to write out a statement.

"What am I supposed to write a statement about?"

"Because of these phone numbers we are arresting you," said the sergeant, waving the address book at him.

Burning with malaria fever, Owens snatched the address book from the policeman. Flipping to one of the entries, he held it up for the officer to see. "Look," he said, "this is the number of my editor at *Tempo*. You can call him. He will tell you who I am. And here, this is the number of the Canadian High Commission. The chap there is named Gerald Ohlsen. You can call him, too. Here is the number of the American ambassador. If you doubt me, you can call any of the numbers in this book. This is a brigadier in the army. If you want to call him, call him. It's in Defence Headquarters. These are my friends. And you are putting me here because you saw foreign numbers in my address book? Is that how you do it? You call them and then we will see what they do to you."

Owens's bravado was extinguished by the appearance of a policewoman who entered to see what the ruckus was about. Owens recognized her immediately – she was an Andoni police officer from Port Harcourt. Not only did she know Owens's real identity, she was aware he was wanted in connection with the murders at Giokoo.

"I know him," she said to the lieutenant. "Why are you holding this man?"

As Owens's blood ran cold, the sergeant explained that Tony Uche was being arrested as a suspected case of 419.

The policewoman shook her head. "No," she said in pidgin, "he no be four-one-nine-oh. Leave my brother go."

"What?"

As the sergeant glowered in indecision, Owens stared at the young woman. Why had she not exposed him? She was Andoni. A police officer. She should have been his enemy.

"He is my brother. We are from the same state. He no be four-one-nine-oh. Leave him go," she repeated. Grudgingly, the sergeant handed back the address book and motioned Owens to leave. Reeling from his illness, he returned to Sam Olukoya's apartment and collapsed on the sofa.

"This is it," he said to Sam. "I have been lucky twice. Now I am going to disappear."

||||||

Ken's military hospital turned out to be just as bad as prison. In a letter smuggled out on May 8, Ken wrote: "For two nights I have not slept a wink, I am being intimidated, harassed and de-humanized, even though I am supposed to be receiving medical attention. . . . I am like Ogoni, battered, bruised, brutalized, bloodied and almost buried."[10]

In a matter of weeks, Ken was discharged from the military hospital and returned to his cell at Bori Camp.

||||||

After years of working toward a degree in political science, Diana was set to defend her thesis at the University of Port Harcourt in early June 1995. As she put the final touches on her dissertation, she dared to venture out only to a nearby business centre to have her handwritten thesis word-processed.

The morning of her thesis defence, she and her father arrived at the business centre to pick up her manuscript, but were informed that a computer operator had hit a wrong key and wiped out the entire 130-page document. It would take three days to retype the manuscript. Diana didn't have three days. Because of her marriage, the birth of her child, and the crisis with Ken and Owens, she had exhausted every possible deadline extension. If she did not defend her thesis that day, she would have to wait another year. As Diana flew into a rage at the woman who had put her in this predicament, Professor Barikor left for the university to fetch a relative who worked as a computer scientist to find if there was a way to retrieve the deleted document. After her father left, Diana sat down with the worker and furiously began retyping. With hard work and luck, she might be able to get to school by day's end.

As Diana was typing at the business centre, Okuntimo's henchmen showed up at the office of the university's dean of student affairs. The secret service had found out when Diana was scheduled to defend her thesis. "We are looking for Mrs. Wiwa," one of the officers said.

"Why are you looking for Diana?" asked the dean.

"It's a security matter. We are here to arrest her."

The dean knew Diana well and was upset to think she was in peril. Giving Abacha's agents just enough vague information to get them out of his office, he found Professor Barikor and told him the secret service was looking for his daughter. "Find a way to get to her so that she doesn't come to school," said the dean. "Security is all over the place. A good number are in plain clothes."

Professor Barikor raced back to the business centre and told Diana to abandon the thesis. She returned to her uncle's house in the back streets of Port Harcourt and hid until her parents delivered Befii at nightfall. Early next morning, she and her son were on the bus to Lagos. Owens's entire family was now permanently underground.

|||||||

In his jailhouse letters, Ken always referred to Lekue Laa Loolo as "Antelope," for indeed, Loolo means antelope in Ogoni.

Ken and Loolo had been friends from their earliest days. Their Ken-Khana families travelled in the same circle; the boys attended the same classes, played on the same fields, shared the same hopes and dreams. Loolo, as physically diminutive as Ken, always seemed to be in Ken's shadow. Their differences increased when Ken went off to university and later sided with the Nigerians during the civil war. Loolo had done neither of these things. His instincts were to stay with the herd, and Biafra seemed to be the direction most young Ogoni men were leaning. Nevertheless, after the Biafran defeat, Loolo was one of the first people Ken sought out and he was one of the first repatriated Ogonis whom he was able to help find a job. Ken introduced Loolo to the governor, who ended up appointing him to a ministerial post in Rivers State. Loolo eventually lost that job – and others as well – but Ken was always there to lend his old friend a hand.

When MOSOP was formed, Loolo had been the organization's first vice-president, but he resigned that post when he decided to run for political office. Out of friendship, Ken helped with Loolo's campaign, donated money, and even lent him his car for electioneering purposes. Loolo won the election but was forced out of office when Abacha came into power. During Abacha's reign, as Ken became more militant, Loolo began to distance himself from his long-time champion. Feeling more comfortable with the so-called moderate Ogonis, Loolo allied himself with Edward Kobani, Albert Badey, Priscilla Vikue, the Orages, and Garrick Leton.

One afternoon, a staff member at Saros International came to see Owens at his hideout near the railway station.

"Someone is looking to see you."

"Who is it?"

"Mr. Loolo."

"Loolo! What does he want?"

"He didn't tell me. He said he had something important to tell you. He said Ken asked him to see you."

Owens thought about this. He knew that Loolo had been one of the elites who had published that damaging apology to the governor after the Willbros incident at Biara – the apology that repudiated Ken as an Ogoni spokesman and called for government troops to quash MOSOP. Yet if Ken had really sent Loolo, should Owens turn him away?

"Are you sure there was no policeman following him, no security agents?"

The Saros employee didn't think so. Taking the gamble, Owens asked him to bring Loolo to the Railway Club.

When Loolo arrived later that day, Owens greeted him with all the respect due a senior. Loolo seemed shocked to find Owens in such a squalid part of the city. "Well, that's life," said Owens. "I hide somewhere around here."

Finding a corner where they could talk, Owens asked what had been going on at home. Loolo admitted he had severed relations with Garrick Leton, Priscilla Vikue, and the rest of the elites. "I told them there was no way I was going to be in that conspiracy to see that Ken – no matter how bad he was – would either be hanged or convicted," said Loolo. "Ken should not be accused of murder. He knows nothing about murder. Ken would never murder anybody. I am no longer following these people. I am now on the side of truth."

"What brings you here? You said Ken asked you to see me."

Loolo glanced around the Railway Club to see if anyone could be eavesdropping. "Before I broke with these people, I have been with them and I saw all the moves they were making. The police and the army and the CIB people have been putting some people in Kobani's house and drilling them as prosecution witnesses."

"Drilling them?"

"Yes. They have some lawyers in to do that. When I used to go to Kobani's house, they were always there. But two of these men, these witnesses against Ken, have had a change of heart. They escaped from Kobani's house and sent people to tell Ken and Ledum about what happened. These men are in much danger, so we have brought them here to Lagos."

"What are their names?"

"Charles Danwi and Nayone Akpa."

"I don't know them," said Owens. "So tell me, what is it they want?"

"They told me that all they wanted was to be safe and to get out of the country. They gave me a letter from Ken, and in it Ken says we should try as much as possible to find out if what they are saying is true and get as much publicity as possible. Ken says that if what they are saying is the truth and the world hears about it, there is no way the government will continue this charade."

Loolo's message was a gift from heaven. Owens arranged for Danwi and Akpa to be taken to the chambers of Gani Fawehinmi for debriefing and the signing of affidavits. On the day of the meeting, Owens sat down with the two at Ken's lawyer's office in Lagos. Fawehinmi brought out a video camera, turned it on, and asked them to tell everything they knew.

Danwi described being arrested on the road by soldiers who questioned him about his whereabouts on the day of the murders. He told them he had been in Eleme, far from Giokoo. When asked about MOSOP, Danwi had said he was a musician and he didn't know anything about MOSOP. Nonetheless, the soldiers took him to the late Edward Kobani's house in Port Harcourt where Nayone Akpa was also being held. Danwi and Akpa alleged that they were each paid thirty thousand naira to say, untruthfully, that Ken Saro-Wiwa had instructed NYCOP youths to murder the four chiefs at Giokoo. Danwi said that, apart from the thirty thousand naira, he was also promised a house, a contract from Shell* and the Oil Mineral Producing Areas Commission (OMPADEC), and that he was placed on a level-five monthly income at the Gokana Local Government without even being an employee of the council. The men testified that it was Priscilla Vikue, I.S. Kogbara, and Edward Kobani's brother Mohammed who supervised the bribery.[11] The men who coached their testimony included the attorney general of Rivers State, several lawyers for the prosecution, and members of the detective team. Danwi and Akpa also mentioned that Bennett Birabi had come by Kobani's house to buy them drinks and congratulate them on the service they were about to perform.

Gani Fawehinmi published these confessions in his tabloid *The Masses* with the blaring headline "OGONI TRIAL SHOCKER: I Was Bribed to Lie against Saro-Wiwa – Says Prosecution Witness" alongside a photo of a dazed-looking Charles Danwi.[12]

The publication of this confession caused a stir across the nation and abroad. In most countries, such revelations would have ended (or at least suspended) any legitimate trial. In the case of Ken Saro-Wiwa's Special Military Tribunal, the publicity exposing the trial to be nothing more than a government-sanctioned lynch mob had the opposite effect. The tribunal's three judges not only proceeded with their kangaroo court but turned

---

* Shell has denied all allegations of bribery, saying, "We have not paid cash, awarded contracts, or used any other means to try to influence events surrounding the cases before the tribunal."

openly hostile to the defence team. Okuntimo's soldiers assaulted Ken's lawyers outside the trial venue, slapping members of the defence team on one occasion and ripping Fawehinmi's jacket on another. During one of these run-ins, Fawehinmi was prevented from attending the trial and was summarily "deported" from Port Harcourt.[13]

Owens then read a newspaper item claiming that the Supreme Court of Nigeria had just ruled that a person sentenced to death for murder could now be executed as quickly as one week after the verdict. Frantic, Owens contacted the Swedish ambassador. He knew that the ambassador played golf with a justice of the Nigerian Supreme Court and perhaps would have some inside information.

"What is this?" Owens asked the ambassador. "Have these people already made their decision about Ken? Perhaps you can find a way to raise this issue with the justice next time you see him."

The Swedish ambassador said no. "I have already raised the issue of Ken with the justice," he said. "The justice told me I shouldn't bother about it. 'It's a done case,' he told me."

"What do you mean?"

"I mean there's nothing anyone can do now."

|||||||

"Mon, do you know this house is sinking?"

"Diana . . ."

"The whole area is sinking! You go to the lower floor, it's always flooded. I'm sure Sam never noticed that. It was one of the first things that caught my attention. And we're right in the plane route. We're so close to the airport, I'm scared that a plane will land on the house! They're flying so low."

"Where would you rather be? With Ken?"

By the time Diana and Befii had come permanently to Lagos, Owens's bank accounts were empty. He had no prospects. His clinic in Bori had stopped producing income; his clinic in Eleme was occupied by Nigerian soldiers. Now that Owens's entire family was forced to hide in Sam Olukoya's small apartment, it was cramped and unpleasant for everyone. Besides the sinking foundations of the drab apartment building and the constant thunder of landing jumbo jets, there was also the stench from the State Abattoir next door and the whiff of charred animal bones from makeshift offal crematoria in the parking lot. Nevertheless, the Wiwas were lucky not to have ended up somewhere even worse.

On Diana's second day as a fugitive, a clandestine courier came to Sam's apartment with a letter for Owens. It was from Ken. The people he had been relying on to run errands from jail had refused to do any more work for him unless he paid them in advance, and he could no longer afford it. Being slowly bankrupted by his legal troubles, Ken was finding it hard even to feed himself and his fourteen co-defendants at Bori Camp. Would it be possible, Ken asked, for Owens to go to the Saros offices on Tejuosho Street and get his staff to scare up more funds?

As Owens readied himself to leave, Diana's bad mood darkened further. She knew that Owens was likely to be gone until long after nightfall. "Mon, don't go," she said. "Befii and I have just arrived. We want to spend time with you. Can you not stay here today, or at least for the morning?"

"No, Diana. Ken needs me to do these things right away. I have to go."

As Owens finished putting on his shoes, Befii woke and began to cry. "Not again!" said Owens. "He's been crying all morning. What's wrong with this boy?"

"I don't know what's wrong with him. He's not normally a crybaby," said his wife. "He's been changed, fed, bathed. He won't stop. I put on the fan and powdered him a bit so he wouldn't feel the heat too much, but he will not stop! The only way he got to sleep was when you held him."

"Well, I can't hold him now."

Diana picked up Befii and handed him to his father. "Here, take him."

Befii laid his head on his father's shoulder and stopped wailing. Owens handed the baby back to Diana but the child began to screech.

"This is not doing too much for my ego," she said. "You're going to have to hold him until he goes back to sleep."

"But I have to go," he said.

"Please, Mon. You have to get Befii to sleep before you leave. Do this for me."

Seeing his wife pout, Owens took Befii once again and lay with him on the bed. As Diana left the room, Owens began singing an Ogoni lullaby his father had once sung to him. By the time Diana returned from the kitchen, both father and son were sound asleep. Neither stirred from nine in the morning until two that afternoon. When Owens finally awoke, he looked at his watch and exclaimed, "Oh my God! What am I doing here?" and bolted out the front door.

Owens was furious with himself, with Diana, with Befii. How could he have slept so long? He rushed to the nearest motor park, one of the city's impromptu transportation depots where *danfos* and *molues* – the wretched and rusty fleet of yellow vans and buses that are Lagos's poor excuse for public transportation – congregate to pick up and disgorge passengers. He clambered aboard a packed *molu* bus and seethed. Alighting near Tejuosho Street, he wended his way through bustling side streets, immersed in his bitter inner monologue.

It eventually occurred to Owens that people were waving at him. Folks on upper-floor balconies were flapping their arms. Shopkeepers peering out of doorways motioned with their hands. "Go back! Go back! Go back, go!" they whispered. Confused, Owens looked around to see if these people might be warning someone else, but all fingers were pointing at him. A food vendor at a *buka* caught Owens's eye and said in a desperate hush, "Na you we de talk to. Make you go back!"

Spooked, Owens retraced his steps and ducked into an alley. Sneaking though the back door of his old hiding spot in the ladies' boutique, he concealed himself behind the rack of dresses and peered out to see what was going on at Ken's office.

Through the glass front door of Saros International, Owens saw unfamiliar faces. For two hours, he watched until the strange happenings had subsided. After a while, one of Ken's employees made a furtive dash and slipped into the boutique.

"What happened?" said Owens.

"Let's just go," said Ken's employee. "Let's leave here straight."

Owens followed Ken's employee out the back of the boutique to a small, secluded tavern far away from Tejuosho Street. Once inside, the man told Owens that a secret courier (no doubt, the same man who had delivered the letter to Sam's place that morning) had come to Saros International and dropped off two letters from Ken. Soon after he left, soldiers and police stormed the Saros offices. The question they kept asking was "Where is Monday Wiwa? Which of you is Monday Wiwa?" Receiving no satisfactory answer, the police arrested two staff members.

"They took the letters Ken sent. One policeman said he had come all the way from Port Harcourt to arrest the brother of Ken and to take him back."

Owens was shocked.

"If you had been there, they'd be taking you to Port Harcourt even now."

"So that's what was wrong with my son," Owens marvelled. "He was saving my life! Okay, let us go back to the office so I can use the phone."

"Mon, are you crazy? They could still be watching."

Owens buried his face in his hands. What was he thinking? Of course returning to Saros would be foolhardy. If he wanted to survive, he must never be seen there again. He thought of Diana and Befii. If the police and the army had been able to trace Owens to the Saros offices, they might trace him to Sam Olukoya's apartment. It was imperative to find a new haven for his wife and child at once.

||||||

When Ken learned about the raid on his Lagos office, he dashed off an urgent message to Owens. "Get out of the country as soon as possible," he wrote. But as long as Ken was in prison, Owens could not bring himself to flee. Instead, he took Ken's letter to Gerald Ohlsen at the Canadian High Commission; Ohlsen agreed that Owens would be safer out of the country and presented him with application forms for immigration to Canada.

Owens couldn't imagine living somewhere like Canada, but he took the immigration papers back to Diana and they decided it was best to fill out the forms. A few days later, they got an appointment to meet with Canadian immigration officials visiting Lagos from their West African regional office in Accra, Ghana.

The Canadians turned the Wiwas' application down. According to immigration regulations, the family simply did not qualify.

Owens proudly insisted he wouldn't have gone even if he had been chosen. Diana and Befii were another matter. They had not been able to come up with a new hiding place so far, but Ohlsen came to the rescue once again by introducing the Wiwas to a young Austrian couple, Marcus Brunner and Doris Danler, who lived not far from the Ohlsens in the upscale neighbour-hood of Ikoyi. The Austrians agreed to take in Diana and Befii; Owens would stay far out of town in his hiding spot in Badagry. Yet even with these precautions, the authorities began to learn more about Owens and Diana's whereabouts.

Shortly after Diana evaded the police on the day of her scheduled thesis defence, the police staged another raid on 24 Aggrey Road in Port Harcourt. This time, they came away with a treasure trove of intelligence.

Ken, being Ken, had insisted that his files be kept in order even while he languished in prison. Every letter smuggled into his prison cell was smuggled back out again and taken to 24 Aggrey Road for his assistant, Apollos, to archive in secret files. Unfortunately, Apollos had been stashing Ken's letters in an unlocked drawer. When the military raided the office, they found all the letters Ken had received in detention. For months, Amnesty International had been forwarding to Ken copies of their correspondence with U.S. senators and representatives, along with the replies. There were copies of letters to Bill Clinton, John Major, and several British MPs. There were letters from Barika Idamkue, who was in Geneva lobbying members of the UN. There were letters from Gerald Ohlsen, Sister Majella, representatives from UNPO, Greenpeace, the Body Shop, PEN International. . . .

But worse, the military found the letters Owens had written to Ken. They read Owens's accounts of embassy visits, of talks with journalists and sympathetic military brass, of his close calls with the law at Hauwa's and at Letam's; of his attendance at the St. Patrick's Day party at the Irish ambassador's house and of the Queen's birthday party with Lord Mansfield. They read about his meetings with Brian Anderson.

Keeping the letters as evidence, the military made a bonfire of everything else in Ken's office. Ken's precious library and his personal files went up in flames. At Bori Camp, they raided his cell and confiscated a laptop computer and transistor radio (items that had been smuggled in to him at great peril), books, personal photos, pens, every scrap of paper.

||||||

Ken's lawyers withdrew from the case in June. It had become obvious that the Special Military Tribunal was following neither Nigerian nor international law. The "evidence" against Ken and the fourteen others was a feeble confection of hearsay, embarrassing retractions, and invective from those who stood to gain from Ken's removal. One of the prosecution witnesses claimed that he attended a MOSOP meeting in which Ken instructed those present to "deal with" a group of Ogoni elders whom he described as saboteurs. (This "witness" was David Keenom, an Ogoni man whom Danwi and Akpa said they saw collect thirty thousand naira in bribes in return for his coached testimony.) Another witness, Limpa Gbaa, swore that he attended a 1994 rally in which Ken told the assembly to "go out and kill the vultures." (According to Danwi and Akpa, Gbaa was also bribed.) As for the other prosecution witnesses, on occasion during the trial, Justice Auta actually told them what to say.

Priscilla Vikue told the tribunal that when she complained to Ken about the destruction of her country home by a group of Ogoni youths who labelled her a collaborator and sell-out, Saro-Wiwa supposedly told her that there was a revolution in Ogoni and that if she didn't join in, she would be swept away. "Heads will roll," Vikue claimed he said.

One of the most damning pieces of testimony came from MOSOP's former president Garrick Leton. "Ken is a very insistent person whose will has to be done," Leton said to the court. "The murders were planned and executed by pre-appointed assassins. . . . Saro-Wiwa must be exposed for what he is: a habitual liar, a person who uses the travails of his people to achieve selfish desires and ambition . . . a person who is prepared to engineer the elimination of his elders . . . a person who in this situation cannot escape complicity in the murder of the four Ogoni leaders."[14]

There were no witnesses for the defence. Following the mistreatment of Jessica Wiwa, Diana Wiwa, and Elfreda Jumbo, and the physical abuse rained on members of the defence team, it was clear that the tribunal had no intention of hearing any testimony on behalf of the accused.

In a televised interview, Nigerian Nobel laureate Wole Soyinka summed up the reasons for the lack of defence testimony: "The military as a body expressed itself as being inimical to the very existence of the Ogoni people," he said. "How can anybody have any confidence or even want to come out and give evidence before a military tribunal – the very military who have been persecuting their very existence? No, I do not think that anybody should look to this tribunal for any justice for Ken Saro-Wiwa."[15]

Another reason for the withdrawal of the defence team was money. Ken was now broke. The government made it no secret that one of its goals was to hold Ken until his personal fortune evaporated. To achieve this, the tribunal's favoured tactic was to adjourn the trial after every two or three days without stating how long the adjournment would last. Each time the trial was halted, Ken's lawyers would fly back to Lagos. Once they had left, the tribunal would suddenly announce that it would be sitting again. Ken's legal team would catch the next flight back to Port Harcourt, only to be told that the trial was again being adjourned. In addition to legal fees, Ken was obliged to pay for plane tickets, hotel rooms, car fares, and restaurant tabs for his lawyers and their associates. His employees in Lagos and Port Harcourt had to be paid, as did his domestic staff in both cities, not to mention his support of Maria and the children in England. With no money coming, in June Ken had to pull the plug on his defence team.

After doing so, he gathered his fellow Ogoni prisoners at Bori Camp and informed them of his decision. Referring to the tribunal judges, Ken said, "These people have made up their mind. There's no need of giving this sham of a trial any credibility by having our lawyers there." He instructed his fourteen co-defendants not to accept any government-appointed lawyers. He proposed that Ledum Mitee, a lawyer, represent each of them individually. Besides containing costs, this would also prolong the trial. Ken reasoned that with Obasanjo and Yar'Adua in prison, and with the rumours of *coups d'état* in the air, Nigeria was so unstable that it was only a matter of time before Sani Abacha was toppled. Once Abacha was gone, the Special Military Tribunal would be disbanded and they would be freed. For Ken, the only strategy that made sense was to play the tribunal's own game, and stall.

Ledum Mitee, however, was of a different mind. He had begun making his own plans.

||||||

With the lives of such internationally renowned figures as Obasanjo and Saro-Wiwa in peril, the United States was taking an interest in Nigerian politics. Though Nigeria was the U.S.'s leading African trading partner, America had largely ignored the Nigerian political crisis.[16] To correct this, by 1995, it had become *de rigueur* for black American politicians to make a tour of the troubled country. Most notable of these African-American visitors were Congressional Black Caucus chairman Donald M. Payne, Florida Congressman Alcee L. Hastings, and the conservative Republican Congressman from Oklahoma, J.C. Watts, Jr.

In late September, Owens received word from U.S. Ambassador Walter Carrington that a contingent of senior congressional staff members had come to Nigeria and would like to meet with him. Owens thought it best for the Americans to meet not just himself, but a whole delegation of Ogoni people. He received permission from Carrington to bring with him Diana and a half-dozen other notable Ogoni professionals living in Lagos.

On the day of the meeting, Owens and the other Ogonis drove in separate cars to the designated venue, a private mansion in Ikoyi. The Ogonis were shown into a large sitting room where Carrington and his political officer had gathered with the congressional staffers. As servants offered food and drink, the Americans welcomed the Ogonis, saying they had been sent by senior officials in Washington to report on the situation in Nigeria; more specifically, they were to get as many details as possible about the Ken Saro-Wiwa issue.

Owens came prepared with a sheaf of documents as well as several of Ken's most notable books. He briefed the team on the history of MOSOP, emphasizing its non-violent nature. The Americans asked pointed questions about the deaths of the four chiefs at Giokoo; Owens and the others told them what they could.

"So those four men were killed apparently by Ogoni people," one of the Americans said. "That means there is some justification for holding MOSOP responsible."

Bristling, Owens said, "Sir, I don't think so. To conclude that would mean that any crime committed at any time in Ogoni would be the fault of MOSOP. Even if those four were killed by Ogonis, and even though many people belonged to MOSOP, that doesn't mean MOSOP was responsible. MOSOP and NYCOP did not organize the killings. Ken is a man of peace."

A political officer from the American embassy looked at Owens coolly. "You claim that Ken didn't know about the murders, that MOSOP was not responsible, that NYCOP was not responsible. So why is it that Ledum Mitee recently informed us that NYCOP was responsible for the killing?"

"Ledum?" said Owens. "Sir, I don't think that what you're saying is correct. There is absolutely no way Ledum could have told anybody from the American embassy such a thing."

"Why is that?"

"Since Ledum, Ken, and the others were detained, none of them has seen any diplomats. In fact, the only person Ken has seen – and it was only Ken who saw this person – was the Nigerian ambassador to the United States who went to visit him in that early time at Afam. What you are saying is just not correct. Ledum has not told you that NYCOP killed those four chiefs."

"Are you calling me a liar?" asked the political officer.

"I'm not saying that, but your information is not correct. Maybe somebody else, maybe Ken's assistant in Lagos, who is not an Ogoni, might have gone to the embassy without knowing what was happening, relying on news that was coming through the Nigerian media and through Komo. Maybe he gave that information, not Ledum. Ledum has not been out of jail."

The political officer smirked. "Owens, you may not know what is happening. Yes, we have spoken with Ledum Mitee recently."

Owens could not be convinced. If it were true, either the Americans had found a way to get into Bori Camp or Ledum had found a way to get out. Either scenario seemed preposterous.

IIIIIII

As September waned, the world awaited the fate of retired General Olusegun Obasanjo. The former head of state had been sentenced to death, but the execution had been delayed as letters poured in from around the globe pleading with Sani Abacha for clemency. On October 1, 1995 – twenty-four hours before he was to face a firing squad – Obasanjo's sentence was reduced to fifteen years' imprisonment. Yar'Adua and his co-defendants also had their death sentences commuted to prison terms ranging from fifteen years to life.[17]

Abacha commuted the sentences after the intervention of U.S. president Bill Clinton, former U.S. president Jimmy Carter, and Archbishop Desmond Tutu (who was sent to Nigeria as the emissary of South African president Nelson Mandela), as well as pressure from many international groups. It was the first time in Nigerian history that coup plotters convicted by a military tribunal were not executed. The commutation of the death sentences was also a good sign because it showed the Abacha regime could be influenced by international pressure.[18]

Ken, who was still managing to get letters into and out of Bori, wrote to Owens: "Well, if [Abacha] can be made to backtrack on Obasanjo, then we have hope."

Unbeknownst to Ken, his hope was being scuttled from within. The only defendant Ledum Mitee chose to represent as a lawyer was himself. Several of Ken's other co-defendants decided to cooperate with the government and accept court-appointed counsel.

A schism between Ken and Ledum had arisen over money, and the rift between them had only widened. Visitors to Ken's cell mentioned the animosity between the two. In letters to Owens, Ken described the difficulty Ledum was causing. The one person Ken felt he could talk to was John Kpuinen, who continued to treat Ken with respect. As for the rest of his co-defendants, Ken felt they were no longer to be trusted.

Some believe that Mitee was allowed out of jail for brief periods during the latter part of his confinement, which Mitee himself sharply denies. Whether or not Ledum was given special privileges, his hatred for Ken is not in doubt. Owens received two letters in Lagos on the same day – one from Ken, one from Ledum. Ken mentioned the "prison blues" Ledum seemed to be going through and his hope that Ledum would soon snap out of it. Ledum, on the other hand, was bitter and mistrustful of Ken, whom he referred to only as "him." Whenever the pronoun was used, it was underlined twice.

|||||||

On October 10, 1995 – Ken and Owens's joint birthday – Ken Junior represented his father at a special ceremony in Belfast, Ireland. Thanks to the work of Sister Majella McCarron and Nobel laureate Mairead Corrigan McGuire, Ken Saro-Wiwa received his official nomination for the Nobel Peace Prize. He had precious little time to savour the honour, however. In late October, the Special Military Tribunal delivered its verdicts.

In his ruling, Justice Ibrahim Auta found that there was sufficient evidence to conclude that Ken Saro-Wiwa had counselled and procured John Kpuinen and Beribor Bera to commit the murders of the four chiefs at Giokoo. No physical evidence linked Ken to the crime, and all the witnesses against Ken had either been bribed or (like Vikue and Leton) had a motive to lie. The prosecution had presented no evidence to suggest that Kpuinen and Bera were anywhere near the crossroads when Ken allegedly issued the order for them to kill. As for Dr. Barinem Kiobel, although he had an iron-clad alibi (he had gone to the police station in Bori to inform the authorities that the murders were taking place), the tribunal concluded that he had planned the very crime he'd reported.[19]

Auta also concluded that since MOSOP was the umbrella organization of NYCOP and that NYCOP members allegedly committed the murders, the leadership of MOSOP was ultimately to blame. "MOSOP and NYCOP laid the foundation for the disaster that happened on May 21st, 1994, carefully planning and carrying out the murders," Auta said.[20] The tribunal did not demonstrate that NYCOP was involved in the killings.

This last finding contradicted many of the judges' final verdicts. Although Ledum Mitee held the second-highest position in MOSOP (vice-president), he was found not guilty along with his Gokana brethren Pogbara Afa, Monday Donwin, Albert Kabrara, Joseph Kpante, and Michael Vizor. Except for Ledum, who defended himself in court, all the acquitted had accepted representation by tribunal-appointed lawyers.

As for Ken and the remaining eight – Beribor Bera, John Kpuinen, Barinem Kiobel, Nordu Eawo, Paul Levura, Daniel Gbokoo, Saturday Doobee, and Felix Nwate – they were sentenced to hang.

A leading British jurist who later published a detailed study of the Saro-Wiwa tribunal concluded: "The judgement of the Tribunal is not merely wrong, illogical, or perverse. It is downright dishonest. The Tribunal consistently advanced arguments which no experienced lawyer could possibly believe to be logical or just. I believe that the Tribunal first decided on its verdicts and then sought for arguments to justify them. No barrel was too deep to be scraped."[21]

From London, Greenpeace issued a statement saying that the blame for Ken's death sentence ultimately lay at the feet of Royal Dutch/Shell "because [Saro-Wiwa] and the Ogoni people have exposed the devastation wrought on their land by Shell."[22] Internationally, it was widely felt that although world leaders could appeal for clemency (and were actively doing so), only Shell held enough sway with the military regime to persuade Abacha to save Ken Saro-Wiwa.

Shell insisted it could not – and would not – intervene. In a published statement, the company said, "We believe that to interfere in the processes, either political or legal, here in Nigeria would be wrong. A large multinational company such as Shell cannot and must not interfere with the affairs of any sovereign state." Shell called for "quiet diplomacy" to resolve the situation.[23]

|||||||

Once more, word came from the underground that Owens had to leave Nigeria; Abacha's boys were looking for him harder than ever. When Owens voiced his concerns to Diana's hosts, the Austrian couple arranged for the Wiwas to find a new hiding spot in the home of Claus-Peter and Marcela Holste-von Mutius, an ex-pat German couple living in the relatively secluded Apapa area of Lagos. The Wiwas were happy to find a place at last where they could be together.

From his new hiding place, Owens heard radio reports of Ken's conviction and of Ledum Mitee's release, and placed a call to Ledum's house in Port Harcourt.

"He is not home yet, but he is coming near," said the person who answered. Since telephone connections are difficult in Nigeria, Owens asked to stay on the line until Ledum arrived. The family member laid the receiver down and left to prepare for Ledum's homecoming.

While he waited, Owens could hear over the phone the happy voices of people gathering at the Mitee house. Music began to play and there was laughter and talk of dancing. Then Owens heard something strange: someone at the homecoming party called out for Priscilla.

Priscilla? thought Owens. *Could that possibly be Priscilla Vikue?*

The party noises grew until shouts began ringing out; Owens surmised that Ledum Mitee had come in the door. Once he was informed that there was someone on the phone for him, Ledum picked up the receiver.

"Congratulations on your release, Ledum."

"Mon? Is that you?"

"Yes. I heard you are now free and I'm very happy. Now that you are out, we have to try to do as much as we can for Ken and the others who have been condemned."

"How do you mean?"

"Well, if you can come and talk with the diplomats here, maybe we can find a way to get you a ticket to go abroad. In Europe or America, if you talk about all that happened, you may be able to draw up world opinion for people to know that Ken is as innocent as you are."

"No," said Ledum. "I can't do that."

"Okay. I understand," said Owens. "You need some rest."

"Where are you calling from?"

"I'm still in Lagos."

"Monday, listen to me. You should leave right away," said Ledum. "But don't go near any of the borders. That's where they're looking for you, at every border post."

"There's no way I can leave Nigeria right now," said Owens. "Ken needs me."

"Listen to me, Mon. Find a way to disappear."

||||||

In desperation, Owens phoned every diplomat he had met. He called all the press houses. He phoned the Archbishop of Lagos, a friend of Abacha. He tried Brian Anderson at his home and his office. When he phoned Nnaemeka Achebe, Shell Nigeria's general manager would say little. Owens contacted many Ogonis living in Lagos and asked them to urge the foreign embassies to put pressure on Abacha to have mercy on Ken. He was shunned at every turn. No one was willing to help. To be linked with Ken was to be in peril.

Gerald Ohlsen, however, remained true. When Owens contacted him, Ohlsen advised him to think of himself. "You've completed all the things you can do," he said. "You've got to get your family out of the country. There's nothing anyone can do for Ken."

Owens refused to heed such counsel. Word was spreading that a hangman had been recruited from the north and that gallows were being erected in Port Harcourt. Owens kept up his frantic campaign, calling contacts in England, the United States, and the Netherlands. Above all, he set his sights on those who were headed for New Zealand.

That week, the Commonwealth Heads of Government Meeting was being convened in Auckland. Earlier, Ken had written from prison to Body Shop founder Anita Roddick, imploring her to send Ken Junior to the meeting to lobby the Commonwealth heads of state for support. Ever since Roddick's first contact with the Ogoni delegates at the UN conference in Vienna, she had become an ally, and now she arranged for her company to fund Junior's trip to New Zealand.

Once there, Junior arranged private audiences with New Zealand prime minister Jim Bolger, Britain's overseas development minister Lady Lynda Chalker (England's self-titled "Mama Africa"), and Canadian prime minister Jean Chrétien. Although the meeting with Chrétien was productive, Junior's confrontational attitude and lack of diplomacy alienated Bolger and

Lady Chalker. After issuing a release and convening a press conference, Junior returned to England demoralized, having achieved little.[24]

In Lagos, Owens returned to the Ikoyi home of Marcus Brunner and Doris Danler, who were entertaining a German journalist from Radio Deutsche Welle. Nearly a week had gone by since the verdict, and no word had come to suggest that Ken's execution might be cancelled. As Owens rocked in panic on the living-room sofa, Marcus offered him a glass of wine.

"I'm not in the mood to drink wine," said Owens.

When the German reporter urged him to calm down, Owens remonstrated, "Sir, how can I calm down? I called Port Harcourt and I was told that they have taken Ken away from Bori Camp to the prison in a Black Maria. They have not allowed him to eat for two days. They even refused him water to bathe. They had put leg cuffs on him. And I am hearing now that the hangman is in Port Harcourt!"

"Would you like to do a radio interview?"

"Well, why not?" said Owens. "I'll do anything to bring attention to this."

The Radio Deutsche Welle correspondent set up his tape deck and began to record a brief speech about how the world needed to put pressure on Sani Abacha and Shell to come to the aid of Ken Saro-Wiwa. The reporter then turned to Owens and asked if he could answer the next few questions as if Ken were already dead.

Thrusting the microphone in Owens's face, the reporter asked, "Owens Wiwa, what is it you want the world to do now?"

Appalled, Owens said, "Sir! There is absolutely no way I'm going to answer as if Ken is dead."

"Well, you know, it's just . . ."

"No! What is it you know that I don't know?"

"Okay," said the journalist, turning off his tape recorder. "Why don't we just forget this for now?"

|||||||

The next morning, November 10, 1995, Owens awoke to find Befii by the bed playing with his father's glasses. Under the toddler's rough handling, the frame had cracked and a lens had fallen out. A bad start to the day, Owens thought. In foul humour, he pried the glasses from his son's grasp, then bathed and dressed. His glasses were hastily repaired with adhesive tape.

Taking a taxi to a nearby business centre, Owens entered a phone cubicle and called a contact at the Agence France-Presse office. Curiously, his friend at the AFP declined to speak with him. Owens phoned another friend in Port Harcourt.

"Monday," said the friend, "I think Ken has been hanged."

"What? This can't be true. Are you certain?"

"No, but that is what I have heard."

With shaking hands, Owens thumbed through his small address book to find the home number of an Ogoni prison guard who worked where Ken was rumoured to have been transferred. When someone answered, Owens said: "Is it true that Ken Saro-Wiwa has been hanged?"

"Who is speaking?"

"I just want to find out if it is true."

"If you hear, it must be true. Why are you asking me?"

Owens hung up. It could not be true. It was just gossip. He had to speak with someone with information he could trust. He called BBC World Service correspondent Janet Anderson, who was in Nigeria reporting on Ken's trial.

"Janet, what's this I'm hearing about Ken and the others being killed?"

"Yes," she said, "I've heard the same thing."

"So is it true?"

"Well – Owens, it's possible."

"So if it's possible, why have you not put it on the BBC yet?"

"I can't put that sort of news on the radio without an official confirmation."

"So what you're telling me is you think it might not be true?"

Anderson hesitated. Her voice breaking, she said, "I think it's true, Owens. I'm so terribly sorry."

The telephone slipped from Owens's hand. In the business centre cubicle, he crumpled to the ground beside the dangling receiver and wept.

CHAPTER**TWELVE**

Numb with shock, Owens went to the front desk to settle the bill for his phone calls. The clerk, seeing him distraught, asked what the matter was.

"I don't really know," he said.

On the trip home, the taxi took him through streets blocked by boisterous crowds shouting, laughing, dancing. According to the driver, the local soccer team had just won an important match. Owens regarded the joyous faces. How could they be happy? Didn't they know Ken was dead? Had the International Football Association, in collusion with the government, timed the match to divert attention from Ken's execution? To Owens, everything now seemed part of some grand conspiracy.

By the time he arrived back at the Austrians' house, it was late afternoon and everyone had heard the news. Diana and their hosts studied Owens in silence, hoping to take their cue from his reaction. He gave them none. He retreated to his bedroom, shut the door, and willed himself into a deep sleep.

Awaking around ten that evening, Owens found Diana and Befii sleeping next to him. He rose silently and stepped into the hallway. From the living room, he could hear Claus-Peter and Marcela in high-pitched, agitated conversation. They were speaking German, something they rarely did in the Wiwas' presence. Noticing Owens in the hall, they asked if they might have a word with him.

"It is clear these people are going to be after you and Diana now," said Claus-Peter. "It might be in your best interest to leave the country."

"You're right," said Owens. "Only the problem is I don't know where we would go. I don't know anybody in Benin or Ghana, anywhere outside Nigeria – apart from maybe Ken's son in London and my cousin Vincent, who is now in the U.S."

"We have taken the liberty of calling a friend in Ghana who will let you stay at his house for a few days. Here is the address and phone number."

Owens thanked Claus-Peter, adding, "That will be very useful."

Claus-Peter looked uncomfortable. Owens did not seem to be catching his drift. "You will have to leave our house as soon as possible. Tonight, in fact. One of the neighbours might already have seen you or Diana. And if they find you here . . ."

"I know. You have been wonderful to keep us, but now we have to make sure that you are safe."

Owens woke Diana and Befii, collected their personal belongings, and led his family out into the Lagos night.

||||||

The Wiwas took a taxi to the home of an Ogoni acquaintance, who turned them away. The same thing happened at another friend's house. And another. And another. Shortly after midnight, Owens and Diana had no choice but to check into a cheap hotel. Consumed with thoughts of Ken, Owens found it impossible to sleep. At dawn he went down to the lobby to get the newspapers. All the headlines were about the executions of "the Ogoni Nine," as they were now called. By the reception desk, employees were discussing the news. Most blamed the government for Ken's demise; one fellow, however, disagreed. "Good for this man!" he said. "The murder he cause it. Now he is dead. Good for him!" Glancing up, he noticed Owens. Owens could tell the fellow recognized him but couldn't quite place him. He nodded a perfunctory greeting and went back to his room.

"Pack your things," he told Diana. "We're going to another hotel."

Over the next two days, the Wiwas checked into six different hotels. Sometimes it would only be an hour or two before a desk clerk would look at them oddly or they'd overhear a bit of conversation and they'd feel too frightened to stay. This was no way to live.

It was time to flee Nigeria.

Both Owens and Diana held legitimate Nigerian passports, but Befii had no travelling papers. The Wiwas, therefore, needed a forger to add Befii's information to Owens's passport. Nigerian passports, however, would be useful only after the Wiwas escaped the country. To cross the border, they would need documents bearing false names. Letam Wiwa knew of a chap named Mr. Ade who, according to Letam, knew how to make things happen.

On Bishop Oluwole Street, Lagos's finest forgers are at your service. For the right price, a person can get almost any document there – a false passport, birth certificate, government document, you name it. Mr. Ade took the Wiwas' legitimate documents and disappeared into the back streets. When he returned, Befii was "officially" listed as Owens's son on his passport, and both Owens and Diana now had ECOWAS (Economic Community of West African States) papers that would let them slip into any country neighbouring Nigeria. Since Owens still had his journalist I.D. card with the name Tony Uche, on their ECOWAS papers he and Diana were known as "Anthony and April Uche."[1]

On the third day after Ken's death, Owens, Diana, and Befii visited Gerald Ohlsen at the Canadian High Commission. Ohlsen placed their Nigerian passports in a diplomatic courier bag to be ferried out of the country. Getting the Wiwas themselves out would be more complicated.

"How do you intend to move?" asked Ohlsen.

"We're going to follow the path that other activists have followed," said Owens. "Soyinka talked about how he escaped, so we'll follow the same path. That's our idea."

"When you get to Cotonou in Benin, go to the Canadian Consulate. There will be someone waiting for you. But you shouldn't remain there. Benin is not safe. The Nigerian secret service is active outside the borders, especially there. You might be recognized."

"Fortunately, our German friends have found us a place where we can stay in Ghana."

"Good. I'll send your passports to the High Commission in Accra. They should be there by tomorrow."

Ohlsen bade the Wiwas an emotional farewell. A car from the Canadian High Commission was waiting to take the family on the first short leg of their escape: to a nearby "motor park," to catch a *danfo* or *molu* for the border.

For the journey, Diana had dressed in the style of a Yoruba market trader with Befii strapped to her back in a fabric sling. Their two suitcases of belongings were packed to resemble goods to be traded. Owens wore his new belt, a parting gift from Marcus and Doris, which contained a concealed compartment. At the motor park, the Wiwas blended into the street market traders and beggars as they searched for transport.

In Lagos, merely taking a bus is a dangerous adventure. When they're not being driven at breakneck speeds by drunken drivers, the wheezing commercial buses known as *molues* (literally "I go beat you") barely chug along. With battered chassis, bald tires, faulty brakes, and noxious trails of black exhaust, the *molues* are little more than affordable death traps. In fact, commuters often refer to them as "moving morgues" because of their frequent, horrific crashes. Bus drivers in Nigeria seldom obey traffic laws or transport regulations. On a *molue*, it is common for the conductor to cram as many as a hundred passengers on board (recommended seating is for forty-four) and even to proffer a hand to a running passenger to help him jump aboard while the bus is in motion.[2] Passengers are crammed in or hang from open doors and windows. As for the smaller *danfo* minibuses, they are meant to seat only twelve. A *danfo* with twelve passengers, however, would be considered empty.

Both the large and small transports are often whimsically painted with religious or pop-culture themes and cryptic slogans. The phrases emblazoned on sides and rears of the buses are often vaguely inspirational or rabidly Christian. *To Be a Man Is Not a Day's Job* is popular, as is *I Am Covered in the Blood of Christ*, and (the unintentionally hilarious) *Oh God Help Me*.

After asking around, Owens found a non-sectarian *danfo* to take them out of Lagos. Owens and Diana loaded their things and climbed onto a bench seat, placing Befii between them. The driver kept adding passengers until the *danfo* was jammed with Yoruba traders and their wares. When not another person could be shoe-horned aboard, the driver pulled out and wended his way through the "go-slows" (traffic jams) and rutted roads of Lagos.

At the junction known as Mile Two, the Wiwas' *danfo* was stopped at a police checkpoint and everyone was ordered to step out. "We will search you," said a policeman to the driver. Owens's attention immediately flashed to the small address book he kept in his pocket. The last time he was searched, he was detained for merely having it on his person. If they found it on him now . . .

In the confusion of everyone climbing out of the van, Owens palmed the diary and jettisoned it on the roadside when the police weren't looking.

It was lucky he did so. As usual, the police performed a thorough body search of all the van's passengers and asked about the contents of their luggage. Owens told the police they were transporting commercial goods that Diana was on her way to trade in Benin. As the van was full of other traders, the police bought the story without question. After the driver gave the police a few naira in *dash*, the *danfo* was allowed to continue on.

On the way out of Lagos, the Wiwas' van was stopped and searched at least ten more times. Not all of these were regular police checkpoints. Many other groups, including the army, immigration officials, border police – even the veterinary police – had set up barricades in order to extort money from passing vehicles.

When the van stopped for fuel in Badagry two hours later, Owens noticed a car driven by a man in uniform pull into the petrol station alongside them. When the driver stepped out, Owens realized it was Monday Abua, the customs officer he'd been hiding with in Badagry all these months. Scrambling out of the *danfo*, Owens went to greet his friend. Abua seemed both pleased and alarmed at Owens's presence.

"What is happening?" asked Abua. "Why are you on this bus?"

"I'm going to Ghana. Befii and Diana are with me. To stay in this country any longer will be our death."

"Look, I am working at the border," said Abua. "Why don't you join me and let Befii and Diana continue alone. It will be easier to get past the border police if you are not three."

"I don't know," said Owens, hesitant to change their plans so quickly.

"Listen man, think! If one of you is caught, at least the others might get through safely."

Abua was right. It was likely that Owens would be recognized, and it would be much safer for his wife and child to cross the border alone. "Okay. I will let them go," he said.

Owens went back to the van and told Diana about his plan. He was carrying about seventy American dollars. After paying off the *danfo* driver, he gave fifty dollars to his wife and hid the remainder in his belt.

"Wait for me at the Canadian Consulate in Cotonou," he said to Diana. "I will join you there."

The passengers for Benin climbed back aboard the *danfo*, and Owens, with a heavy heart, watched his wife and baby son drive away, uncertain if they would meet again.

|||||||

Monday Abua was stationed at one of the major crossings to the Republic of Benin, at a border town called Sime. He parked in a big dusty lot on the Nigerian side, home to one of the area's open-air markets.

"I'm going to check in my office," said Abua. "I'll come back and see you soon."

"I'm hungry," said Owens. "I'll just go over to one of those *bukas*."

"I'll join you there," replied Abua, and handed Owens a bundle of naira. After Abua walked away, Owens saw that his friend had given him enough money to pay for transportation from Sime to Cotonou and even on to Ghana. It was a stirring display of generosity.

Owens went to the first small *buka*, a thatched-roof hut with picnic tables in front, and ordered a bowl of pepper soup. He sat at one of the outdoor tables, alert for security personnel. Looking around, he noticed that the international border was nearby. Extending from the sophisticated complex of immigration buildings was a simple fence of chain link and barbed wire that separated the two countries. Less than fifty metres from the official border crossing, the people of Sime had torn a gap in the fence. Traders from the Benin side, wishing to avoid the harassment of the Nigerian border police, were using this breach to cross freely back and forth to trade their wares.

The owner of the *buka* set Owens's soup before him. As if in a trance, Owens paid her no mind. Heart pounding, mind racing, he stood and placed a few naira on the table. Then, eyes on the barbed wire fence, he began to walk slowly, nonchalantly. After a few steps, unable to restrain himself, he broke into a sprint. Without looking back, he ran for his life. He did not stop until he was well clear of the fence on the Benin side. Racing up to a group of buses idling in a car park, he shouted breathlessly, "Cotonou! Cotonou! Cotonou! Which car? Which car?"

"Come! Come! We are ready!" called out several drivers. Choosing a bus at the head of the queue, Owens leapt aboard.

Once the bus had pulled out of sight of the border, Owens drew a calming breath and thought about his family. What had happened to Diana and Befii? Did they cross? Were they safe? Did they have enough money? There was nothing he could do except wait and worry.

A couple of hours later, he arrived in Cotonou, Benin's de facto capital. At first glance, this flat, smoggy, sprawling metropolis seemed like a smaller version of Lagos, but the differences soon became apparent. The contrast was most noticeable in the streets themselves. Radiating out from pleasing, well-ordered traffic circles, the wide boulevards of Cotonou exuded an ineffable European charm. In place of Nigeria's wheezing *molues* and *danfos*, Cotonou was abuzz with motorcycle taxis called *zemi-johns*, piloted by drivers in yellow shirts with numbers stencilled on the back.[3] And, unlike the still very British Nigeria, Benin was French. On the streets, traders had baskets of crusty baguettes; the sidewalks were lined with French street signs, and the skyline was filled with French billboards. The gutters were relatively free of litter. Not only did the traffic lights work, they were obeyed. After the crackpot chaos of Lagos, Owens found Cotonou's orderliness inexpressibly foreign.

After passing through two police checkpoints, the driver of the *danfo* dropped Owens off at a car park where he hailed a taxi to the Fonds canadien d'initiatives locales on a street named rue 238. Entering through the front gates of the consulate, Owens did not see a soul. It was almost six o'clock, twilight was descending, and the consulate offices were closed for the day. There was no sign of his wife and child.

It was almost impossible not to be consumed with worry. Where were Befii and Diana? Were they simply delayed or had ill fortune found them? If they were in trouble, where would he even begin to look for them?

As he stood there, crestfallen, in the inner courtyard, Owens heard a tentative voice. Out of the corner of his eye, he saw movement.

Diana and Befii emerged from shadow. With joy and relief, Owens and his wife raced toward each other.

|||||||

The trip to Cotonou had been eventful for Diana and Befii as well. Although her paperwork, manner of dress, and cleverly wrapped parcels identified her as a poor Yoruba market trader, her American-accented English and her inability to speak Yoruba had caused the border guard to grow suspicious. Squinting back and forth between the young mother and her skilfully forged ECOWAS passport, the guard asked, "You're a trader? You sound like you went to school outside Nigeria. Why must you sell goods in Cotonou?"

"Times are hard," she said bluntly.

They locked eyes. Her expression betrayed nothing. After a tense moment, the guard handed back her papers and said in Yoruba: "*Ó dàbò*." *Goodbye.*[4]

Now, in Cotonou, before the Canadian consulate, she turned to her husband and asked, "What do we do now?"

Owens had no answer. Even though the Canadian consulate was closed, the Wiwas managed to find someone working late. No one at the consulate seemed to know anything about the Wiwas. Confused and exhausted, Owens and Diana lugged their bags to a cheap hotel on a dark side street.

Yet even out of Nigeria, Owens could not sleep. Every noise, every footfall in the hallway, was the sound of Abacha coming. "Benin is not safe," Gerald Ohlsen had said. It would be best to put Benin behind them as soon as possible.

The next morning Owens began looking for a way to get to Ghana, two international border crossings away. The Republic of Benin is a narrow sliver of a nation, less than a hundred kilometres wide along its seacoast, and Togo, the next country to the west, is even smaller. The first step of their remaining journey was easy. Cotonou to the Togo border is a short car ride. Without luggage, one could even get there in a *zemi-john* in little more than an hour. Taking a bus, the Wiwas arrived at the border in late afternoon. At the crossing, passengers were required to disembark and pass through customs and immigration. Worriedly, Owens whispered to Diana, "Will any of these people recognize me? What if they ask a lot of questions?"

"Mon, why are you worried?"

"Because this story is going around how Abacha's agents are all at the borders because of the pro-democracy activists who have been escaping."

"We have our false papers and new names," she said. "Calm down. There is no way they will stop us."

Diana was right. The Togo border police let them pass with little scrutiny. Boarding another bus, the Wiwas began a torturously slow trip across the Togolese countryside. Although the road leading from border to border along the Atlantic coast was barely seventy-five kilometres long and in good repair, it took nearly the entire day to creep that short distance. The many stops and police checkpoints en route gave Owens and Diana much time to nervously contemplate the palmy villages and quiet lagoons passing outside the bus windows. It also gave Owens time to agonize over what they would face at the next border crossing.

Since the early 1990s, tensions between Togo and Ghana had run high and there was extra security on both sides of the border.[5] No doubt Abacha's spies would be there, too. The border between the two nations was notorious for its frequent and unexpected closures. As Owens feared, when they arrived at the crossing, the border to Ghana was officially impassable.

The bus offloaded its passengers at the *gare routière* at Akodessewa, a little town about three kilometres from the capital city of Lomé, and the Wiwas were left at the station.

Lomé is no place to be stranded. Violent crime is endemic among the crowded, narrow streets, with tourists like Owens and Diana bearing the brunt of attacks by armed robbers. Waves of political-inspired violence and military shooting and looting were commonplace in Togo. Out in the country's lush countryside, dark voodoo practices lurked.

As darkness fell, Owens and Diana were befriended by two Togolese women who said they knew a secret passage to Ghana. Hoisting the Wiwas' luggage atop their heads, the women led them on foot out of the city and into back-country bush.

Exhausted and terrified, Owens and Diana followed their guides for hours in the jungle darkness, sometimes carrying, sometimes leading by the hand their silent toddler son. The only break they took during the night was in a village where they stumbled on a secondary school riddled with bullet marks from a recent military clash. Staring at the bullet holes, Owens shuddered at the memories of Kaa, Kpean, and Oloko. Crossing a small stream, the guides led them another five kilometres into the night, bypassing military checkpoints and shrines filled with voodoo fetishes.

In the wee morning hours, the Wiwas reached the Ghanaian border town of Aflao. Through the dark jungle, they had crossed the international boundary without even knowing it.

‖‖‖‖‖

The home of the Holste-von Mutiuses' friend in Accra, Ghana, was a comfortable yet unprepossessing villa in a new estate. The Wiwas' new hosts were out of the country, but their servants had been told to expect visitors. The housekeeper welcomed them warmly. Thankfully, the staff had been given no personal information about the Wiwas, and Owens and Diana kept their horrific tales to themselves.

After the euphoria of escape and the tension of flight, this safe place opened the door to the reality of Ken's execution. Owens became depressed and irritable. His moodiness and excessive smoking left the house staff bewildered and uncomfortable. Diana, unable to explain her husband's conduct, chastised his insensitivity. In her anger, she took his prized possession – his cigarettes – and threw them in the garbage.

"What is wrong with you?" he cried. "Here's the time I need a cigarette!"

"Pull yourself together," she commanded. "The time is not right for you to act this way."

Through his bitterness, Owens had to admit that his wife was right. Rage and depression were luxuries he simply could not yet afford.

In Ogoniland, the unaffordable luxury was mourning. When the Ogonis heard the first radio reports announcing the executions, multitudes spilled into the streets, wailing with grief. Within hours, the Abacha regime deployed four thousand troops to flog and arrest anyone mourning the Ogoni Nine in public. Officially, it was illegal to express grief over the death of Ken Saro-Wiwa.[6]

In an interview with a *Newsweek* reporter, an Ogoni teacher described life in the days following Ken's death. "If the military sees two or three people gathering, they may imprison you," he said. "If you wear black, they may beat you. If you carry newspapers, they will seize them. Our headmaster was arrested last week as a warning to us not to discuss Ken in the classroom.

Pastors were arrested because they prayed for Ken Saro-Wiwa. They take away people every day."[7]

Around the world, the executions triggered international condemnation. The United States, Canada, South Africa, and all fifteen members of the European Union withdrew their ambassadors from Nigeria. At the request of South African president Nelson Mandela, the organization of former British colonies temporarily suspended Nigeria from the Commonwealth.[8] Western nations restricted travel by government officials and cut off sales of military equipment to Nigeria. Harsher, more meaningful sanctions – such as an international embargo on Nigerian oil or the seizure of Nigerian assets abroad – were not considered.[9]

||||||

On Owens and Diana's third day in Ghana, Owens visited the Accra offices of the United Nations High Commissioner for Refugees (UNHCR). A protection officer at the UNHCR gave him a sympathetic ear but little other help.

"Look around you," said the UNHCR officer. "Look around this building. You see the people outside? I cannot guarantee you that some are not Abacha's agents. I cannot guarantee your security in Ghana at all."

At the end of the meeting, Owens offered to give the UNHCR officer his address and phone number in Accra so that he might be contacted in case any options arose.

"No! Don't let me know where you're staying," said the officer. "Please keep it to yourself. It's better that way. Just take my advice. Get out of Ghana as soon as possible."

Assuming that Gerald Ohlsen had forwarded their documents to Accra, Owens next paid a call on the Canadian High Commission and asked to see the political counsellor. Owens introduced himself and told him what had brought the Wiwas to Accra. "Mr. Ohlsen sent our papers here in the diplomatic bag from Nigeria," he said. "Or at least I hope he did."

"Yes," the counsellor said. "I believe we do have them here." After a quick check, he located the documents.

Relieved to have his family's passports again, Owens felt emboldened. "I have just come from the UNHCR," he said. "They said we were not safe here in Ghana and should leave as soon as possible. I would like to seek asylum in Canada."

The political counsellor directed Owens to the office of the consular officer. When the consular officer heard Owens's request for asylum, she stared incredulously, then laughed. "If you think you'll be going to Canada, you're mistaken," she said. "If you want to be a refugee, go back to the UNHCR."

"But they said they can't help us. The UNHCR officer told us to get out of Ghana."

"Look," she said pointedly, "you're not going to Canada. It's not going to happen."

||||||

Dejected, Owens returned to the place where they were staying and asked the housekeeper if he might be permitted to make a long-distance call. His options in Ghana and Canada extinguished, Owens dialled the number of Ken Junior in London. After expressing his condolences to Junior for the loss of his father, Owens confided that he, Diana, and Befii were trapped in Ghana and needed a way out. Junior promised to see what he could do.

Before long, a call came through. It was Gavin Grant, a spokesman for the Body Shop in London. At Junior's request, Anita Roddick's company was calling to say it would be honoured to help.

"We're sorry to hear about your brother," said Grant.

"Thank you," Owens replied. "And thank you for all you've been doing in London. I know we all tried our best for Ken."

Grant asked Owens if he would speak with Polly Ghazi, a journalist from *The Observer* who wanted to do a story. It was an intriguing proposition. Owens hadn't thought of turning to the media. Perhaps the Body Shop was right: a little publicity might be useful. An hour later, Ghazi phoned Owens and conducted a long interview. He told her about his meetings with Shell's Brian Anderson and the conditions Anderson had placed on MOSOP before the company would consider intervening in Ken's case. This news about Shell's connection to Ken's execution was sensational stuff: next morning, Ghazi's article appeared on the front page of *The Observer*.

Gavin Grant called back later that day. "So Owens, do you have your passports?"

"Yes. Why?"

"Take your wife and child and documents and go to the British embassy in Accra. They've been notified to provide you with a six-month visitor's visa to the U.K. My colleague David Wheeler is making arrangements with KLM Airlines for your transportation."

"But I cannot afford to pay for plane tickets."

"Don't worry. It's been taken care of."

Owens was amazed. The day before, he and Diana had been facing a bleak future with their baby. Now, Grant was informing him that the Body Shop had persuaded KLM to give them complimentary seats on the November 21 flight from Accra to London. Not only that: Owens, Diana, and Befii would be flying to England *business class.*

||||||

The Wiwas sorted through their belongings, keeping only what they thought would be essential for the trip and jettisoning the rest. Diana gave her Yoruba market trader's outfit and other pieces of clothing to their hosts' house staff. In the taxi to the airport, Owens thumbed through their pile of travelling documents, both falsified and genuine.

"We still have our ECOWAS passports," he said to Diana. "And my I.D. as a journalist. If they search me and see this, which has a different name than my Nigerian passport, there will be trouble. I think we should throw it away."

Diana shook her head. "You never know. Let's tuck them somewhere away for an emergency."

"No, I'm not taking any chances. Maybe Abacha's people are here in the airport. I understand they're everywhere." At the airport, Owens stealthily tossed all their false I.D.s into a trash bin.

As promised, business class tickets were waiting for the Wiwas at the KLM counter. The staff seemed well informed about the family's situation and handled their fugitive passengers with skill and courtesy. There was no need to tell the crew that the Wiwas' identity must be kept secret; their names were never spoken aloud, and KLM listed the Wiwas on the manifest under false names. Any apprehensions Owens had about fleeing Royal Dutch/Shell on KLM Royal Dutch Airlines evaporated.

When it came time to board, Owens kept looking left and right. "What's wrong?" asked his wife.

Incredulous yet suspicious, he whispered: "Are we really going to go?"

After everything he had endured, it seemed too easy. Nevertheless, the flight crew showed them to their seats, where they were pampered. When Befii became excited by the unfamiliar sights and sounds and started to cry, the crew brought things to soothe and distract him.

Just after the doors were shut and the plane pushed back from the jetway, the pilot announced over the PA: "Ladies and gentlemen, we've had a slight change of itinerary. En route to Amsterdam, we'll be making a short stop in Kano, Nigeria, to pick up more passengers and fuel. We regret any inconvenience this may cause."

Inconvenience? Good God! Owens flew into a panic. They were going back to Nigeria. And Kano! Abacha lived in Kano. Owens knew that Diana was thinking the same worried thoughts, but they said not a word. The ninety-minute flight to Kano was almost unendurable. After so much grief, to be inadvertently delivered back into the arms of the enemy – what cruel irony.

On landing, Owens's worst fears seemed to be realized. The pilot informed the passengers that Nigerian security forces would be boarding the plane to conduct a search. The forward hatch was opened, and three soldiers bearing automatic weapons entered the plane. As two of the uniformed personnel wandered up and down the aisles of the aircraft, their squadron leader studied the passenger list. Thanks to the forethought of the KLM staff, neither Owens's nor Diana's names appeared.

"Keep your head down," whispered Diana, terrified.

But Owens did not. When the soldiers looked at him, he gazed back at them defiantly. He had made up his mind: there was no way he was going to allow himself to be arrested again. If the soldiers recognized him, he would do anything, no matter how desperate, to save his wife and child. "If they get me, they will hang me," he whispered to his wife. "They won't take me, not alive."

After an hour on the ground, Nigerian passengers were added to the flight and the soldiers disembarked. When the doors were sealed again and the plane began to taxi, Owens hugged Diana and Befii.

"Thank God," he sighed. "London, here we come!"

|||||||

At Heathrow, Body Shop representatives Gavin Grant and David Wheeler were waiting to welcome the Wiwas to England. After claiming their single piece of luggage, Owens and Diana headed out to the parking lot and the first shock of their new lives: cold weather. The Wiwas were wearing light-weight tropical clothing. It was November in England and they owned no coats, scarves, gloves, or hats.

Grant and Wheeler drove them to Ken's home on the outskirts of London. Junior and his mother, Maria, met them at the door. The two were polite if not warm – an understandable reaction, Owens thought. No doubt Ken's wife and son were still reeling from the executions. Owens and Diana once more expressed their condolences, saying they had tried their utmost to save Ken but perhaps God knew best. Junior helped his uncle's family with their bag and Maria showed them to an upstairs guest room. Before long, Owens's brother Jim showed up along with his wife, Linda. There was much excitement over the article in *The Observer*. Not all the excitement was joyous, however; Jim told Owens that many members of the small Ogoni community in London were unhappy.

"Why should they be upset?" asked Owens.

"It's politics," said Jim. "They feel that they are the Ogoni spokesmen in the U.K. You have not even come yet and you're talking with the press."

Owens shrugged. "On the plane, I decided my goal was to do everything I can to make sure everybody hears what has been happening to Ogoni and what happened to Ken." He told his brother he also intended to campaign for a worldwide embargo on Nigerian oil and a boycott of Shell products. Owens figured Abacha was in power thanks to his ill-gotten oil wealth. If the world cut off his income, his regime was bound to fall.

"What about you?" asked Jim.

"What do you mean?

"Now that you're here, how are you going to survive?"

Owens stared at his brother blankly. Moment-by-moment safety had preoccupied him. They had arrived at Ken's house, and in Ogoni culture one was welcome to stay indefinitely at the home of a senior brother. Their U.K. visas were good for six months. There would be time to get their lives in order. Perhaps in the coming months Abacha would be toppled, leaving Owens and Diana free to return to Ogoniland. It was no problem.

||||||

Owens spent the next two days meeting with expatriate Ogonis. He described to them the forces that had led to the murders of the four chiefs and the execution of the Ogoni Nine. "There are still nineteen Ogonis in detention," he told them. "What we have to do is make sure that they are released and continue the campaign to see that Ken's death was not in vain."

The London Ogonis, however, had mixed feelings about Owens's presence in England. Some were afraid that he intended to upstage their efforts by using his family's genius for publicity. Others were relatives and friends of the Giokoo victims. At best, their welcome of Dr. Monday Owens Wiwa to England could be described as lukewarm.

On the third day of Owens and Diana's stay at Ken's house, Maria Saro-Wiwa delivered some difficult news. They would have to leave, she said. One of her twin daughters was moving back from school and her son, Gian, who had been away in hospital, was also coming back. Maria needed Owens and Diana's room vacated right away. "The Body Shop people are coming to take you," she said. "You should be ready."

Owens was shocked. Turning to Diana, all he could do was murmur, "Okay, let's pack."

Gavin Grant came to collect Owens, Diana, and Befii. Despite what Maria had intimated, the Body Shop had no plan for Owens's family. On such short notice, the only thing Grant could think of was to take them to his own house south of London in the coastal town of Littlehampton, Sussex. Although Grant's home was small, it had a cramped spare bedroom. Owens thanked him profusely, but he knew their time at Grant's house would be brief. Littlehampton was in the middle of nowhere. How could he remain so far from London, when he had an anti-Abacha/anti-Shell campaign to wage?

But Owens wasn't as far removed from the centre of things as he assumed. Littlehampton was Anita Roddick's hometown and the international headquarters of the Body Shop. Next morning, Grant introduced the Wiwas to the Body Shop staff, and Owens and Diana had their first formal meeting with the remarkable Anita and T. Gordon Roddick.

Small and fidgety with a head full of auburn curls, fifty-three-year-old Anita Roddick was at the time perhaps the world's most famous environmental capitalist. The daughter of Italian immigrants, she and her husband, Gordon, a handsome Scotsman, opened their first Body Shop cosmetic boutique in Brighton, England, in 1976. Two years later, the Body Shop had gone international and the company was a successful world brand, championing social responsibility and environmental sustainability through no-nonsense, cruelty-free beauty products made from natural ingredients. When the company went public in 1984, the Roddicks became instant multimillionaires. At the height of the Body Shop's success in 1992, the Roddicks' company had an estimated worth of US$1.3 billion.[10]

The Roddicks gave Owens, Diana, and Befii a welcome befitting heroes. Owens returned the greeting in customary fashion: "I want to thank everybody at the Body Shop. I appreciate all the energy and the money and the effort that you put into our cause and particularly want to thank you for making it possible for me, Diana, and Befii to leave Ghana. Since my first contact with Gavin Grant, things have changed quite rapidly for us. We went from a period of uncertainty and despair to one of opportunity. Here we are safely in London. Thank you."

Owens told Anita that in detention Ken had composed a poem for her. It read:

> Had I a voice
> I would sing your song
> Had I a tongue
> I would speak your praise
> Had I the time
> I would live for you
> But here, gagged and bound
> To the floor and injustice
> And waiting for death
> I can only wonder
> How you have breathed life
> Into the thought that we give
> Meaning to being
> Only when we share and care.[11]

Touched, Anita said, "I never met Ken, but he must have been a great man." She asked what the Wiwas would be doing next.

"My goal is to see that the others who are still in detention are freed and those in Ogoni have some level of safety."

"But what about you personally? Where are you staying?"

"We're here in Littlehampton with Gavin Grant. Ken's family was unable to accommodate us any longer."

Anita turned to Grant and said, "We need to make arrangements for these people. For a start, where are they going to live? How are they going to eat? How are they going to get around?"

These were all the practical details Owens had so far neglected. Anita's concern for his family's welfare moved him beyond words. Perhaps this woman's earth goddess persona – the untamed hair, the natural-fibre tribal-sewn clothing, the socially conscious shampoo company – was more than a marketing gimmick. Anita Roddick seemed genuinely empathetic. When Gavin said he had no grand scheme for the care and feeding of the Wiwas, Anita and Gordon agreed that, for the time being, it would be best if Owens, Diana, and Befii moved into the Roddicks' elegant townhouse in Central London.

CHAPTERTHIRTEEN

The weekend after Owens and Diana arrived in England, the *Sunday Times* ran the front-page headline: "Body Shop helps Saro-Wiwa's brother flee Nigeria to Britain."

The article chronicled the Wiwas' flight from Lagos to London and lionized the company for its corporate heroism. "The dramatic escape of Owens Wiwa was organized by Body Shop International, the cosmetics retailer, which lobbied the Foreign Office and airlines to secure his safe passage," reported the *Times*. (This was at best a half-truth; while the company was responsible for getting the family from Ghana to England, it had nothing to do with the Wiwas' escape from Nigeria.) "Anita and Gordon Roddick, the Body Shop founders who have repeatedly criticised Nigeria's military regime, were kept closely informed throughout. They met Wiwa and his family, who fled with him, soon after they arrived in Britain."[1]

With the Body-Shop-friendly coverage in *The Observer* and the *Times* creating a buzz worldwide, the cosmetics company kicked its PR machine into high gear. After moving Owens's family into the Roddicks' townhouse, the Body Shop launched them on a press junket. For a week, Gavin Grant came to the house at eight every morning to take the Wiwas to the company's London offices, where Owens and Diana would sometimes sit for five-hour-long television, radio, and newspaper interview sessions. Reporters came from Japan, Sweden, Holland, and the United States – and all were obliged to schedule their interviews through Body Shop spokespersons.

On December 1, 1995, the Body Shop, in conjunction with Greenpeace and the Ogoni Community Association of the United Kingdom (OCA-UK), staged a protest outside Shell International headquarters. Hundreds of people marched to the Shell building in central London carrying picket signs, coffins, and a large crucifix. Owens and Diana wore oversized black T-shirts emblazoned with the words "BOYCOTT SHELL" above a Body Shop corporate logo. Ken Junior, his mother, his siblings, and most of the London Ogoni community also attended. As protesters massed before Shell headquarters, a squad of Greenpeace activists affixed large anti-Shell banners above the building entrance, and someone handed a microphone to Owens and encouraged him to address the crowd.

Owens stared mutely at the assembled protesters, police, reporters, and passers-by. No one had told him he would be giving a speech. He had never given a speech before. His public speaking experience had so far been limited to statements at MOSOP steering committee meetings or crowd control at disasters. "Oh my God, what is this?" he whispered to Diana. "I can't stand on a podium and talk to people at rallies!"

"Why not?"

"That was the work of Ken and NYCOP and those sort of people. I was always in the background."

"Just say something, say anything," she said. "Think of Ken. How would Ken address a crowd like this? What would he have said to Shell?"

At the thought of Shell, Owens gripped the microphone and the words came pouring out. He enumerated for the crowd the reasons why he felt Shell was responsible for the deaths of Ken and the others. He talked about Shell's ties to the Nigerian army, the company's involvement in the bribing of Ogoni people to give false witness against Ken, about Brian Anderson's offer to help free Ken only on condition that Ken and MOSOP retract their statements about Shell. "I hold Shell responsible for the death of Ken," Owens told the crowd, "and I vow that I will make sure everybody around the world hears about the atrocities in Ogoni unless they pull out of the Niger delta and stop supporting the military dictatorship."

Overcome with emotion, Owens handed the microphone to Diana, who gave her own eloquent speech. At the end of the rally, Owens and Diana were besieged by journalists. As media stunts go, the "Boycott Shell" protest was an unqualified success.

For another week, the grind of interviews did not stop, and the strain of being cross-examined about deeply traumatic matters was often unbearable. As Owens sat beneath harsh lights with the cameras trained on his face, he began to feel more interrogated than interviewed. Still, he cooperated, hoping to provide incriminating testimony against Shell and Abacha. As for more tangible evidence, among the few things Owens had smuggled out of Nigeria in his suitcase were some letters Ken had written to him from prison. Owens had these letters typed and gave copies of them to David Wheeler for safe keeping.

Along with the media interviews, the Body Shop subjected Owens to several debriefing sessions that lasted for hours. During one of these sessions, Owens was delivering his well-rehearsed testimony rationally and un-emotionally. Perhaps hoping to pull a better performance from him, the Body Shop's Gavin Grant said, "Okay, let's watch a film." He put on a video-tape of Ken Saro-Wiwa giving a rousing speech on the first Ogoni Day. Since Ken's execution, Owens had not allowed himself to look at images of his brother in newspapers or on television. Now there was Ken right before him.

"In recovering the money that has been stolen from us, I do not want any blood spilled – not of an Ogoni man, not of any strangers amongst us. We are going to demand our rights peacefully, non-violently, and we shall win!" cried Ken triumphantly.[2]

Grant rewound the tape and played it back in slow motion. Owens watched his brother's face glide across the screen, alive and vibrant. It was too much. Owens's tenuous grip on his composure faltered and he collapsed in uncontrollable sobbing.

The Body Shop then handed Owens over to Greenpeace UK, which per-formed an exhaustive interview of its own and then widely disseminated the video and transcripts of his testimony. Next, Owens recounted his story in detail for the British Human Rights Caucus and was invited to attend a session of Parliament and have photo ops with high-level members of the British government, including Liberal Party leader Lord Ashdown and oppo-sition Labour Party leader Tony Blair.

Gavin Grant also arranged for Owens to be the subject of a segment on the American television news magazine *60 Minutes*. During the first week of

December 1995, Morley Safer and crew arrived in London. In the pre-interview, however, Safer became concerned. Although Owens was educated, well-informed, and articulate, he spoke English with a pronounced Nigerian accent that Safer had trouble understanding. To Grant, Safer said, "I don't know how I'm going to present this guy to the American public." He then turned to Diana and asked, "How come you speak with an American accent?"

When Diana told him that she had lived in the United States as a girl, Safer suggested that she should "translate" what Owens was saying for the TV audience. Arranging Owens and Diana side by side, he started the interview by asking Owens a question. Owens answered in English, and then Diana repeated what he said word for word. Owens found this awkward exercise insulting, and his discomfort showed in his answers and in his sullen demeanour.

Next, Safer turned to Grant and asked, "Where are the documents?" Owens had no idea what Safer was referring to and was astonished to see Grant handing over copies of the letters Ken had written from prison. Peering intently at the letters, Safer could not make out Ken's rather difficult handwriting and asked Owens to read them aloud to him. Owens balked at the suggestion. The letters were highly prized and deeply personal. How could he be expected to read them aloud? Still, for the good of the campaign, he did as he was asked.

Safer was displeased with what he heard. Ken's letters from prison did not contain the smoking gun linking Shell with the atrocities in Ogoniland. Safer terminated the interview and instructed his crew to pack up.

That was that. There would be no *60 Minutes* segment on the Ogoni crisis.

||||||

The Roddicks' handsomely appointed London residence was a narrow row house in the city's fashionable West End. At the time, Anita and Gordon were using one of their other homes and left the London place entirely at the Wiwas' disposal. Owens and Diana found it a marvel that a house that old would have a spectacular twenty-first-century bathroom – and the smallest spiral staircase one could possibly imagine. On the downside, the house was devoid of food.

A new London friend who had come to their aid, a woman named Christina, arrived with a gift of groceries and Diana used some of the little money they had left to buy further provisions. Christina was shocked, however, that Diana would even consider going outside. It was December, and the Wiwas' light clothing was woefully inadequate for the bitter climate.

Christina alerted Anita Roddick to the problem, and Anita immediately had an employee take Owens and Diana shopping and treat them to whatever clothing they needed. Owens felt uncomfortable accepting such charity; he bought only a wristwatch worth about five pounds. Diana chose only a grey cloth coat for herself, but did buy some warm clothes for the baby.

The telephone at the Roddicks' house was another issue. Understandably, the staff at the Body Shop would not divulge the Roddicks' private number. Calls for the Wiwas had to be made to Gavin Grant, who took a message. Even Ken Junior, who had a long-established relationship with the Body Shop, had to go through Grant to speak with his uncle.

Before long, the flush of excitement over the Wiwas' arrival in London and the heady whirlwind of media attention began to fade. Gavin Grant no longer came by to pick up Owens every morning. He was now off on assignment, and his replacement, former UNPO staffer Richard Boele, lived far away and did not have a car. Although Owens was reimbursed by the Body Shop for his daily commute by taxi to its downtown office, he chose to travel the distance on foot instead. Without a winter coat, it was a long, brisk walk through the streets of London.

One morning after Owens walked to the Body Shop offices, he ran into Glenn Ellis, the filmmaker of the Ken Saro-Wiwa/Shell documentaries *Delta Force* and *The Drilling Fields*. When Ellis saw how Owens was dressed and that he was carrying his paperwork around in his bare hands, Ellis emptied his own shoulder bag and offered it.

"Oh no! I don't want to take your bag," Owens protested.

"Please, take it. You can't keep walking around like this. Tell me, do you have any money?"

"No."

"Then take this as well." Ellis reached into his wallet and handed Owens three hundred pounds. "Go buy yourself a coat."

To Owens, three hundred pounds was a small fortune. Even so, he could not bring himself to spend the money on a winter coat. Instead, he bought some badly needed items for the baby, then tucked the remainder away. The little cash they had left Nigeria with was long gone, and with no prospect of more income the three hundred pounds would have to last a long, long time. Owens and Diana did, however, allow themselves one luxury: they took Befii to a fast-food restaurant that evening, where they ordered hamburgers.

|||||||

At the end of two weeks, the Body Shop began to search out alternative accommodation for Owens, Diana, and Befii. To be fair, it had been made clear that the Roddicks' invitation to the Wiwas was never intended to be permanent. Utilizing the refugee network in London, the company found the family some temporary lodging. A local family agreed to host the Wiwas for two weeks or so, until their daughter returned home at Christmas.

When Roger Diski received the call from the Body Shop asking if he might put the Wiwas up for a while, he expressed surprise that a company with assets in the billions would be looking for free housing. Nevertheless, Diski said he'd be glad to help. He and his wife, Judith De Witt, both worked for a firm that produced television documentaries. They lived with their four children in a three-storey Victorian house in the working-class neighbourhood of Islington. Diski shared custody of a teenage daughter by a previous marriage, and it was this daughter's bedroom that would be unused until she came to spend the holidays with her father. The basement bedroom was of modest size and decorated like a typical teenage girl's room, but there was a double bed and space for a cot for Befii. The youngest of Roger and Judith's children was a year older than Befii, so the house was child-friendly. For the first time in his life, Befii would have playmates. There were toys he could share and hand-me-down clothes to wear. The household also had two kitchens, so Owens and Diana were able to keep house for themselves.

The problem remained of how they were going to continue to pay for groceries, diapers, and transportation. Even with free accommodations, the little money they had wouldn't last long. What was Owens to do for income? He had no way of earning a living, yet he was still being sought out for media interviews for which he was obligated to pay his own transportation, photocopying, and faxing.

Before he could figure out the answer to this problem, the Body Shop informed him that UNPO would like him to go to the Netherlands for a speaking tour. In addition, a Dutch television crew was coming to England to shadow him on the journey. Thus it was with decidedly mixed emotions that Owens agreed to make the trip to the Netherlands – or, as he called it, "Shell Country."

|||||||

A month after Ken's execution, Royal Dutch/Shell ran a full-page advertisement in the *New York Times*. It read: "Some campaigning groups say we should intervene in the political process in Nigeria. But even if we could, we must never do so. Politics is the business of governments and politicians. The world where companies use their economic influence to prop up or bring down governments would be a frightening and bleak one indeed. Shell. We'll keep you in touch with the truth."[3]

The ad failed to mention that Shell was at that moment "keeping in touch" with some decidedly unwelcome truth. In December 1995, news reports began to surface about a purge in management at Shell Nigeria. The company was reportedly considering the dismissal of twenty employees after an independent auditor's report revealed that whole sections of Shell's Nigerian operations were mired in graft and corruption. According to the report, much of the nearly US$20 million per year Shell allotted for community development had wound up in the pockets of military officials, Ogoni "vultures," and Shell employees. A Shell executive requesting anonymity told London's *Sunday Times*: "I would go as far as to say that we spent more money on bribes and corruption than on community development."[4]

Asked for his response, Shell Nigeria's managing director Brian Anderson admitted, "It's like a black hole of corruption, acting like a gravity that is pulling us down all the time."

Documents obtained by the *Times* indicated that Lieutenant Colonel Paul Okuntimo – the man responsible for "sanitizing" Ogoniland with his campaign of murder, rape, and torture – had been on Shell's payroll during the atrocities. Confronted with this allegation, Anderson expressed surprise that Shell might be connected with Okuntimo's reign of terror. "From what I hear of his recent past, he is a fairly brutal person," said Anderson. "I'd like to know if we were involved with somebody like that so we could stamp it out."[5]

In the delta, it was widely known that the head of Shell Nigeria's eastern region public and governmental affairs department, was Shell's main link to Okuntimo. Okuntimo himself made no secret of his dealings, mentioning their relationship in conversations to (among many others) British environ-mentalist Nick Ashton-Jones and Nigerian human rights lawyer Oronto Douglas. Okuntimo went so far as to tell a reporter for London's *Sunday Times* that he was being paid by Shell to crush the Ogonis' anti-Shell protests. "Shell contributed to the logistics through financial support," Okuntimo told the *Times*. "To do this, we needed resources and Shell provided these." (He later denied having made these comments.)

Despite Okuntimo's statements and the documentary evidence un-covered by the auditor and the *Sunday Times* (not to mention Shell's own official statements admitting the hiring of Okuntimo's men for protection during the deadly Biara and Korokoro incidents of 1993), Shell denied ever having had financial dealings with the Nigerian military. "Shell Nigeria has not authorized any financial support for the military," reiterated a Shell spokesman in London in late 1995. "If there is any evidence that it has hap-pened, we will gladly look at it."[6]

||||||

On the approach to Schiphol Airport, Owens gazed down at the lights of Amsterdam. *So this is Shell Country*, he thought. *This is what they use all the money for that they bring from Nigeria.* From the air, the city appeared to be an orderly jewel. If the Dutch could create such a place from an area for-merly below the sea, he wondered, why couldn't they make the Niger delta beautiful as well?

Owens spent his first day in Amsterdam in media interviews and that night stayed at the home of the head franchisee of the Body Shop Netherlands. Owens had been unaware that the Netherlands arm of the Body Shop had been deeply involved in the campaign to free Ken Saro-Wiwa, and it was the Dutch franchisee who had contacted the president of KLM to arrange the Wiwas' evacuation out of Ghana – not the Body Shop's London office.

The main goals of the Netherlands trip were to lobby the Dutch gov-ernment to accept Ogoni refugees, to campaign for the withdrawal of Shell from the Niger delta, and to press for the Dutch government to support an international oil and arms embargo on Nigeria. As UNPO was the organizer of this mission, Owens made it a point to visit the headquarters of UNPO at The Hague.

From outside, the UNPO building looked like all the other Dutch homes on the languid boulevard named Javastraat. The only distinguishing feature of the four-storey townhouse was a small brass plaque beside the front door engraved with the letters UNPO.

When he was escorted up to the boardroom, Owens noticed the organization's grand display of flags. Hung about the room were the banners of fifty groups of unrepresented peoples from around the globe who were struggling to regain their lost countries, preserve their cultural identities, protect their human and economic rights, or safeguard their environments. There among the flags from Abkhazia to Zanzibar was a banner of six red stars floating on a background of blue, yellow, and green stripes – the Ogoni flag.

Seeing the flag made Owens realize that political autonomy was one aspect of the struggle in which he had not been involved. He had been mainly concerned with the environment and health, but Ken had understood the importance of political autonomy. Without Ken, that part of the struggle would be lost forever. In front of the Ogoni flag, Owens felt the tears well once more.

During an emotional meeting in the boardroom, the staff of UNPO told Owens about the work they had done to rally support for Ken and the others in prison. Owens reminded them that, although the world had failed to save Ken, there were still nineteen Ogonis in detention charged with the same crime, and their fates lay in Abacha's hands. These men desperately needed the support of groups like UNPO to escape the gallows.

In the Netherlands, Owens was invited to attend a session of the Dutch Parliamentarian Committee on Human Rights and spoke with several parliamentarians and other representatives of the Dutch NGOs. Owens spelled out the alliance between Shell and the Nigerian military dictatorship and argued that if that alliance were not broken, democracy would forever be a pipe dream in Nigeria.

Owens was summoned to a photo op with Hans van Mierlo, Dutch minister for foreign affairs. Afterwards, Owens withdrew with van Mierlo and an UNPO representative to the minister's chambers. There, Owens spoke about the Ogoni refugees trickling into the Republic of Benin, as well as the thousands of others who were trying to escape Nigeria. For his part, van Mierlo explained that Royal Dutch/Shell had tried to lobby Sani Abacha for the release of Ken Saro-Wiwa. Essentially, this effort consisted of a fax from Shell to Abacha pleading for clemency – a fax sent after the official confirmation of Ken's death sentence.

"I suppose it could be true that Shell sent a fax," said Owens, "but did Abacha see it? Why didn't Brian Anderson speak to him personally? He even speaks the Hausa language, not English, to Abacha! That is what he told me. It would have been the easiest thing to say, 'If you go on with that execution, we are going to reconsider our investment in Nigeria.'"

Van Mierlo informed Owens that the chairman of Royal Dutch/Shell wanted to have dinner with him in Amsterdam to discuss the matter further.

"Well, thank him for the invitation," said Owens, "but I don't have permission from the Ogoni people to have that sort of discussion with the chairman of Shell."

The UNPO representative had said little during the meeting, but sat quietly taking notes. On the way out, she told Owens that he was only the second person from an UNPO-member group that the minister of foreign affairs had deigned to see. "The only other UNPO member he ever gave an audience to," she said, "was the Dalai Lama."

||||||||

Back in Britain, Owens immersed himself in his campaign to spread the news about the plight of the Ogoni. He became obsessed with seeing that Shell be shamed, that democracy be restored in Nigeria, and that Nigerian oil be boycotted. Immersed in this campaign, he gave little thought or attention to those around him.

At the home of Roger Diski and Judith De Witt, Diana spiralled into depression. A few months earlier, she had been an independent young woman on the brink of receiving a degree in political science. She and Owens had beautiful homes in both Port Harcourt and Bori. Her mother and a full-time nanny shouldered much of the burden of Befii's care. Even with the horrors in Ogoniland, being part of the struggle – especially as a member of the illustrious Wiwa family – was dangerous and exciting. Now, in London, she found herself reduced to poverty, sleeping in a child's bedroom in a house of strangers in a foreign land. The care of Befii was her responsibility alone. Not even Owens was there to help mind the child.

The entire family's sense of gloom was only exacerbated by the arrival of a letter from David Wheeler stating that the Body Shop had decided not to offer the Wiwas further assistance. Wheeler wrote that the Body Shop was aware that – like his brother Ken – Owens was a proud person and that any more help would seem paternalistic. Owens later learned that the Body Shop's motivation might have partly come from the looming possibility of a lawsuit from Shell, which claimed that Owens's statements regarding the oil company were defamatory and accused the Body Shop of aiding and abetting the slander. The Body Shop had also been coming under pressure from the Ogoni community in London to drop their support for Owens and Diana, whom they felt were receiving unfairly preferential treatment.

As Christmas approached, Owens asked his sister Comfort in Nigeria to sell off his personal belongings. Comfort wired her brother back to say that if he needed immediate cash, she had a friend in London who would advance him money, which she would repay. Comfort's friend came through with a sorely needed five hundred pounds.

But five hundred pounds was not going to get them a place to live. The Christmas deadline at the Diskis' was nearly on them and Owens had yet to find his family's next accommodation. There had been an offer of a spare bedroom elsewhere, but the prospective hosts retracted their offer on Christmas Eve.

Owens telephoned Jim for advice. Jim, his wife, and their four children lived in a tiny flat far outside London. "You had better come here," said Jim, "even if you all have to sleep on the couch." He told Owens and Diana he'd pick them up on Christmas morning.

Christmas came and the Wiwas were packed and ready to go. Expecting to have Christmas dinner at Jim's house, Owens and Diana didn't eat. When Roger Diski saw the Wiwas sitting glumly beside their suitcase, he asked what the matter was. "My brother is coming to pick us up," said Owens. "We shall have to sleep on his couch."

Diski had a few words with his wife, "Look, Owens, I'll talk to my daughter and ask her to spend Christmas and New Year's with a friend. So you don't have to go. Why don't you stay and use this period to look for a place to live. Maybe something will crop up."

Owens was relieved, yet despondent. "Look at me!" he said. "A few months ago, I was providing for my family. I was looking after my parents. I had money to give to a lot of people. When we came to England, I believed the world was going to make Abacha leave. I thought we'd be in England for two or three months and then Abacha would be off and we'd go home. Now I don't think that will happen. At least not soon."

Over the holidays, Roger Diski spoke to a psychotherapist friend who lived down the road. She and her husband, a white South African psychiatrist, had a furnished apartment in their basement. For £250 per month, the Wiwas could have it. The one-room flat, tiny and grim, became their new home.

|||||||

One afternoon in January, Owens received a phone call at the flat. "If Shell doesn't get you," the man whispered, "we will."

"Who is speaking?" Owens demanded.

"Remember Ojukwu," the man said, referring to the former Biafran secessionist leader, and hung up.

The threatening calls came regularly, and a recent event made them hard to ignore. A couple of weeks earlier, the London residence of Dr. Barinem Kiobel had been firebombed. Kiobel was one of the men hanged with Ken Saro-Wiwa; like Ken, he had earlier relocated his wife and children to England. In the attack, Mrs. Kiobel and her three children sustained serious burns. The Wiwas were terrified that something similar would happen to them.

Soon after the calls began, Owens returned exhausted from an early-morning BBC News interview. As Befii sat on the floor playing with his toys, Owens stretched out on the bed while Diana stepped into the bathroom. To the sound of water running for Diana's bath, Owens nodded off. He awoke with a start at Diana's voice: "Where is Befii?"

Owens looked around. "Wasn't he with you?"

The front door was ajar. The two froze for a moment. Owens leapt off the bed and Diana dashed up the basement stairs and onto the front lawn shouting for Befii. Owens ran about checking the bushes, the alleys, the street. There was no sign of their son. They phoned the police. They phoned the child welfare department. They phoned Roger Diski to see if Befii had wandered back to their friend's house. No one had seen the child.

For two hours, Owens and Diana did not speak to one another. As they combed the neighbourhood, each blamed the other. Their greatest fear, of course, was that Befii had been kidnapped or had wandered down quiet Huddlestone Road to the busy intersection at the end of the block.

When the police phoned to say that a small black child had been brought into the station, Diana was out the door, hailing a cab even before Owens could hang up. "Diana!" called Owens. "You don't have any money! Who is going to pay the taxi? You'd better wait for me!"

The authorities were not amused that the Wiwas had allowed their two-year-old to wander away and threatened not to return him. In ways Owens and Diana could never have foreseen, London had turned into a nightmare.

|||||||

At the start of 1996, Owens was invited by the European Parliament in Brussels to give a presentation at the January 23 EU Conference on Development and Cooperation. He was broke, but the Body Shop agreed to cover the cost of a train ticket to Brussels. Unexpectedly, the biggest hurdle he had to face was obtaining a Belgian visa. Owens showed embassy officials the invitation from the European Parliament, but they did not believe that an unemployed Nigerian would have received such a summons. They suspected that he was either a criminal or an asylum seeker. When the European Parliament heard about the visa denial, they phoned Owens and instructed him to take the train to Brussels, guaranteeing to have all the proper arrangements made at the port of entry.

Owens travelled with Oronto Douglas, one of Nigeria's leading environmental human rights lawyers.[7] He and Owens caught the train from Victoria Station the day before the meeting, arriving at the Belgian border at two in the morning. As instructed by the Belgian minister, Owens gave the visa officer his Nigerian passport along with the minister's name and telephone number. He explained the European Parliament had made all the necessary arrangements for his visa.

Despite this, Owens was immediately arrested and jailed.

Behind bars again, Owens flew into a panic. If the Belgians deported him to Nigeria, they would be delivering him to the gallows. In desperation, he beseeched the border control officers to call the minister in Brussels who could straighten out the mess, but the officers were reluctant to place such a call in the middle of the night. Owens insisted, however, and after forty-five minutes, they phoned and woke the minister, who told them to release Dr. Wiwa at once.

Owens and Oronto arrived, sleepless, twenty minutes before the start of the conference. Outside the EU building, Greenpeace had arranged a small demonstration against Shell that Owens joined long enough to say a few words to the press and have himself photographed holding an anti-Shell picket sign.

Entering the meeting room, Owens saw representatives from Royal Dutch/Shell already seated. The first person he laid eyes on was Brian Anderson's second-in-command, Nnaemeka Achebe, director and general manager of Shell Nigeria. As Owens passed, Achebe rose and extended his hand. Owens pointedly refused to shake it. The snub was a visible embarrassment for Achebe.

Owens's speech centred on the underlying cause of Shell's actions in the Niger delta. "What they are doing amounts to developmental racism," he said. "The fact is that they are doing in Ogoni what they do not do – or cannot do – in the West." He pointed out that, in Ogoni, Shell ran its pipelines over the ground, flared gas dangerously close to habitation, expropriated farmlands without permission or compensation, left oil spills untouched, and gave virtually no employment to locals. He contrasted this with Shell's record in Europe and North America, where the company was respectful of the environment and the "stakeholders." "The military loots oil funds from Nigeria and puts them in Western banks. I call on you to put in a regulation that will make the acceptance of such ill-gotten money an offence in your countries. In that way, we may be able to at least stop this greed and the theft of our money from the developing world to Europe."

Afterwards, Owens and Oronto met with several parliamentarians and the European minister for development. The following day was filled with media interviews, but as the daylight waned, Owens excused himself and hastened for the train station. His Belgian visa was valid for only forty-eight hours. He was not about to stay in Belgium a minute longer than necessary.

||||||

The day after returning to England, Owens flew to New York to begin a marathon media tour. This time, a half-dozen environmental and human rights organizations had joined forces to present him and the Ogoni problem to North America.

Owens was met at JFK Airport by Juliet Berriut of UNPO USA, Steve Kretzmann of Greenpeace, Stephen Mills of the Sierra Club, and Adotei Akwei of Amnesty International. It was like being welcomed by old friends. Each of these activists had worked with Ken Saro-Wiwa either in Nigeria or during Ken's travels to the United States, and Kretzmann had been in contact with Owens during his time in hiding in Lagos. It was late January, and New York was sporting a sooty crust of snow. As usual, Owens was underdressed, wearing his brother's borrowed sports coat.

On the cab ride into the city, they discussed the next day's itinerary. Fifteen media events had been planned, so Owens was advised to rest. The group had booked rooms at the Pickwick Arms, an unpretentious midtown hotel. The next morning, when it appeared that Owens had gotten no rest at all, Juliet Berriut asked, "What's the matter? Was there something wrong with the room?"

"Not really," said Owens. "There's just something about it that makes me uncomfortable."

"That's odd. That's the same room we always put Ken in whenever he came to the United States."

First stop next day was an early morning meeting with the *New York Times* editorial board. The *Times* had just run a story about Owens's escape from Nigeria ("Unholy Alliance in Nigeria: A Doctor Fails to Save His Brother from Execution"),[8] and the editorial board wished to explore the Ogoni issue further.

Owens was then taken on a round of the networks. Peter Jennings of ABC News interviewed him for half an hour and later ABC sent an entire film crew to the Niger delta for an in-depth report. NBC interviewed Owens next. Unlike ABC, NBC News opted to hire an Ogoni resident of New York to take a camera back to the delta and report from the scene freelance. Unfortunately, NBC's Ogoni freelancer was promptly arrested by Okuntimo's men and spent nearly six weeks in detention.

After radio interviews in Harlem, Owens was taken downtown to a New York City Council meeting where he asked the council to pass a resolution to avoid buying either Nigerian or Shell petroleum products. Newspaper and magazine interviews lasted until eleven at night. Owens, exhausted, slept in the cabs between interviews. "I've given my presentation on the Ogoni and Shell so many times," he joked, "I can practically do it in my sleep!"

The frenetic pace was kept up for two days. Owens was miserable. It was the coldest weather he had ever experienced. On their way to a long meeting with the editorial board of *The Wall Street Journal*, Steve Kretzmann gave Owens the winter coat off his back as a gift.

The next stop on the North American tour was Washington, D.C. On January 30, 1996, Owens and his group met with Colonel MacArthur Deshazer, the National Security Council's director of African affairs, and three assistant secretaries of state. Owens boldly asked Deshazer why the U.S. government had done nothing to stop Ken's execution. "On October the second, I talked to three congressional staffers who came to Lagos," said Owens. "In the presence of the American ambassador to Nigeria, I was assured that there was no way your government was going to let Ken be killed. What happened?"

The room was silent.

Tears welling, Owens continued, "Ken told me after he visited America in 1992, that the answer to the Ogoni problem would come from America. He said the United States used more Nigerian oil than the Europeans and that the people of the U.S. have a conscience. He said the people who came to America were people who were persecuted in other places and cared about people who are undergoing the sort of treatment that we are undergoing. And when I was told that there was no way the United States was going to allow Ken to die, I really believed it. But what happened? What happened?"

Deshazer and the assistant secretaries of state had no answer.

As Owens and the activists left the meeting, Steve Kretzmann and Juliet Berriut took Owens aside. The day before Ken was executed, they said, they had phoned the White House to request President Clinton's intervention and were directed to Colonel Deshazer. They asked him to persuade the president to pressure Abacha to cancel the execution. Many other activists also called Deshazer that day. A few hours later, Steve and Juliet said, Deshazer called them back to say that Clinton had made the call to Abacha. The next morning, however, after the news of Ken's execution was broadcast, they said Deshazer phoned them again to say that the information he had given them earlier about Clinton calling Abacha was not correct.

In Washington, Owens was interviewed on National Public Radio's *All Things Considered*, then stopped in to visit the U.S. Senate building, where he met with a few senators and held a news conference in the Senate Press Gallery. Reporters kept Owens answering questions for nearly five hours. That evening, he attended an event given by *National Geographic* at which his late brother was one of the guests of honour. The party celebrated Penguin Books' recent release of Ken Saro-Wiwa's latest publication, *A Month and a Day: A Detention Diary*.

The book is Ken's chronicle of thirty-two days in prison starting in June 1993 after he was arrested for urging the Ogoni to boycott the presidential election. In *A Month and a Day*, Ken used his detention as a framing device for an analysis of the Ogoni situation. It is a moving memoir, at times wry, at times harrowing. Sadly, it was Ken's execution that helped to make the book an international bestseller.

Owens next flew to Toronto, where Jeanne Moffat, executive director of Greenpeace Canada, met him at Pearson International Airport. Having left his entourage in Washington, Owens was to be chaperoned alone by Moffat. As they drove from the airport to her home, Moffat found it amusing to see Owens staring out the car window in wonder, like a child at Christmas.

"I haven't seen this sort of thing before," he exclaimed. "There are no trees; there are no leaves, just mounds and mounds of snow! I cannot believe you are living like this."

When they pulled into Moffat's driveway, Owens ran into the house and asked to be directed to the fireplace. He had heard that people in Canada had big stone fireplaces and was disappointed to learn that central heating was the best she had to offer.

The next day, Moffat took Owens to lunch and asked about his situation in London. He told her about the gloomy basement flat he and Diana could not afford, the lack of money, clothing, and food, the inability to earn an income, their temporary status in Britain, Diana's depression, the threatening phone calls.

"Do you want to come to Canada?" Moffat asked.

Owens glowered. "I've wanted to come to Canada since we left Nigeria, but the Canadians did not want us." He then described his experience with the Canadian High Commission in Ghana, and the immigration official who told him he "must be joking" for even thinking of coming to Canada. Shocked that Owens should have received such treatment from her countrymen, Moffat offered to explore other possibilities of getting the Wiwas to Canada.

Moving on to Ottawa, Owens met with Canada's foreign minister, Lloyd Axworthy, and had lunch with Flora MacDonald, the intrepid former Canadian external affairs minister. Owens was delighted to meet familiar names he and Ken had corresponded with over the years. In Nigeria, of course, Flora MacDonald, had seen to Diana's welfare during her cloak-and-dagger adventures at Kaa.

Owens continued on to San Francisco and to Chicago. As long as Abacha remained in power, Owens was not going to allow himself a day of rest. Back in London, he was the special guest at an event sponsored by One World Action, a British NGO. There he met Britain's overseas development minister, "Mama Africa" Lady Lynda Chalker, who was still smarting from her unpleasant interview with Ken Junior in New Zealand.[9] At the One World Action dinner, Lady Lynda asked, "Do you actually think you had any success on your trip to the United States?"

"Yes, of course," said Owens. "Most of the people I spoke to – very powerful people – were interested in the idea of an oil embargo and targeted sanctions."

"Well, I have been seeing the same people," said Lady Lynda, "and I want you to know it's not going to happen."

She explained that the Clinton administration was reluctant to add Nigeria to the list of oil producers the United States was already boycotting – Iran, Iraq, and Libya. American corporations, especially those with investments in Nigeria, were also strongly opposed to sanctions. "It won't happen," she repeated.

Lady Lynda's certainty was demoralizing, and recent newspaper reports seemed to bolster her bleak assertions. One month after the execution of Ken Saro-Wiwa, SPDC had signed a contract to build a $3.8-billion natural gas project, the largest single investment Shell had ever made in Nigeria.[10]

In Ogoniland, too, little had changed since Ken's death. The area was still sealed off with military roadblocks. Young men were still being hunted for the murder of the four chiefs at Giokoo. The people still lived in fear and poverty. Ledum Mitee had assumed the mantle of acting president of MOSOP and was alleged to be in negotiations with Shell to resume operations in Ogoni. It appeared Ken had died for nothing.

|||||||

As spring approached, it became Diana's turn to represent the Ogoni cause. In May, she travelled to Brussels to speak to the European Union and afterwards met Owens and Befii in Copenhagen for an anti-Shell protest. Her next stop took her halfway around the world when she was invited to represent the Federation of Ogoni Women's Association at a conference in Tokyo.

While Diana was in Japan, a woman named Raga, one of Diana's new London friends, offered to help Owens with Befii's care. When Raga first arrived at the basement flat in Islington, she was aghast to find the refrigerator empty. Owens had spent his last cent. He didn't even have enough to buy milk for the baby.

"This cannot be happening to you!" Raga exclaimed.

To Owens's embarrassment, she went to Amnesty International to ask for emergency financial assistance for the Wiwas. Thanks to her, three days later Owens received a cheque from Amnesty for three hundred pounds. With this windfall, he'd be able to pay rent for one more month and have fifty pounds left for food.

Like Raga, Roger Diski was concerned about how Owens, Diana, and Befii were going to survive. As it happened, one of Diski's closest friends was the British novelist Doris Lessing. At her home, a handsome three-storey Victorian in West Hampstead, Lessing maintained a furnished one-bedroom apartment that she kept available for visiting friends and artists and the occasional individual in need. When Lessing heard about the Wiwas' plight, she offered the apartment to them.

The move to Lessing's house lifted Owens and Diana out of despair. Lessing lived on a short, crooked lane of single and semi-detached houses. The many trees and tranquil gardens surrounded by walls of brick and stone gave the street an unexpectedly rural feel.[11] The Wiwas' new flat was bright and well-kept and had access to a cheerful back garden.

When Owens and Diana met Lessing, they liked her at once. Charming and direct, the seventy-six-year-old earth mother had a slender, iconic face and centre-parted hair pulled back in a bun. Her house was warm and comfortable with tables and shelves overflowing with books. She not only offered them her flat rent-free for as long as necessary, but said she would keep their kitchen stocked with food.

"No, thank you, Miss Lessing," said Owens. "We have our own food." He explained that they had discovered where in London to purchase traditional African staples. Lessing, having spent many years in southern Rhodesia, understood.[12]

For the Wiwas, good fortune, like bad, seemed to come in waves. Once they had established themselves in the relative luxury of Lessing's home, they received more uplifting news. One morning as Owens sat down to open the mail, he found a letter from John Waddington of Bloor Street United Church in Toronto. His church, Waddington wrote, would be honoured to sponsor the Wiwas' immigration to Canada.

Jeanne Moffat of Greenpeace Canada had come through. With the help of Bob Fugere of Toronto's Interchurch Action and the always-reliable Flora MacDonald, she had found a way to restore, in small part, Canada's reputation for being a compassionate society and a haven for the oppressed.

CHAPTER FOURTEEN

When Owens immigrated to Canada, he was informed that his Nigerian medical qualifications were insufficient for him to practise in his new country. At the age of forty, he found the prospect of repeating medical school (along with the requisite internship) impractical, as he would have no way of supporting his family in the interim. Fortunately, while still in London, he had landed a full scholarship at Johns Hopkins University in Baltimore, Maryland. While his wife and child remained in Canada, he lived for a year in the United States, working toward a master's degree in public health and sending home the little money he earned as a student teacher. After graduation, Owens returned to Canada, where he accepted a position as a Research Fellow at the Centre for Addiction and Mental Health, a WHO-affiliated centre in Toronto. The job didn't pay much, but it was enough to provide for his family.

In his newly adopted country, Owens became the executive director of the African Environmental and Human Development Agency (AFRIDA) and the coordinator of MOSOP Canada. For the Sierra Club, he was elected to serve on the U.S. Sierra Club international committee and was elected as a director of the Sierra Club's Eastern Canada Chapter, later becoming vice-chair. Owens was also elected secretary of Africans in Partnership Against AIDS, an organization for Africans in Toronto living with HIV.

At dozens of worldwide events – from the WTO protests in Seattle to the International Conference on Globalization in Belgium – Owens campaigned for increased corporate responsibility, sharing the stage with the likes of Bill Clinton, Kofi Annan, Desmond Tutu, Vaclav Havel, and Nelson Mandela. Owens found it hard to admit that, like his brother, he had a talent for public speaking. Nevertheless, in a very short time, he earned a reputation as a tireless and charismatic crusader for environmental and human rights. The thing that audiences found so compelling about Owens was that his crusade remained deeply personal. There seemed to be no moment when the unjust execution of Ken Saro-Wiwa was ever far from his brother's thoughts.

In 1996, the UN sent a fact-finding team to Nigeria to investigate the circumstances behind the trial and execution of the Ogoni Nine. It reported that it was clear that even by Nigerian standards, the trial and executions were illegal, and recommended that the government pay compensation to the families of the Ogoni Nine for their loss. With Abacha still in power, any hope of this occurring was slim to none.

Then, two years later, something quite unexpected occurred.

At dawn on June 9, 1998, Nigeria's military council named the relatively unknown General Abdulsalam Abubakar the new head of state. In a televised statement, Abubakar revealed the reason for his sudden elevation: "General Sani Abacha passed away in the early hours of this morning," he announced. "May his soul rest in peace." Officially, the fifty-two-year-old dictator was the victim of a heart attack. Unofficially, his death was much more interesting.

Reportedly, Abacha had a voracious sexual appetite and enjoyed the company of prostitutes imported nightly for his pleasure from Eastern Europe, Russia, and India. Tangentially, Abacha was also renowned for his ravenous consumption of a certain popular erectile dysfunction medication. At four in the morning on that fatal day, as he frolicked in the arms of two Indian prostitutes and one Nigerian girl, Abacha broke into convulsions and expired. There was immediate speculation that someone had placed cyanide in the dictator's Viagra.[1] Whether Abacha was poisoned, died of natural causes, or succumbed to debauchery will never be known; he was buried the next day without autopsy or funeral.

A few weeks after the dictator's death, his widow, Maryam Abacha, declared she wished to mourn her husband by visiting a few "Muslim holy places" in Saudi Arabia. At Kano airport, police became suspicious of her thirty-eight pieces of luggage, and on inspection discovered most were stuffed with cash.[2] Though she claimed that her husband did not steal any money ("He merely put away the funds in some foreign accounts for safe-keeping," she said), Maryam Abacha was detained under protective custody, she was placed under twenty-four-hour surveillance by the police, and her passport was impounded. Not long thereafter, she was put on trial for embezzlement of public funds and human rights abuses.

It has been estimated that during Sani Abacha's four and a half years as head of state, he stole more than $4 billion.[3] The Nigerian government reclaimed from the Abacha family 34 mansions, 54 luxury cars, and US$2 billion in local and foreign currencies. In Switzerland, another US$645 million in Abacha family bank accounts was frozen, as were accounts in Luxembourg and Liechtenstein that contain hundreds of millions more.[4]

After Abacha's death, there arose the problem of what to do with Moshood Abiola, the man elected Nigerian president in 1993. Although he had been in prison for four years, millions of his countrymen (and Abiola himself) still considered him the legitimate head of state. Because the military felt that releasing him would destabilize the country, General Abubakar brought in UN secretary-general Kofi Annan and Commonwealth secretary-general Emeka Anyaoku, as well as a delegation from the United States, to try to convince Abiola to drop his claim for the presidency. The evening before his scheduled release, Abiola was speaking with visiting U.S. undersecretary of state Thomas Pickering when he started feeling "uncomfortable." He was taken to hospital and pronounced dead ninety minutes later. U.S. officials said that they had no reason to suspect an unnatural cause of death, but Abiola's family was convinced of foul play.[5]

They were not alone in their suspicions. As news of Abiola's death spread, mobs of Lagos youths began destroying shops belonging to Hausa businessmen. (Lagos has a largely Yoruba population while the country's military elite is predominantly Hausa.) Rioting was also reported in Ibadan and Abiola's hometown of Abeokuta. Witnesses in Lagos said angry crowds murdered at least eight people in the Hausa neighbourhood of Agege in revenge for the death of Abiola. Parts of Lagos were almost deserted; many businesses were closed, and smoke filled the air from burning cars at hastily erected barricades.[6]

||||||

One afternoon in March 1999, Owens received a telephone call from the wife of his elder brother, Goteh James, in London, informing him that Jim had died that morning.

The news came as a shock. The last time he had seen his brother was in June 1997 when Owens was on his way through London after a speaking engagement in Finland. At the time, it was Owens, not Jim, who had been ill. Owens assumed he had contracted a bad case of amoebiasis on a trip to visit the Ogoni refugee camps in the Republic of Benin. Because Owens was so gaunt and weak from sudden weight loss, the doctors suspected colon cancer. Owens had seen Jim in London before the results of those tests came back and the brothers talked about the possibility of Owens's having cancer.

"There's something I need to tell you," Jim had said.

"What?"

Jim paused and said, "Well, maybe not now. I should wait until you get the results of your tests back."

"What is it?"

"It's nothing to worry about. I'll tell you later."

Jim never did tell his younger brother what was wrong. Owens's cancer tests came back negative; his intestinal malady turned out to be an easily treated parasitic infection.

Owens flew to London for Jim's funeral. After the ceremony in the church and the interment at the cemetery, Jim's co-workers at Nokia held a wake at a local pub. Owens had never attended a wake and found the evening cathartic. It took a while for Owens to work through his hurt that his brother had never confided in him that he was dying. At the same time, he felt greatly relieved that the family and community had honoured Jim's life and buried him with dignity.

Jim's death only exacerbated Owens's obsession with the horrible end Ken had suffered and the unknown manner of his disposition. There had been no official reports of what had been done with the bodies of the Ogoni Nine. The only clue came from a newspaper interview given by a former inmate of Port Harcourt prison who claimed prison officials threw Ken and the others into a common grave and poured acid over the corpses.[7] Now that Sani Abacha was dead, there was hope that whatever was left of these bodies could be located and a funeral could at last be granted for Ken.

A few weeks later, in a telephone interview with the Nigerian tabloid *Tell*, Owens announced: "No matter what these people do, my brother is going to go back to Ogoni and Ogoni will see the biggest funeral and festival ever. It will happen. Even if it is only one piece of his bone, he is going back to Ogoni."

||||||||

By 1998, Nigeria had struggled under military rule for fifteen years. Sani Abacha's replacement, General Abdulsalam Abubakar wasted no time in instituting reforms to return Nigeria to democracy. One of his first actions was to free Olusegun Obasanjo from prison. Obasanjo, the retired army general who served as military head of state from 1976 to 1979, promptly began campaigning to become the country's leader once again, and, in 1999, was elected president of Nigeria.

Shortly before the election, Owens had a phone discussion with Obasanjo, who gave his personal assurance that should he be elected, he would be in favour of releasing the remains of Ken Saro-Wiwa and the rest of the Ogoni Nine. In June 1999, with Obasanjo now in power, Owens returned to Nigeria to meet with the new president and begin the official process of reclaiming the body of his brother.

Owens prepared for the presidential meeting by rereading Ken's letters and one of his last short stories, "On the Death of Ken Saro-Wiwa." In this story, Ken predicted that the Nigerian state and the multinational oil companies would cheat him even in death, as they had been cheating the people of the Niger delta for decades. He predicted that they would deny him his six feet of earth in Ogoniland because they wanted all the area for oil drilling.

As he arrived in Lagos, Owens had doubts about his own safety. Almost all the military officers involved in the Ogoni tragedy remained in positions of influence. How would they view Ken's long-overdue funeral? Certainly the chief of staff of the Nigerian Army – who had announced Ken's death sentence on national television – wouldn't like it. Nor would Justice Auta, who, after sending the Ogoni Nine to their death, had been promoted to serve in a federal high court. The current governor of Rivers State, Peter Odili, had been deputy governor during the Ogoni-Andoni "wars." Recently, Odili had visited British foreign office minister Peter Hain to discuss the return of Shell to Ogoni. People who had been part of the conspiracy to silence Ken were not going to like his younger brother's returning to Nigeria literally to dig him up again.

But many people were overjoyed to see Owens. At the airport in Lagos he was met by his younger brother, Letam, who escorted him around the city for a morning of happy reunions. The next day Owens attended a rally Hafsat Abiola held in memory of her father, Moshood Abiola. Owens's picture and stories of his return were plastered all over the Nigerian papers.

Shortly after the Abiola rally, Owens received a phone call from Gbenga Obasanjo, the president's son, whom Owens had met at Johns Hopkins University in Baltimore. Gbenga said that Owens should get to Abuja right away – that afternoon if possible. He and Letam were out the door and on the next flight. Gbenga met the Wiwa brothers' plane and took them directly to the presidential palace at Aso Rock.

In the president's office, Owens and Letam marvelled at the louche splendour before them. The late Sani Abacha had crammed Aso Rock with plush furniture, tasteless art, ornate draperies, garish rugs, and state-of-the-art electronic equipment, leaving the palace looking more like the den of a cocaine dealer than the official residence of a president.

President Obasanjo entered and greeted the brothers before seating himself behind an imposing desk. Owens thought to himself: *So that is the desk where Abacha sat when he made those horrible laws and gave his orders for people to be raped, detained, tortured, shot, or hanged.*

Obasanjo offered his condolences to the Wiwas and told Owens he remembered Ken fondly. He and Ken had been together during the civil war, he recalled, and in later years Obasanjo often came to Ken's house for an evening of chess. Owens reminded him that, when he himself was a boy, Ken would order him to clean Obasanjo's shoes when he came for these visits. This made the president smile.

Owens handed Obasanjo a letter from Ken Junior requesting the return of his father's body, as well as a half-dozen letters from Canadian NGOs pleading for the dignified return of Ken's remains. Like his uncle Owens,

Ken Junior had moved with his wife and child from London to Toronto, where he continued in his father's footsteps as a journalist.

Obasanjo read the letters and said that he recently had received confirmation about the location of the graves of the Ogoni Nine. Telephoning one of his assistants, he gave instructions for a presidential order to be drafted, releasing the remains of Ken Saro-Wiwa and the other eight men to their families. In deference to Ogoni culture, the order was addressed to Junior, Ken's eldest son.

Gbenga arranged for Owens and Letam to stay at a nearby hotel. There, Owens placed a call to Ken Junior in Toronto to give him the good news. He then phoned Port Harcourt to let his parents know that he was on his way home to Ogoniland.

|||||||

Owens was unprepared for the reception that awaited him on his arrival at Port Harcourt International Airport the next day. As he walked off the airplane onto the tarmac, Owens could hear the sounds of singing and drumming from the parking lot just beyond the terminal. The melody was familiar – an Ogoni song. The moment he stepped into the baggage claim area, a boisterous scrum of reporters hustled Owens into the airport press room and subjected him to endless questions about his return home to Nigeria after four years in exile.

"Sir! What does it feel like coming back to Rivers State again?"

"Can't you hear the singing outside?" Owens answered. "It's very good to be in the midst of Ogoni people once again."

Another reporter held up a piece of paper and said, "But here is a press release that says MOSOP is not aware of your coming and that Ogoni people should not come out to welcome you."

Owens looked at the man askance. "What do you mean? Those people out there, where are they from?" The reporters all burst into laughter. "That answers your question," said Owens.[8]

In front of the airport, Owens was greeted by ten thousand cheering people who surged forward[9] and lifted him up, carrying him shoulder-high. "Great Ogoni people! Great! Great Ogoni people! Great!" many chanted, while others burst into the song "Arise, Arise, Ogoni People Arise."

Owens gave a brief speech, and then others gave welcoming speeches, praising Owens for his courage and his promise to continue the struggle for which his brother had laid down his life.[10] Owens was escorted to a car, which became part of a five-kilometre-long convoy of vehicles led by a motorcycle with a siren and an Ogoni flag.

At the Abacha Road junction, the convoy stopped. People emerged from their cars, buses, and flatbed lorries to sing an Ogoni song and offer a minute of silence for Ken and the others. Owens gave another speech while standing on the mock coffin of Sani Abacha. Leaders of several different Niger delta tribes declared that Abacha Road would now be known as Ken Saro-Wiwa Street. Erecting a new signpost, they called on the government of Rivers State to pass their road-renaming declaration into law, warning that any attempt to replace Saro-Wiwa's name with another would be met with resistance. This statement elicited a huge roar of approval.

The convoy continued on to Ken's long-shuttered offices at 24 Aggrey Road in Port Harcourt. The whole of Aggrey Road by that point had become an impromptu pedestrian mall with people dancing in the middle of the street from one end of the long thoroughfare to the other. Through dogged perseverance, Owens's driver managed to make it fairly close to 24 Aggrey Road, where Owens, Oronto Douglas, Joy Nunieh Yorwika (the self-titled "Esther of Ogoni"), and others made yet one more round of speeches before the cheering throng.

"I am very proud that you have continued the movement despite the human rights abuses and that you've kept the struggle alive," said Owens to the crowd.

As Owens ceremoniously snipped a ribbon draped across the front door of 24 Aggrey Road, Ken's offices were declared open once again after more than four years of abandonment. Because Mrs. Yorwica was an acquaintance of Governor Odili, after the ribbon-cutting ceremony, Owens handed her a copy of Obasanjo's letter concerning the remains of the Ogoni Nine for her to deliver to Odili at Government House.[11]

The next stop was the Eastern Zonal headquarters of Shell Nigeria in Port Harcourt. When the security staff caught sight of the approaching mob, they locked the front gates, but the effort was in vain. A group of Ijaw youths scaled the fence and opened the gates for two thousand dancers and drummers who defiantly cavorted in front of the Shell building. Holding a megaphone, Owens reaffirmed that Shell remained *persona non grata* in Ogoniland, and that the people of the Niger delta would struggle non-violently until the company was gone.[12]

Late in the afternoon, the convoy turned toward Ogoniland. At Eleme, Owens was received with a dance and masquerade performance, along with a formal welcome. Scooping sand into his fingers, Owens said to the people of Eleme, "This is real Ogoni sand. Some people thought I would never come back to this land, but that has not happened. I am back. And I have come with some news which I am sure the Ogoni people will like. Tomorrow, after I have informed the other people who are concerned, I will give the announcement at the big rally we are having in Bori. Please, I invite all of you to come!"

In the Kingdom of Tai, thousands of people lined the roads around Nonwa to cheer Owens, and even more were gathered at Kira Junction, where they unfurled an enormous banner reading, "Dr. Owens Wiwa and other exiled MOSOP activists, the Tai people welcome you back to Ogoni from exile." At the Local Government Council headquarters in Saakpenwa, Tai, another mammoth crowd had been waiting all day.

But when the convoy turned into Gokana Kingdom, no reception awaited. Ledum Mitee and the Gokana branch of MOSOP had dissuaded the people from coming out.[13] Owens later learned that a formal complaint had been filed with the commissioner of police of Rivers State Command, alleging that Dr. Wiwa was coming home to "cause confusion."[14]

Whatever animosity Owens felt while passing through Gokana was washed away the moment he crossed into Ken-Khana Kingdom. In Bori, the horde was so thick that the car could only creep along. Wearing a MOSOP Canada T-shirt and turquoise and orange MOSOP visor, Owens opened the sunroof of his vehicle and stood on the back seat, waving. The crowd screamed and stretched to touch him.

Owens stopped briefly at his clinic. There, he embraced his loyal nurse Zor and congratulated the doctor and the other nurses for their unfailing courage in keeping Inadum Medical Centre open.[15] Many of his former patients offered gifts of yam, *gari*, and fish.

By five in the afternoon, Owens finally made it to his home village of Bane. At the Marian High School gate, he was met by the local FOWA chapter; people danced and sang, and as he approached St. Dominic's Catholic Church, the elders welcomed him with a two-gun salute.

|||||||

Owens arrived at his father's compound just before twilight. When Owens saw his eighty-two-year-old mother and ninety-four-year-old father, he burst into tears.

Embracing him, Jessica said, "Don't cry, Mon. Why are you crying?"

"I'm sorry, Mama," he said. "I'm sorry that I could not save Ken. I thought that no matter what, I would be able to see that he wasn't killed. I just want you to know that I tried my best."

"What has happened has happened, and we are very happy you have come back alive."

Owens told his parents about his meeting with Obasanjo and the presidential order. Jessica told him she had visited Ken often in prison and described being beaten by Okuntimo's men and taken into detention simply for trying to attend her son's trial. Jim said that the police came looking for them after Ken was executed and that he and Jessica had had to live in the bush for two weeks.

"I should not be in Canada," said Owens. "I should be here taking care of you."

"No," said Jessica. "It is better you are not here. Every time you would go out, I would be asking everybody whether they've killed you, too. If anything happened to you, I don't know how I would live. I will not have any tears for you because all my tears have dried up. I have given them all to Ken."

That evening, Jessica cooked for her son. It was the first real Ogoni meal he had eaten in years. Owens could see the sadness in his mother's eyes. She had lost weight and some of her memory as well. She stared blankly, her cheek resting in her hand, and Owens knew that nothing could ever compensate her for Ken's loss.

Next morning, Owens had his father's driver take him to Mogho, Gokana, to meet with the families of the eight men executed with Ken. The families had gathered in the house of Michael Vizor,[16] an Ogoni who had been incarcerated with Ken but was acquitted by the tribunal. Like Owens, Vizor had ended up in Canada.

"I have been to Abuja and I have seen the president," Owens told them. "Obasanjo has given the permission for our loved ones to be brought back and buried in our villages."

The relatives of the Ogoni Nine were pleased, but Owens could sense that some of them clearly expected more. In a word: money. Rumours had been going around Ogoniland that the Wiwas had been given a substantial windfall in compensation for Ken's execution. Some said the Canadian government had given them millions of dollars. Some said it was the UN. It was hard for the poorer Ogonis to forget that for most of the past century the prosperous Wiwa family had been "above the crowd." To many Ogonis, it did not seem possible that the Wiwas had been financially wiped out or that Owens was not living the life of a millionaire in Toronto.

Just after this meeting with the families, Governor Peter Odili sent word to Owens that he wished to speak with him personally. At Government House in Port Harcourt, Owens greeted the governor respectfully and said, "Thank you for agreeing to see me. As you can see, I am wearing a MOSOP T-shirt and a MOSOP hat. There was a time when if you would wear this, you'd be arrested – not to talk of wearing it to the office of the governor of Rivers State! I wore these things purposely. It indicates that a lot has changed."

"I understand the emotions you and your family must be going through," said Odili. "I want to say that Ken was a great person, and he will be accorded a burial befitting a great man."

He told Owens that Ken's grave had been located, adding that Ken and his companions were in nine separate graves, and their bodies had not been defiled by acid. This was joyous news indeed.

"Come back tomorrow," said Odili, "and we will take you to the graves."

The following day, Owens was instructed to go to police headquarters on Moscow Road. The first person he saw when he entered the station was the superintendent of police, Hasan Abila, a former member of the Rivers State Internal Security Task Force and second in command to the sadistic Major Paul Okuntimo.

"You!" Owens shouted. "What are you doing here? How can you be here? You killed my brother and abused all these women and violated people and you are still here in this state?"

Owens then realized that most of the officers around him were the same men he used to plead with to release Ogoni prisoners, who tortured and extorted money from their captives, who raped Ogoni women and beat the elderly. The commissioner of police came out of his office to see who was making the fuss. Owens managed to calm down and apologize.

"I'm sorry I behaved like that," he said. "I did not realize these officers would still be in Rivers State."

The commissioner told him that the staff at the Central Police Station had not changed much since Abacha's reign. Hasan Abila, on the other hand, had been transferred to Lagos but had recently been brought back especially to locate the gravesite of the Ogoni Nine. As they were talking, Government House telephoned to say that the governor wanted Owens to come and speak with him immediately.

"But we were supposed to go and see the grave."

"Apparently, there has been a new development," said the commissioner.

Governor Odili had in fact changed his mind and decided to disclose the location of the graves only after Owens returned with the pathologists who would be doing the exhumation. Odili reasoned that if the location of the gravesite were revealed beforehand, the news would spread like wildfire and the site would be desecrated. Therefore, it was in everyone's best interest if the Ogoni Nine's resting place remained secret a while longer.

This seemed reasonable to Owens, though it would be hard to break the news to his father. Pa wanted to see Ken's grave that very afternoon.

|||||||

On June 26, 1999, a selected group of Ogoni leaders met at the Conference Hall in Bori. The meeting was intended to explain why MOSOP International had suspended Ledum Mitee as acting president and to give him a chance to respond.

The hall was packed and noisy, the tension thick. His Royal Highness Mene Gbarakoro sternly requested that anyone with weapons should leave immediately. (Ledum Mitee's supporters had arrived bearing arms.) No one budged. To avoid any unrest or violence, the chairman of council, Senator Cyrus Numieh, decided to move the meeting to the smaller, more private venue of his own offices. Even with the thugs removed, the atmosphere was charged with animosity.

The chairman opened by saying, "There is no way MOSOP at home can succeed without MOSOP abroad – and vice versa. We depend on assistance from international communities for the sustenance of the struggle."

Owens then stood and held up a magazine article with a large photograph of Ken Saro-Wiwa. He scanned the faces in the room as he brandished the portrait of his brother. "Someone is dodging from the picture," he said. "This is a sign of guilt in him."

Placing the magazine on the table, Owens continued: "I came back to Nigeria to achieve four things. One, I am here to secure the release of the remains of the Ogoni Nine. Two, I am here to see my aged parents, who have been weakened by this struggle. Three, I am here to meet with the elders to try to seek ways to resolve the crisis in MOSOP with a view to working as a team for good results. And last, I have come to see to it that the MOSOP office at 24 Aggrey Road was reopened for MOSOP activities."

On the issue of the crisis in MOSOP, Owens narrated what had transpired between Ken and Ledum in detention. He related how American diplomats had told Owens that Ledum alleged NYCOP had been responsible for the murders of the four chiefs at Giokoo and that this statement so tainted Ken Saro-Wiwa in the eyes of the Americans that President Clinton had opted not to press Abacha to stop the executions of the Ogoni Nine.[17]

"The image of MOSOP has gone into the dust as a result of Ledum Mitee's anti-MOSOP activities both at home and abroad," he said. "I call on all of you to ensure that MOSOP regains its vision and comes out of the crossroads."

Ledum Mitee was called on to defend himself. Rather than rebut the charges, he went on the attack. He accused Ken of refusing to hand over the money from the Right Livelihood Award to Ledum, who was acting as their lawyer during the military tribunal. He accused Ken of pocketing large sums from foreign film companies optioning projects based on the events in Ogoni. Ledum did admit that he and Ken had personal differences in jail and that their relationship had become "strained."

After all the allegations and counter-allegations, an Ogoni Methodist bishop present at the meeting said, "I am truly shocked by what we have heard. We never knew the fundamental nature of the issues, which are deeper than we thought them to be." He suggested that COTRA form a committee to examine the issues involved. When the floor was opened for nominations, squabbles turned to pandemonium and the meeting was halted.

With the exhumation of Ken's body now on hold and the sorry state of MOSOP unsolvable at the moment, there was little left for Owens to do in Nigeria. The next day, downcast, he returned to Canada.

|||||||

One of his first tasks on returning to North America was to contact Hafsat Abiola, daughter of the Nigerian opposition leader who had died mysteriously in 1998. Owens knew that Moshood Abiola's family had asked for an autopsy to be performed by a team of international pathologists, and that the team from Canada and Britain had concluded that Abiola had died of "natural causes as a result of his long-standing heart disease."[18] Owens called Hafsat to ask how she had arranged for the pathologists. She told him she had called the U.S. State Department and they had put her in touch with Physicians for Human Rights (PHR).

Owens had joined Physicians for Human Rights while at Johns Hopkins and was acquainted with some influential members. After contacting PHR, Owens soon received word that the group would agree to perform the exhumation and identification of the bodies in January 2000. With this news, the Wiwa family began putting plans in motion for a quiet family funeral for Ken Saro-Wiwa in Bane on the Monday after Easter, April 24, 2000. Ken's estranged wife, Maria, returned to Ogoniland to meet with the other eight families and keep them informed. Everyone was happy that the exhumation and proper burials would take place.

Physicians for Human Rights gave the Wiwas a list of things it needed from all the families of the Ogoni Nine to perform the DNA analyses. The group also requested that the exhumation be videotaped. Owens contacted British documentary filmmakers Glenn Ellis and his wife, Kay Bishop. They had produced the acclaimed television documentary *Delta Force* chronicling the rise and fall of Ken Saro-Wiwa. Ellis and Bishop agreed to accompany Owens to Nigeria, and it was a relief for Owens to know he'd have someone at his side. Owens's safety in Nigeria was still a huge concern. A white English couple with video cameras would, he felt, provide a measure of security.

Two days before he was to return to Nigeria, Owens received a call in Toronto from Physicians for Human Rights. Their question was simple: did the other eight families know that the exhumation was taking place?

"Of course," said Owens.

"We have information that this is not so," said the PHR representative.

"Who would tell you something like that?"

"We have been contacted by a Mr. Newsome. Do you have any evidence that the other families have been contacted?"

"Who is Mr. Newsome?" Owens asked.

"Do you have any proof that the government has given you the proper clearance to proceed?"

Owens explained that he had received an official declaration from President Obasanjo and would fax it to them right away. After receiving the fax, Physicians for Human Rights called back to say they had been notified a second time by Chris Newsome that the other eight families did not wish to proceed. The only thing PHR knew about this Mr. Newsome was that he was a New Zealander telephoning from London. Owens took Newsome's number and dialled him immediately.

"What is happening?" he asked Newsome. "Who gave you the right to call the Physicians for Human Rights about this issue? Do you have any of your relatives dead whom they are about to exhume? You're from New Zealand! Who gave you the right?"

"I was acting on behalf of Ledum Mitee," he said.

Newsome explained that he was a former employee of the Body Shop in New Zealand who was now working in Ledum Mitee's London office of MOSOP.[19] He offered Owens his apologies, saying he assumed he had been doing the right thing. After Owens threatened him with a lawsuit, Newsome wrote a formal note of apology to Physicians for Human Rights, saying if he had acted improperly he was doing so only as an employee of Ledum Mitee.

But the matter did not die. PHR contacted Owens again, saying it had been faxed a press release from the other eight families indicating that they did not wish PHR to come. PHR had also received a call from Bishop J.B. Poromon in Port Harcourt who claimed to be the head of the Ogoni Nine burial committee.

"Look," said Owens, "I don't know of any burial committee. You should check the signatures on that press release."

But the signatures were genuine. Ledum Mitee's younger brother Batom, the national coordinator of Ledum's faction of MOSOP, had gathered representatives from the eight families and persuaded them not to participate. He said the Wiwa family had been paid compensation by the federal government, and the rest of the families would not receive money if the burial continued. The families were now insisting that the government first pay them all compensation as recommended by the UN special report, and that they wanted a common burial in Bori for all of the nine, including Ken Saro-Wiwa.[20] The Mitees encouraged Bishop Poromon in Port Harcourt to phone Physicians for Human Rights to say that he was now the head of the official burial committee for the Ogoni Nine and the families would like the doctors to postpone their trip for six months.

PHR was in a bind. Just before Owens was to fly out of Toronto, the doctors informed him that they were sending only one pathologist to Nigeria to try to solve these conflicts. Until then, there would be no exhumations.

Exhausted, bitter, but full of resolve, Owens met Glenn Ellis and Kay Bishop at Heathrow and returned to the internecine battlefields of his homeland. Perhaps Ken's exhumation could still be achieved.

|||||||

In Lagos, as Owens's driver made his way through the choked city streets, a young newspaper peddler darting in and out of traffic rapped on the back window of the car. There was a large picture of Ken on the front page. Owens rolled down the window and bought an assortment of Nigerian dailies.[21]

Stories of the exhumations were splashed all over the headlines. "We Won't Identify Saro-Wiwa's Grave," read the headline of the *Nigerian Tribune*; "Saro-Wiwa Burial Blocked: No Honour For Him" read the *P.M. News*.[22] One newspaper gave a graphic description of Ken's face after he was hanged.

All the newspapers carried the story of the Methodist bishop in Rivers State who was trying to stop the exhumation, reporting that two pathologists from Physicians for Human Rights had already cancelled their trips after speaking with him. Fortunately, one pathologist, Bill Haglund, was still scheduled to arrive, and Owens's meeting with the governor was still on; perhaps all was not lost.

Owens and the film crew made their way to Port Harcourt and on to Ogoniland. When Owens broke the bad news to his parents, they were disconsolate and angry that a bishop they had never heard of was claiming to represent them.

"Who is he?" asked Jim. "If he is a member of a burial committee, he should have met me!"[23]

Interestingly, the bishop in question was Johnson Poromon, who had led the worship service on the eve of the first Ogoni Day march in 1993. Ken had been impressed because the bishop was from Gokana but delivered his sermon in flawless Khana.

"Papa, we have to go see this man," said Owens.

Next morning, Owens and his parents went to the bishop's residence in Port Harcourt. The meeting, also attended by several Ogoni academics and community leaders as well as ten women representing FOWA, was contentious yet restrained. Jim Wiwa asked, "Who gave you the authority to preside over the burial arrangement of my son without asking me?"

Bishop Poromon, a balding man with wide, heavy features, respectfully replied, "MOSOP appointed me head of the burial committee."

"Ledum Mitee, you mean," said Owens.

"And why have you not bothered to visit me or send words of condolence to me or to Ken's mother?" Jim demanded. "It is five years after the death of my son!"

"I have been ill," the bishop said. He also explained that during the past five years he had often been away – in Ghana, in Abuja – so he had no opportunity to come to Ogoniland. The bishop maintained that he was preventing the exhumations at the request of the members of Ledum Mitee's burial committee and the families of the eight activists executed with Ken. Because Ken was killed for the struggle, he said, his body rightly belonged to MOSOP and the Ogoni people, not to the Wiwa family.

The Wiwas found this idea appalling and lambasted him for claiming ownership of Ken's remains. In the end a meeting between the burial committee and the families was scheduled for that evening.

||||||

As Owens was leaving the bishop's residence, a little man with an elfin face and a receding hairline approached him. He introduced himself as an Ogoni, one of the gravediggers who had buried the Ogoni Nine.

The nine, he confirmed, had not been dumped in a common grave, as reported, and neither had anyone poured acid over the corpses. To Owens's astonishment, the little gravedigger described how he and his co-workers had been surrounded by policemen with automatic weapons as they worked, how the gravediggers carried the bodies packed in coffins from an open truck and placed them in nine separate graves, and how the lid on Ken's coffin had fallen off, allowing the gravedigger to see Ken's face. After the grave was filled in, the gravedigger surreptitiously placed an identifying mark on a nearby tree to identify Ken's plot.

"The other gravediggers will back my story," he said.

In the evening, Owens attended another meeting with the bishop and representatives of the other eight families. The relatives described how they had been persuaded to sign the document opposing burial and wanted to know the truth about the stories of UN compensation payments. Owens tried to convince them that no one in his family had received a cent in compensation, nor was any money to be forthcoming. At the end of the meeting, they agreed that the exhumation could proceed.

But all was still not well. Anthropologist Bill Haglund, PHR's International Forensic Program director, who was supposed to be helping Owens, had not yet arrived and no one seemed to know where he was. Haglund had extensive experience in the exhumation and examination of mass graves. Previously, he had exhumed bodies and testified before international tribunals in Rwanda and the former Yugoslavia. Owens hoped that he had not been put off by the bishop and his burial committee.

The morning brought good news. Haglund phoned to say he had arrived in Lagos and was catching the next flight for Port Harcourt. Owens went to the airport and awaited his arrival. Haglund, it turned out, was easy to find in the crowd. One of the few white faces, he looked like a typical American academic – fiftyish, with short brown hair and full beard gone grey, a necktie, and a brown corduroy jacket. On the drive back from the airport, Haglund listened as Owens explained the situation.

"It's encouraging to know that the bodies were at least buried in caskets," said Haglund. "Even if they are not still in good condition, it isolates and it makes the recovery of the skeletons more efficient."[24]

They discussed an article that had appeared that morning in the *Boston Globe* that stated that Ledum Mitee's faction of MOSOP wanted the body of Ken Saro-Wiwa as a money-generating tourist attraction in Bori.

Regarding Ledum's "burial committee," Haglund said that since the Ogoni Nine were executed as criminals, their bodies legally belonged to the state. Therefore PHR did not need permission from the families to exhume. He agreed with Owens, however, that it would be preferable if everyone were onside.

Haglund wasted no time scheduling a discussion with the bishop and the families of the Ogoni Nine to see if he could smooth the water and possibly get DNA samples. While he was away, Owens summoned the courage to visit Ken's house in Port Harcourt.

So far, Owens had avoided going back to 9 Rumuibekwe Road. The last time he had been at Ken's house was in the early morning hours of May 22, 1994, just after Ken's arrest. Now, as Owens walked through the iron gates into the front yard, raw emotion overcame him. He had lived in this house for eight years. It was filled with memories of late nights discussing politics and listening to Ken's monologues about Shell, injustice, his Ogoni people, and the future. Although the house was freshly painted, it was otherwise as it had been. The staff had kept the lawns and flowers well tended. In the garage, Ken's car was still covered with a tarp.

While Owens was looking around the house, a messenger arrived with good news: the governor wanted to meet with him on Monday. Hearing this, Owens decided to go to Bane to tell his parents and to see the burial site the family had chosen. Before his death, Ken had told his family that he wanted his plot be located in a certain clearing behind the village where he was born. It was a tranquil place, close to the rain forest, where the only sounds were of insects, birds, and children playing in a stream.

||||||

While Owens was in Bane, residents of Gokana came to tell him that an old Shell pipeline had ruptured near B-Dere, despoiling farmland that fed about a hundred families. When Owens went to check out the spill, he looked at the faces of the farmers and knew that none were likely to live to see their land reclaimed. Disease and poverty were their future now, with their oil-sodden land impairing their quality of life and their life expectancy.

Bill Haglund was also having a depressing time. When he had gone to visit one of the widows of the Ogoni Nine, a man introducing himself as the MOSOP chief of that village announced that there would be no exhumation unless the government paid compensation. He said the Nigerian government would also have to buy all the victims' families plots of land and build them each a house before Haglund could begin digging. When Haglund visited a family in Bomu, they also insisted that the UN must pay them before they would allow an exhumation. They even threatened to sue if Physicians for Human Rights touched the grave of their loved one. And so it went with every family Haglund visited that day. Though all had promised to cooperate only two days earlier, the eight families now refused to give DNA samples. Some were unwilling to meet with Haglund at all.

Owens decided to inform the Ogoni people about what was happening. At a town hall meeting in Bori, he explained the role of the bishop and the burial committee and Ledum Mitee's bid to stop the burial. Most of the Ogoni people seemed sympathetic, a response that left Owens feeling reborn by the end of the meeting. "This was the way it used to be," he said to Haglund. "Ken's vision is still alive."

Next day, Jim and Jessica Wiwa accompanied Owens to see Governor Peter Odili. When they arrived, the governor had changed his mind and refused to see them. The Wiwas were told they had to go back to the bishop and his committee to negotiate the exhumation. Owens decided to meet Poromon one final time in a last-ditch attempt to work together.

At Poromon's church, Owens and Haglund were escorted upstairs to an empty meeting room with yellow walls and blue doors and windows. The bishop entered wearing a scarlet cassock with a white band collar and a crucifix hanging from a golden chain. As Owens listened, bemused and disbelieving, the bishop said that his committee needed six months to get ready. Brandishing a large, official-looking book, Bishop Poromon explained, "We are the intermediary between the nine families and the . . ."

Owens cut him off: "You cannot represent – *cannot* possibly represent Ken Saro-Wiwa!" he exclaimed.

"Well," said the bishop, "that is your particular emphasis, which is not our emphasis."

Owens said that his elderly parents had been waiting five years to bury their son. The experts from the Physicians for Human Rights were volunteers whose time was precious. In six months, his parents could be dead and the PHR doctors unavailable. The president, the government of Nigeria, MOSOP, and the other eight families had known about the exhumations since last summer. No one raised an eyebrow or offered to help – and now Owens was being told that they all needed more time!

Owens called Gbenga Obasanjo. "It's at the point where only the governor will have to make a decision now and he is stalling," he told Gbenga. "He is not even obeying your father's order. Can you get through to your father? I understand he's in France."

Gbenga promised to contact his father in Europe. From France the next day, President Obasanjo made a statement to *The Nigerian Guardian* reaffirming his order for the Ogoni Nine to be exhumed and returned to their families.[25] In response, Governor Peter Odili's secretary reiterated that the exhumations would not be allowed until the governor received yet another directive from the president. Owens concluded that the Rivers State government had no intention ever of returning Ken's body.

En route to the airport to begin his journey home, Bill Haglund waxed philosophical. "I was set to bring my spade," he said. "The dead lie patiently while the living anguish over the supposed state of their spirit. But the dead are very patient – they'll wait."[26]

||||||

In Ogoni culture, death is not the end of life, but a phase in which the soul passes into the spirit world. For a soul to complete its journey, certain rites must be performed during the burial ceremony. Since Ken and his colleagues had not been given these rites, their souls were stuck between the two worlds.

Owens's training as a physician put him at odds with these beliefs. Still, he felt that these cultural rituals were important for his parents and for Ogoni society. Unfortunately, without the cooperation of the state government, there was not much he could do. For now, he could at least see where Ken and the others were buried.

After seeing Bill Haglund off at the airport, Owens instructed the gravedigger to take him to the cemetery. The little gravedigger led Owens along the dirt path through the municipal graveyard to the spot where he remembered burying Ken. At first glance, that particular section of Port Harcourt Cemetery did not much resemble a graveyard. With no apparent headstones, markers, or memorials, the area looked like nothing more than a patch of untamed rain forest in the city. Owens and the gravedigger walked along a dirt path with tall trees to one side and six-foot-high grass to the other. When they reached a certain spot, the gravedigger stopped and pointed at an iroko (African teak) tree about fifty metres from the path. It was the tree with the special mark. They had reached Ken's grave.[27]

Owens gazed at the forlorn patch of weeds and underbrush. So near his goal, yet so far away. It was as if he and his brother were separated not only by death but also by a perverse power whose victory was measured by the torture it inflicted on those who insisted on justice and human dignity. Still, Owens reflected, the Nigerian government and Ledum Mitee had not won entirely. Ken's burial was still scheduled for April 24. The funeral service was going to go ahead, with or without a body.

CHAPTER**FIFTEEN**

In December of 2001, a year and a half after Ken's symbolic funeral, Owens again raised the issue of exhumation during a Christmas trip to Bane. The main obstacle last time had been Rivers State governor Peter Odili. Knowing that an election would be held in 2002, Owens toyed with the idea of using his own influence in the state to ensure that Odili was not re-elected. As it turned out, political pressure became unnecessary when Odili had a change of heart. Inexplicably and unexpectedly, he invited Owens for a meeting at Government House on December 29, 2001.

Owens thanked the governor for giving him an audience and apologized for anything divisive he might have said in the past.

"You had reason to be angry," said Odili. "If it were my own brother, I would have behaved the same way. The truth is, I was misled." The governor explained that members of MOSOP's leadership and other interested individuals had misinformed him about the situation regarding the remains of the Ogoni Nine.

Owens nodded. "I have not talked to the other families this trip, but the issue is does anyone have the right to keep us from having Ken's body? The president's letter was very clear."

"It's true," said Odili. "Nobody should stop the other person from having the remains. The remains are there. Those who want their relatives, they should have them. Those who do not want them, they can return the bones to the grave. I'm going to instruct the attorney general to make this possible for you."

"May I have that in writing?"

The governor drafted a letter to the attorney general of Rivers State, Aleruchi R. Cookey-Gam, promising to give "moral, financial, and security support" for the exhumation of the Ogoni Nine. Six weeks after Owens returned to Canada, Cookey-Gam instructed the police and the State Security Service to cooperate with the exhumation process.

Owens contacted Bill Haglund at Physicians for Human Rights. Satisfied that it had government authorization to proceed, PHR made plans to re-commence the exhumation on June 15. David Sweet, a lecturer in the Department of Oral Medical and Surgical Sciences at the University of British Columbia, offered to perform DNA identification on eight of the recovered bodies for free. (The ninth – Ken's – would be financed by the Wiwa family.) For the first time in years, Owens had a glimmer of hope.

The politics of bones, however, remained rife with hidden hobgoblins. PHR's work required a three-person team, so Haglund invited forensic archaeologist Melissa Connor and forensic anthropologist Federico Costa to assist him. Haglund and Connor got visas from the Nigerian embassy in Washington, D.C. Costa, on the other hand, applied in Athens and was turned down.

When Owens phoned the Nigerian embassy in Greece, he was told that the visa officers there had been ordered by the permanent secretary of the Nigerian ministry of Foreign Affairs to block any attempt for scientists from the Physicians for Human Rights to come to Nigeria. "Your case is very sen-sitive," the official told Owens.

"But the president has given us an executive order. I can fax it to you. We also have a letter from the state attorney general instructing the police and security agents to give us all the necessary assistance."

"I can only act on what my superior says."

To further complicate matters, police officials in Port Harcourt stated they could not be ready for an exhumation on June 15. The police claimed they had no idea where the Ogoni Nine were buried and were contacting prison officials to identify the sites. Not surprisingly, the prison had not responded. Therefore, the police required another three weeks.

Haglund rearranged his schedule once again and set a new exhumation date. "I'll be free to hop into Port Harcourt about the fifteenth of July," he told Owens. "Melissa will only be able to stay for six days because she has a conference." As for Federico Costa, his participation remained in doubt.

Unlike the scientists from PHR, Owens was unable to postpone his trip. Arriving in Lagos on June 16 – now a month before the scheduled exhumations – he put his time in Nigeria to good use. In Port Harcourt, he sought out the man his contacts identified as the biggest obstacle to Ken's exhumation, the assistant commissioner of police, Bayo Agileye. He turned out to be a pleasant, welcoming fellow, apparently eager to help. "I'm still waiting for prison officials to reply to my letter," he said.

"The prison in Port Harcourt?" Owens asked. "I can walk there right now. Give me a copy of the letter you sent and I'll go and see the controller of prisons myself."

Picking up a thin file folder from his desk, Agileye said, "Let me tell you, I sympathize with your family. I am a Yoruba man. The Yoruba people suffered under Abacha. In fact, I think we suffered the most under Abacha. Most of our big men were sent in exile. Obasanjo was in prison. Abiola was in prison. Those of us in the force were demoted or taken to one distant place or another."

"Yes, Abacha was a terrible man."

"I am saying this so you will know I am telling you the truth," said Agileye. "There is no file on Ken Saro-Wiwa in this police station." He opened the folder. Inside were only two pieces of paper. "All I have here is the letter from the attorney general and the copy of the letter we sent to the prison. There is no document to show Ken was executed. No document to show he was ever arrested. There's nothing for me to work with."

Owens shook his head. As far as the Port Harcourt police were concerned, Ken Saro-Wiwa had never existed.

||||||

In fact, it turned out the prison had responded promptly but police dispatch had left the message sitting undelivered. Relief turned to disappointment when Owens read the prison's response. Ken, apparently, had never been their prisoner. On December 10, 1995, the prison merely opened the gallows for the army and security personnel who delivered the condemned. The prison was not involved in their hanging or the burials; thus, it had no idea where the bodies might have been disposed of.

"There's nothing more I can do," said Agileye. "In 1995, I was posted to Lagos. I thought the prison people would know where your brother is, but they say they don't know. There's nothing more I can do."

"But I know where the grave is," Owens said. "I know one of the gravediggers. I can ask him to bring gravediggers who were there that day. They can show you the place independently. I also know the police officer who supervised the burial: Hasan Abila. The last time I came here, in June 1999, he was in this police station. I even shouted at him! I used to treat him for free and he was the one who dealt with Ken's hanging and burial. I know he is around. I hear Mr. Abila is in Lagos now."

Agileye thought a moment. "Okay," he said.

|||||||

Until the modern era, Ogonis traditionally did not use cemeteries; they buried loved ones in their family compounds beside the house. Certain individuals, however, were not accorded this honour. If the cause of death were considered taboo (for example, by drowning, accident, or violence), the corpse was considered tainted and was disposed of far away, usually in an evil forest where people feared to go. Ogonis who died with unnatural bodily protrusions, such as hernias or goitres, were also shunned in death.

One of the biggest taboos was death by hanging. No wonder, then, that many of the families of the Ogoni Nine wanted no part of the exhumation. (The corpse of Ken Saro-Wiwa was an exception; it was not considered tainted because of his achievements.)

Modern, Christianized Ogonis had little trouble with the concept of welcoming the return of their martyrs, but the traditional beliefs of the animists needed to be respected. More than anything, however, the burials presented a financial hardship for all the families. Funerals in Ogoni culture are elaborate shows of social status. It is not unusual for Ogonis to keep a family member stored at a mortuary for a year as they struggle to collect enough money to hold an enormous ceremony and feast. Family fortunes have been wiped out on the death of a wealthy patriarch. Unlike the Wiwas, the families of the other eight men had no way of paying for the funerals that many felt these martyrs deserved. It did not help, of course, when Ledum Mitee's faction of MOSOP spread rumours that the Wiwas had received a windfall from the UN because of Ken's death. This was why several of the families insisted that they would not accept the remains until the UN (or the Wiwas) gave them money.

Politically, the problem of the exhumations lay with the families of the four chiefs killed at Giokoo, for they remained powerful and had access to sympathetic men in elected office. One of Governor Odili's commissioners was, in fact, the son of one of the slain chiefs, and Odili had given political appointments to several other Giokoo victims' relatives. In newspaper interviews, the children of the slain chiefs stated that since they had not recovered the remains of *their* fathers, it was only fair that the government should not cooperate with the exhumation of the Ogoni Nine.

The mere act of exhuming the remains of the Ogoni Nine threatened to have an enormous legal impact. As Bill Haglund pointed out, when a criminal is executed in Nigeria, the corpse becomes the property of the state. By giving the bones of the Ogoni Nine back to their families, the government would in essence be admitting that the trial and executions were in error, opening the way for lawsuits. The exhumation would also stir old resentments against those in government who'd been close to Abacha. It was also felt that the remains of Ken Saro-Wiwa would act as a sort of fetish object that would rekindle anti-Shell sentiment in the region.

Owens discussed all this with his father in Bane one Sunday morning. In his upstairs parlour, Pa Wiwa stared forlornly at a photograph of Ken. "I'm desolate," he said. "I am ninety-eight years old. This is the time my son should have taken care of me. All this was for him, and now" – he waved his cane in disgust at the view from his window – "the government of Nigeria is bad, Shell is bad. They will not give Ken back. You are going to fail again."

||||||

The little gravedigger who had sought out Owens in early 2000 was named Friday. As it turned out, he was also Owens's first cousin, born out of wedlock, and the son of Jessica Wiwa's elder brother. When Owens asked Friday if he knew the whereabouts of the other gravediggers on duty the day the Ogoni Nine were buried, he said that three were still employed at the cemetery, three had retired and returned to their villages, and three had died. Two others were not from Ogoni; Friday had no idea what had become of them.

Owens asked Friday to take him to speak to the three Ogonis who still worked at Port Harcourt Cemetery. Their names were Friday, Thursday, and Saturday, and they remembered the day of Ken's burial in detail. As they gathered around Owens, another man approached and introduced himself as a public health official. He was also on hand the day Ken Saro-Wiwa was buried.

Owens listened carefully, asking questions and seeking detail, as the five men told him the following story. At seven in the morning on November 10, 1995, a Port Harcourt local government engineer came to the municipal cemetery to ask Friday, the overseer, if he had graves available. Yes, said Friday, he had some graves already dug. How many were needed?

"Nine," said the engineer.

Surprised, Friday said, "Sir, what will you do with nine graves? The local government has never asked for nine graves before! Normally it's one grave or two to bury a pauper that died on the road or an unidentified body."

The engineer was so overcome with emotion that Friday could not understand his reply. Friday told the teary man that he had one exceptionally wide and deep grave dug for a white man who had just died, but he also had twelve smaller, shallower graves at the ready. The engineer told him to reserve nine of the ordinary graves and left the cemetery to visit the city council public health official.

"I need to buy nine coffins," the engineer said to the official.

"Nine coffins!" he exclaimed. "Who would they be for?"

The engineer became emotional again and said that Ken Saro-Wiwa and the others were going to be killed that morning. He would need nine coffins and one of the city's flatbed garbage trucks for transport to the cemetery.

When other employees at the health office heard that Ken was about to be executed, many closed their offices and went home. The few who stayed hung their heads in silence. The official suggested to the engineer that they go to Marine Base to purchase the coffins, and that they should buy the type of basic wooden boxes used for paupers. "And we'd better bring a tarpaulin to cover the coffins on the truck," he warned. "If the people on the street see nine coffins, they will know it is Ken and there is going to be trouble."

Both the public health official and the city engineer took one of the city council trucks with the tarpaulin along with them to Marine Base, bought the coffins, and asked the driver to take them to the prison to wait.

A few hours later, at the cemetery, the assembled gravediggers were astounded to witness the arrival of a contingent of police, army, and air force personnel escorting a municipal garbage wagon transporting a large load covered with a tarp. After ordering all other civilians out of the cemetery, the police instructed the workers to unload the nine wooden boxes and stack them under a nearby mango tree. The gravediggers were told to cut a path through the tall grass and shrubs to the open graves. With machetes, the gravediggers opened a trail in the underbrush from the mango tree to the graves, a secluded section of the cemetery where two tall termite mounds rose under the shade of an enormous iroko tree.

Around two in the afternoon, police sergeant Hasan Abila came to supervise the nine burials. As the gravediggers laid the coffins into the ground, Owens's cousin Friday stumbled and fell into one of the graves atop a coffin. The lid of the casket cracked from the force of his fall and shifted, revealing to the frightened Ogoni man the unmistakable face of Ken Saro-Wiwa. With Abila shouting angrily at him, Friday replaced the coffin lid as best he could and scrambled out of the grave.

By seven that evening, the graves were filled in and the mounds flattened and covered with grass. The Abacha regime was determined that the final resting place of Ken Saro-Wiwa be kept secret for all time. All that remained was the matter of what to do about the witnesses.

In the deepening twilight, the police trained their guns on the gravediggers and commanded them all to strip. Terrified, they complied. Standing naked, there were told to line up near the gaping hole of the large white man's grave. For reasons not entirely clear, the slaughter of the gravediggers did not take place. After an interminable wait, the men were forced into a police wagon and transported to a detention centre outside Port Harcourt. There they were held without food or visitors. After four days, the gravediggers were ordered back to their villages and told never to return to their jobs at the cemetery or to set foot in Port Harcourt again.

They were all forced to sign documents swearing that they would not reveal what had happened or disclose the locations of the graves. Port Harcourt Cemetery was then closed for six months. When rumours began circulating that the police had thrown the Ogoni Nine into a mass grave and doused the bodies in acid, the gravediggers said nothing, not even to their wives.

Owens listened to the tale of the gravediggers silently. It was strange how such a horrible tale could provide some small measure of relief.

|||||||

Two things still stood in the way of Ken's exhumation. Federico Costa was still being denied entry into Nigeria, and the whereabouts of Officer Hasan Abila were still a mystery. Abila had supervised the execution and burial of the Ogoni Nine and was the one person in authority who could say with certainty where the bones of Ken Saro-Wiwa could be found.

The last Owens had heard, Abila had been transferred from Port Harcourt to Lagos. Letam Wiwa, still with Army Intelligence, ran a check through official records and discovered that Abila was teaching at the police college in the remote northern city of Jos. If Owens could convince the police to summon him to Port Harcourt, Abila would perhaps be able to corroborate the gravediggers' tale.

By July 5, there was still no word from Nigerian visa officials concerning Costa, the police in Port Harcourt had not sent a message to Abila in Jos, and no official attempt had been made to notify the other eight families that the exhumation was soon to take place. With only ten days left before the arrival of the team from Physicians for Human Rights, Owens decided he would have to do the government's work for them.

He returned to the office of assistant police commissioner Bayo Agileye, who had taken no action regarding Abila since their meeting two weeks earlier. Owens himself drafted the official communiqué to summon Abila to Port Harcourt. He also composed for Agileye letters to be sent to the families of the Ogoni Nine, and letters to inform the city council of the upcoming exhumation. When Owens asked Agileye to distribute the letters to the families, Agileye declined, saying that he had no idea of the names of the families or their villages, and no budget to pay for the letters to be sent. Exasperated, Owens arranged for the delivery himself.

It looked as if Owens would have to deploy the same tactics to resolve the issue of Costa's visa. On July 8, he flew to Abuja, then made the four-hour drive from the capital to Jos. When he arrived at the police college with the official letter summoning Abila to Port Harcourt, he was told that Abila was in a meeting. Owens insisted that his case was urgent. The person on duty went to fetch Abila, who came into the hallway and glowered at Owens quizzically. "Abila, it's me. It's Owens."

"Ah!" gasped Abila. "I didn't recognize you!"

Owens had made up his mind to treat Abila differently than the last time; he apologized for screaming at him in the police station in 2000 and reminded them that they had once been friends. Abila said he understood how Owens must have felt that day. Then, taking Owens to a place where they could talk privately, he assured him that until the morning of the hangings he had no idea that Abacha was intent on killing Ken. He had assumed that Abacha would bow to international pressure and commute Ken's sentence. Abila also considered Ken a friend and was horrified to have to supervise his execution.

"Well," said Owens, "I didn't come to talk about that. I'm here to ask for your help. I need you to come to Port Harcourt on the twelfth to identify the grave of Ken. We want to exhume."

"I will come," said Abila. "If it's the last good thing that I'm able to do, then I will do it."

Owens returned to the capital city of Abuja to tackle the Ministry of Foreign Affairs. There he was met with unwavering resistance. The permanent secretary refused to assist in the matter of Costa's visa unless Owens received yet another directive from the president. It was a fool's errand; there was no way Owens would be able to see the president of Nigeria on such short notice. Besides, Obasanjo's existing order was clear and valid.

|||||||

Hasan Abila met Owens in the lobby of the Hotel Presidential. Owens had asked him to be in Port Harcourt on the day before the pathologists arrived.

"I want to know about Ken's last day," he said. "What happened on November tenth? I think I am ready to know."

"Are you sure you want to know everything?"

"Yes."

Abila gathered his thoughts. At 7 a.m. on November 10, 1995, he said, he'd got a call from Lieutenant Colonel Komo instructing him to take a detachment of mobile police to Port Harcourt Prison. When Abila asked why, the governor told him that Ken Saro-Wiwa and the eight men convicted with him were to be executed that morning. Abila had never believed the government would carry out the death sentences; now Abila found himself mandated to supervise the unthinkable.

He did what he was told to do. Knowing that the executions carried the potential for riots, he posted uniformed personnel at intervals along Aggrey Road. At 11:30, a Black Maria carrying Ken, John Kpuinen, Dr. Kiobel, and Beribor Bera came rumbling through the arched entrance of Port Harcourt Prison.[1] When the wagon came to a stop, Ken was the first to step out, his wrists in chains, his favourite curved-stem pipe clenched between his teeth. Catching sight of Abila, Ken gave a wry smile and said, "Well, Abila, I told you that it's going to end here. I've always believed that I'm going to spend some time in this prison. I have a feeling I'll be here until this government is done."

Abila stared wordlessly. No one had told Ken that this morning would be his last.

Kpuinen, Kiobel, and Bera followed Ken out of the Black Maria. All were put in leg cuffs and marched single file to the prison records office. On the walk there, Ken was bored and bemused, seeming to enjoy being away from his cell at Bori Camp. His expression changed, however, the minute he walked into the records office. A priest was waiting in the room.

A prison official read aloud the death sentence to each man individually. After hearing his warrant pronounced, Ken calmly made a brief statement that if he were to be killed for Ogoni, the struggle would continue to its logical end. The four were then taken to Cellblock C, the holding area for the condemned. Their five compatriots were already there, kneeling together in chains. At the sight of Ken, Nordu Eawo burst into tears. "So are we to die like this?" he sobbed.

"Don't cry," said Ken. "It is an honour to die for Ogoni. We will live forever in the minds of those we are leaving behind."[2] As he spoke there came the sounds of voices singing, "Glory, glory be to God, amen, alleluia" – the melody sung by the guards of Port Harcourt Prison before an impending execution.

Each of the Ogoni Nine was put in a separate holding cell before being summoned to the gallows. The gibbet had been set up inside a shack on one side of the prison courtyard. Approaching the fearful hut, Ken indicated to his captors that he wished to go first. Before he went to his death, however, he made a brief, ringing speech to the men assembled in the courtyard. "I have always been a man of good ideas," said Ken, "and whether I be killed, my ideas will live forever and Ogoni, for which I am dying, will one day be emancipated from the shackles of oppression."[3]

Placing his pipe back in his mouth, Ken allowed Abila and the guards to march him into the hut. Inside, the pipe was taken away and a hood placed over his head. The executioner manoeuvred Ken to the trap door, looped the noose about Ken's neck, and then activated the mechanism.

Nothing happened. The contraption did not activate and the floor did not give way. To give the executioners time to fix the problem, Ken was released from the noose and taken back outside.

The condemned men were astounded to see Ken Saro-Wiwa emerging alive from the execution hut. Whatever joy they might have felt was soon extinguished. The problem was quickly addressed.

Unexpectedly, John Kpuinen was then seized and trundled to the hangman. The apparatus worked as intended. After his body was dragged from the hut and discarded against a prison wall, Barinem Kiobel was taken next. Those outside heard the terrified man cry out, "God, I am innocent! God, I am innocent!" A few minutes later, his corpse was flung atop Kpuinen's.

The guards came for Ken once more and tried to frogmarch him back into the hut. Ken upbraided them for treating him disrespectfully; proudly and calmly he walked back into the hut. As the hood was placed over his head a second time, he called out: "Lord, take my soul, but the struggle continues!" and the executioner tripped the switch.

As if by some cruel joke, the gallows malfunctioned again. The floor gave way, but not in a swift, clean drop; Ken was left dangling but not dead. The executioners let him swing for as long as it took for him to stop thrashing about. Once he grew still, he was lowered to the ground and a doctor came to check his pulse.

He was still not dead.

Frustrated, the executioners roughly bundled him up, reset the contraption, and hanged him a third time.[*]

One by one, the remaining six men were dispatched. After the doctor certified that all were dead, inmates were conscripted to place the bodies into the nine coffins the engineer had at the ready, and to stack them onto the waiting flatbed garbage truck. The engineer covered the coffins with a tarp, and a military convoy escorted the truck to the cemetery.

Lieutenant Colonel Komo arranged for a cameraman to videotape the entire execution. This tape Komo hand delivered to Sani Abacha at the presidential palace in Abuja. The dictator is said to have cheered like a sports fan when he saw Ken drop through the trap door for the third time.

||||||

When Hasan Abila finished telling his story, Owens asked if Ken had given him anything before he died. Abila said no; he'd found it difficult even to look at Ken that day.

"Well, I thank you," said Owens, standing up. "I hope I will see you on the fifteenth at ten o'clock."

"That is Monday morning?"

"Yes, the morning after the pathologists from PHR arrive. We'll come to the police station."

Owens and the Physicians for Human Rights had by this time given up all hope of Federico Costa's assistance. Bill Haglund and Melissa Connor landed in Port Harcourt on Sunday, July 14, having run the usual gauntlet of graft, extortion, and corruption at Murtala Mohammed International Airport

---

[*] As stated, this is the version of Ken Saro-Wiwa's execution as recounted by Hasan Abila to Owens Wiwa. Some additional details were taken from Ken Wiwa, Jr.'s book *In the Shadow of a Saint*. Some witnesses to Ken's hanging have said that Ken was the first of the Ogoni Nine to be executed, while other witnesses have insisted that all the others died before him and that he was hanged ninth. Some of these witnesses claim that it took the executioners five attempts over the span of twenty minutes to kill Ken. Despite the differences between these remembrances and Abila's tale, all the accounts do agree that the execution of Ken Saro-Wiwa was botched, was attempted several times, and was cruel beyond imagination.

in Lagos. Haglund suggested they travel directly to Ogoniland to pay a call on Owens's parents and the families of the Ogoni Nine.

The visit with Jim and Jessica Wiwa was brief and cordial. When the doctors from PHR went to Gokana, however, representatives from all eight other families were gathered at the home of John Kpuinen's brother, a school principal, for the naming ceremony of a newborn child. The Kpuinen family compound was crowded with locals dancing, singing, drumming, and feasting. With difficulty, Owens and the Americans finally got the eight families sequestered for a private discussion. At once, the bickering began.

The chief complaint seemed to be coming from Blessing Eawo, the widow of Nordu, who was threatening the Wiwas and the government with lawsuits if they dared touch the grave of her late husband. Blessing still harboured intense resentment toward Ken Saro-Wiwa for starting the political movement that had swept her husband to his execution.

"Ken started MOSOP and my husband died for MOSOP!" she exclaimed.

"Blessing," said Owens, "there are times when I am also very sad. I was a doctor with two clinics and properties and money. Now I have to live in a foreign country where I have nothing. The thought sometimes occurs to me that it was Ken who made me come to be this destitute. But then I remember that Ken did not come to my house or to my hospital to tell me that I must follow him. I was the one who came to MOSOP when I saw that the things Ken was talking about were the things I had to fight for, the Ogoni environment and Ogoni survival. I am very, very sure that your late husband also made that decision on his own. The person who killed your husband was not Ken. It was Abacha."

Someone asked if Ledum Mitee had been informed of the exhumations.

"No, Ledum is not involved," answered Haglund. "We are here at the invitation of the government. The Wiwa family is paying our expenses and my friend David Sweet has agreed to do the DNA testing for free for you eight families." After more discussion, Owens agreed to help with the families' funeral expenses by offering to buy all the coffins for the eight men executed with Ken.

By the end of the meeting, six members of the eight families agreed to donate blood. Two families remained adamantly opposed.

||||||

When the day of the exhumations arrived, Owens, Bill, and Melissa gathered at the Port Harcourt Central Police station for their scheduled 10 a.m. rendezvous with Hasan Abila.

Abila failed to show up.

Unable to wait, Owens and the pathologists thought it best to continue to the cemetery. Perhaps Abila had got confused and gone straight to the graveyard.

He was not there, either. The four Ogoni gravediggers waiting in the cemetery said that the only people who had come by that morning had been the health supervisor for the local government and a special assistant to the chairman of the local government council.

Haglund and Connor spent the remainder of the morning listening to the gravediggers describe events in the graveyard on November 10, 1995. After the testimonies, the pathologists separated the men and asked each individually to show them where he remembered burying the Ogoni Nine. All four led them to the same location – under the large iroko tree.

Friday, the supervisor, took Owens aside. "We have to talk," he said. "You can see the whole place is overgrown and we have to cut a new path. We will have to get workers in but these workers have not been paid for the last nine months."

"Before we do anything, I want the policeman here to also point out the graves," said Owens.

"Look, we are the people who did the burial. We work here. There is no way he is not going to show you the place that we have shown you."

It was now two in the afternoon, and Bill Haglund and Melissa Connor were impatient to begin. Supervisor Friday added some stipulations before his workers would lift a finger. "This was a powerful man!" said Friday. "In order to exhume him, for him not to affect us, we will have to make some sacrifices. You will have to buy a goat and to buy some yam. You will need to buy some schnapps – that is the Dutch gin – and we will need money, two thousand naira."

Owens bristled. "Look, I am the brother of this man buried. If I leave here and go to the waterside, the Down Below, and I tell the Ogoni people that I want to exhume Ken and I want labour for those who are going to cut the grass and clear the place, I will get a hundred people to do it for free!"

Holding up his hands, Friday said, "That is why we are not wanting a big amount. The money we need is the minimum to protect us spiritually and culturally. Okay, since it is Ken, we won't need a goat. You should get four chickens and four yam and one bottle of schnapps. And instead of two thousand naira, eleven hundred."

Knowing that such negotiations in Nigeria can be stretched out for days, Owens grudgingly capitulated and handed over sixteen thousand naira (about US$125 at the time) to cover the wages and sacrificial groceries. As soon as the money changed hands, Friday's cemetery crew sprang to work, clearing a path through the tall grass and clear-cutting the land over the graves.

As the workmen prepared the site, Owens returned to the police station to see if he could learn what had happened to Hasan Abila. Oddly, Abila walked in the door at almost the exact same moment. It was three o'clock.

Owens took Abila to the cemetery and ordered the workmen to stop clearing the brush. It was time for Abila to tell the Physicians for Human Rights where the graves of the Ogoni Nine were.

Abila went straight to the mango tree where the coffins had been offloaded from the garbage wagon. He saw the path that the gravediggers had opened and said, "Yes, this is the path." Walking toward the iroko tree, Abila hesitated. Haglund cast a worried glance at Owens, but said nothing.

"There was a church opposite where these graves were," said Abila. "I remember because people came out to the balcony of the church, and we had sent police to go and arrest them when we asked them to leave and they didn't leave. Major Obi – that's the army major who was with us – he wanted to shoot at the people who were watching. But where is the church?"

Seven years had passed since the executions. There had been a brush fire in the cemetery that had caused the undergrowth to grow back thicker than ever. Abila fought his way through it and finally caught sight of a steeple and balcony beyond the cemetery perimeter wall.

"Yes, that is it," he said. "That's the church. And see? This anthill was here – and that one." Abila tapped his toe on the earth. "Okay, this is where the first grave was. This is where we buried Ken, because Ken is the first person we put here."

It was almost the same spot that the four gravediggers had pointed out. When the workers finished cutting the grass, at least ten coffin-sized depressions could be plainly seen.

"Why should the ground look like that?" asked Owens. "Shouldn't the earth be up in a mound?"

"No," said Haglund. "Over graves there's usually a depression. When a body decomposes and the coffin collapses, the ground goes down a bit."

Haglund placed a small wooden stake at the foot of each potential grave and affixed a small paper flag with the numbers one to nine. The four corners of each grave he delineated with more stakes. At grave 9, Haglund instructed the gravediggers to begin their dig carefully.

The gravediggers had said they'd prepared one exceptionally large grave for a white man, digging it six feet deep and six feet wide as per his family's instruction. All the other graves that day were of a normal, coffin-sized length and width, three feet in depth. When the gravediggers reached the two-foot six-inch mark on grave 9, Haglund noticed a change in the quality of the soil. If a basic wooden coffin had been laid in a three-foot-deep grave, one would expect the coffin lid to appear at approximately this depth. Haglund had the workers stop the digging and asked them to turn to the excavation of grave 1. According to the witnesses, this was the burial spot of Ken Saro-Wiwa.

By 6 p.m., only two feet of soil had been removed from grave 1. Night was falling and no one wanted to continue digging in a graveyard after dark. "Okay, let's go," said Haglund to the gravediggers. "We'll start again tomorrow."

IIIIIII

Owens did not return to the cemetery the following morning; he did not think he could bear to look on the bones of his brother. Busying himself with other matters, he stayed away until early afternoon. When Owens appeared at the cemetery gates to see how things were proceeding, Haglund and Connor were not happy; the gravediggers were in a state of despair.

"What's wrong?"

"Either we have not been shown the graves or we're in the wrong place," said Haglund. "In grave 1, so far I've counted nine bodies. It's a mass grave."

Had the inflammatory stories about the Ogoni Nine being dumped in a common pit and covered in acid been true? Had the government dug up the Ogoni Nine at a later date and dumped them all together? Had they tried to conceal Ken's grave more recently by placing the bodies of others on top of him?

The gravediggers insisted they had no idea who was in the mass grave. "I cannot remember when we buried so much people here," said Cousin Friday. "They only have mass graves for paupers, and then not so many at one time."

Thursday said, "Maybe this is that grave for the white man. I don't think we started from this place. Why don't we go the next one? I think the next one will be the first one for the Ogoni Nine. I don't remember this grave."

A public health official who was standing by shouted, "Oh! No, no! I remember! I remember this mass grave. These are the bodies of armed robbers and paupers that we put in here."

Bill Haglund shot the man an exasperated look. "Are you sure?"

The health official stammered, "I'm, I'm not very sure, but . . ."

If what he said were true, the day had been wasted. Haglund decided to return to the partially excavated grave 9 to see what it contained. The day before, the first two and a half feet of soil had been removed. Any buried body would be in the next few inches of earth.

To everyone's amazement, grave 9 was empty.

No one had any idea what to make of it. Urging the gravediggers on, Haglund asked them to open grave 8. It was empty as well.

It was beginning to look as if the bodies in the mass grave really might be the tangled remains of the Ogoni Nine. Still, Haglund thought it best not to jump to conclusions until the rest of the presumptive graves had been opened. With only an hour of daylight left, the gravediggers moved over to grave 7 and removed the first two and a half feet of soil.

"Well, this is something!" exclaimed Haglund.

Coming closer, Owens peered into the hole. In the fading light he could make out a pattern in the earth in the distinctive shape of a coffin lid.

|||||||

Next morning, Owens again stayed away, preferring to let the scientists perform their grisly task by themselves. In the afternoon, he found a very different reception waiting for him.

"We have found it!" cheered the gravediggers as they ran to greet him. "We found it! It is done!"

Haglund and Connor had exhumed the remains of an adult male from grave 7. Continuing through graves 6, 5, 4, and 3, the scientists had unearthed four more skeletons. All appeared to be male, and all had been buried in nearly identical simple wooden coffins.

Grave 2, however, presented a problem. The workers had found nothing in the first four feet. At five feet, nothing. After removing another six inches of soil, they finally came to a coffin lid that matched the ones in the other graves. There was also evidence that grave 2 had originally been wider than the graves beside it. It appeared that this was the extra-large grave for a white man.

Owens noticed that, unlike the jubilant gravediggers, Haglund did not seem delighted at all. "Bill, everyone's happy!" said Owens. "Why aren't you?"

"I still don't know if we're getting the right graves. I haven't done much physical examination yet, but in the grave Ken was supposedly buried in, grave 2, the femur we found looks too long for someone of Ken's height. These might not be the graves."

"What are you going to do?" asked Owens.

Haglund shrugged. "I suppose we'll continue," he said. "Melissa has to leave tomorrow, so we haven't a lot of time."

Going back to the grave sites, Haglund pulled up his flagged stakes and renumbered the plots. Since the grave formerly labelled 1 had been a mass grave of paupers and robbers, the large grave dug for the white man now became the new grave 1. The original nine graves identified at the beginning of the exhumations had yielded only six skeletons. They seemed to belong to the Ogoni Nine. But where were the three other graves?

"Remember that day?" said Cousin Friday to his fellows. "There were three we put over there." He pointed to a patch of weeds several yards away and a couple of the others nodded, saying, yes, they did remember burying three of the coffins a little removed from the rest.

Haglund said to Owens, "All these stories keep changing; I'm hearing different things from different people. I have no idea if these are the right set of graves. First they said the graves were all in a straight line. Then somebody remembers three graves being somewhere else. It's been seven years and memories have changed. I think they're making things up."

"Bill, you can't say that. There are areas of agreement here. The story about the anthills, the iroko tree, the church. I do not think these are all false."

Haglund drew a breath. "I'm a scientist. I don't go by what people say; I go by what I see."

Encouraged by Owens, Haglund flagged the three new suspected grave sites as numbers 7, 8, and 9, and had the gravediggers start digging.

New grave 7 did indeed contain another male skeleton that had been in a simple wooden coffin. Grave 8 was the same. It seemed that Cousin Friday had remembered correctly after all. With the opening of grave 9, however, the cemetery was once more thrown into turmoil.

The skeleton in 9 was wearing a brassiere. The gravediggers gave out a yelp and jumped back from the grave. A juju! Magic had changed the body of Ken Saro-Wiwa into that of a woman. Others began speculating that this might once have been Ken's grave, but the government had later disinterred him and placed someone else in there. Haglund simply sighed and said, "Well, I'm sure there were no transsexuals among the Ogoni Nine. This coffin is totally different, more ornate. It's obviously the wrong grave." He ordered the gravediggers to fill it back in and leave the unknown woman in peace.

|||||||

By the end of Thursday, only eight bodies had been discovered. For the time being, that would have to be enough. The important thing now was to determine, if possible, before going through with the expense of DNA testing, if these really might be eight of the Ogoni Nine.

Besides the bones and fragmentary coffin remains, personal items were recovered from the graves. Haglund was keeping his findings secret, but Owens noticed that he was asking questions about what Ken might have been wearing at the time of his death.

"Would he have worn a watch with a plastic strap?" he asked.

"No way," said Owens. "Ken either wears leather or chain."

Haglund asked again about a cheap plastic watch as well as Ken's belt and clothing. Since Owens had not seen his brother for more than a year before his death, he could be of little help. He did know who might be able to provide such information: Hauwa Madugu.

Owens placed a long-distance phone call to Hauwa in San Francisco. Ken's mistress had fled to the United States with baby Kwame after the Abacha regime subjected her to detention, physical abuse, and public humiliation following the executions. During the Ogoni Nine's final months in prison, she had travelled from Lagos to Port Harcourt in order to visit Ken almost daily. Owens asked her what Ken had been wearing when he was taken to the gallows.

"I didn't see Ken that day," she said. "I saw him the day before, so I don't know. All I can tell you is, unless they mutilated his body, there is no way that Ken's remains would not be with his ring."

Why had Owens not thought of it? Ken always wore a large ring on his left hand. In fact, he had worn the ring for so many years it had become impossible to remove.

As Melissa Connor was packing to leave for her conference, Owens stopped by her hotel room and asked, "Did you find any rings among the remains?"

"Yeah," she said. "There were a couple of rings. Why?"

"Ken's ring will be very distinctive."

"Well, I didn't do any of the final cleanings, so I really can't say. You should tell Bill about it, though."

After dropping off Connor at Port Harcourt Airport, Owens returned to the graveyard to find Haglund working to finish the exhumations of the eight bodies. When Haglund crawled out of one of the graves to take a break, Owens asked him, "Did you find a ring?"

"Yes, there were about three rings."

"Well, of these men, only two were likely to wear rings of any good nature – Ken and Dr. Kiobel. Ken's ring is very distinctive."

Haglund rummaged through his pockets and pulled out a pen and a scrap of paper. "Draw me a picture of the ring," he said.

A poor artist, Owens awkwardly sketched out a large ring with a square face surrounded by Xs representing small gems. He said he thought the stones were green and asked if he had come across anything like it.

"I don't want to tell you much right at the moment," said Haglund.

One thing he did let Owens know was that with every new piece of information, he was becoming increasingly confused. For example, Ken's driver, Sonny, had recently been by the graveyard and had remarked that he recognized Ken's shoes being lifted out of grave 4. But this was not the grave Cousin Friday had recalled falling into during the burials in 1995, the grave in which he saw the face of Ken Saro-Wiwa when the coffin lid cracked open. And neither of these graves had contained skeletons wearing rings. Still, there was something about Owens's description of Ken's ring that intrigued Haglund. He suggested that they speak to Ken's widow.

That afternoon, Owens phoned Maria Saro-Wiwa in London and told her a member of Physicians for Human Rights wanted to speak with her. Taking the phone, Haglund said, "Maria, I wanted to ask you about Ken's wedding ring. Could you describe it for me?"

As he listened, a look of surprise came over Haglund's face. He handed the phone back to Owens. Maria was in a rage, shouting into the phone. After listening to her rail for a few minutes, Owens explained to Haglund what was going on. "She's upset because you said wedding ring," he said. "Ken never wore a wedding ring."

Owens explained to Maria that Bill had misspoken. He had meant only to say "ring." Somewhat mollified, Maria calmed down and said that Ken had always worn a signet ring.

After hanging up, Owens asked Haglund what a signet ring was, and Haglund drew him a little sketch.

"I think this is Ken's ring!" said Owens.

"Well, this is different than the picture you drew for me, Owens. Now you're looking at my drawing and saying it's Ken's ring, too. It's been seven years since you've seen it. Just like the gravediggers, your story keeps changing."

Owens suggested that they ask Sonny, Ken's driver, to see what he remembered about Ken's jewellery. Sonny roughly described the same sort of ring that Owens and Maria had remembered Ken wearing. When Haglund asked what Ken had been wearing the day he died, Sonny didn't know. He was in Lagos at the time of the hangings. A person who might know, suggested Owens, was Ken's cook, Deebom. When telephoned, Deebom said: "White on white. White shirt and white trouser."

Lowering the receiver, Haglund said to Owens, "You see? Here's another one: he said white shirt and trousers. The body with the shoes wasn't the body with the ring, and the body with the ring wasn't wearing white pants."

"Ken never wore white on white," said Owens. "I never saw him wear such a thing, and I don't think he ever became that Christianized."

Peeved, Owens took the phone from Haglund and sternly asked Deebom in Khana, "What did you tell this white man?"

Flustered, Deebom explained that two days before the execution, Ken said he no longer wanted to wear a Nigerian prison uniform, and he ordered

Deebom to arrange for a tailor to sew an outfit for him out of white brocade. When he was presented with the loose-fitting shirt and trousers, Ken sent the trousers back because they were too long. Deebom took the trousers home, but Ken was executed before any alterations could be started.

When Owens explained this, Haglund suggested they go to Ken's house to take a first-hand look at the returned pants. If the fabric of the pants matched the fabric of a tunic taken from one of the graves, that would lead them closer to a positive identification.

But things weren't that simple. When Owens and Haglund showed up at Ken's house and asked Deebom to fetch the white brocade trousers, Deebom said they weren't there. "You see?" said Haglund. "The story keeps changing. He said he brought the trousers back to the house and now he says they're not here."

"Where are they?" Owens snapped at Deebom.

In Khana, Deebom explained that many of Ken's private effects had been lost during renovations of the house prior to his symbolic burial.

Now that Deebom mentioned it, Owens remembered that many of Ken's things had gone missing at that time. Haglund then asked if there might be any photographs of Ken that he could look at. Owens led him into the living room and asked Deebom to go upstairs and bring down some of Ken's photo albums.

As they were waiting for Deebom, Haglund noticed Ken's large portrait photograph hanging on the living-room wall. It was the same picture in the golden frame that the family had used in Ken's April 2000 symbolic funeral. Haglund stood astonished before the photo. There was Ken Saro-Wiwa in a grey pin-stripe suit, a white shirt, smart necktie, and pocket square. He was shown from the waist up in three-quarter view, seated, with his left forearm resting atop his lap. The photo clearly showed that on his left wrist he wore a gold-plated Longines watch. And on his left ring finger? A signet ring with a square face surrounded by small stones.

When Deebom arrived with the albums, Haglund flipped through them, looking for more pictures of Ken's hands. "Here's the ring again. Yes," he said, regarding a photo. "It's clearer in this one." His eyes then lit on another photograph and he stopped. In the picture, Ken was wearing a wristwatch with blue hands.

"The body we found wearing the white brocade shirt was the one we found wearing this kind of ring and watch," said Haglund. "Now we're getting somewhere."

As they left Ken's house, Haglund asked Owens to go to a local shop and purchase eight "Ghana-must-go" bags, the pidgin name for a type of square woven nylon satchel first used by Ghanaians fleeing Nigeria during a period of mass deportations. Cheap, lightweight, and rugged, these bags had since become the poor man's luggage of choice in Nigeria. Ghana-must-goes seemed the perfect solution for carting human remains around Port Harcourt without attracting attention.

At sunset, Owens and Haglund returned to the cemetery. While Owens waited at the gates, Haglund went back to the opened graves to pack the eight skeletons and their artefacts, placing the salvaged contents of each grave in its own carefully labelled Ghana-must-go. As twilight deepened, Haglund returned to the gates and told Owens that they would have to wait a little longer. "We have to take the bags out of the cemetery at night," he explained.

"All those bags with skulls and everything, where will you take them?"

"Back to my hotel room," he said. Seeing Owens's look of alarm, Haglund chuckled and said, "It's no problem. I like bones. They're my friends."

Over the next few days, Haglund turned his hotel room into a veritable charnel house, cleaning, cataloguing, and stacking all the bones he had pulled from the graveyard. On the dresser and the desk, on the chairs and on the television, rows of skulls and femurs, ribs, and vertebrae gleamed. For positive identification, the bones along with their associated rings, belts, shoes, and rotten fabric remnants were not enough. DNA evidence would still be needed.

To get blood samples from willing family members, Owens had drawn up a short consent form: "I agree to give a blood sample to be used for matching skeletal remains for DNA identification purposes." Haglund and Owens went to the Inadum Medical Centre in Bori where they collected sterile syringes, tourniquets, and cotton balls. A doctor from Inadum then followed the two out to the Wiwa compound in Bane. After Jessica Wiwa was read the consent form and appended her thumb print to the copy, Owens drew his mother's blood sample. The doctor from Inadum then drew Owens's blood.

While Owens went with his parents to pray in Pa's living room, Bill Haglund left with the doctor to go to Gokana to collect blood samples from the other families. Haglund knew that this part of the job would be tough.

The Kpuinens would be the most difficult of the Ogoni Nine families, certainly – especially if Ledum Mitee had continued to lobby them not to cooperate. To the doctors' surprise, when they showed up at the Kpuinen compound, John Kpuinen's brother sighed and said, "Look, I'm tired of all this. MOSOP or no MOSOP, I want the remains of my brother back."

John Kpuinen's mother came forward. "And what has MOSOP done for me? I want the remains of my son, too."

Owens arrived just as the Inadum doctor was taking blood from Kpuinen's mother. The happy look on her face was the same as the one Jessica Wiwa wore when she gave her blood to identify her son.

||||||

In all, Physicians for Human Rights had retrieved the remains of eight of the nine slain activists, and gathered DNA samples from seven of the families. It seemed certain the body of Ken Saro-Wiwa had been reclaimed. The identity of the missing remains was still a mystery, but with luck it might turn out to be one of the men whose families did not provide a DNA sample.

After Haglund and Owens finished giving interviews to the BBC and the Associated Press at a hotel in Lagos, Haglund took Owens aside, sat with him in a quiet corner, and handed him a large envelope.

"What is this?"

"It's for you," said Haglund. "Open it if you want."

Curious, Owens lifted the flap of the envelope. Inside were the rotted remains of a man's belt. It did not seem familiar. But then, Ken usually wore loose, untucked African shirts so one rarely saw his belt. Owens then withdrew from the envelope a signet ring. Time had indeed played tricks with his memory. The small stones were not green, they were diamonds. The face of the ring was a little different than he remembered, but it was Ken's ring, no doubt about it.

The last item in the envelope was a wristwatch.

Holding it to the light, Owens examined it, puzzled. It was a cheap watch with blue hands and a plastic strap, unlike anything a person as fastidious and well-to-do as Ken would normally wear.

Inspecting it closely, Owens saw on the face of the watch a distinctive logo and a slogan:

"UN Human Rights 1993."

NOTES

Prologue

1. Absolute poverty rate comparisons were adjusted to reflect purchasing power in 1985 U.S. dollars.
2. Virginia Postrel, "Widening Rich-Poor Gap Is Based on Misleading Data," *International Herald Tribune*, August 16, 2002, p. 12.
3. http://www.ratical.org/corporations/OgoniFactS.html.
4. David Wheeler, et al., "Paradoxes and Dilemmas for Stakeholder Responsive Firms in the Extractive Sector: Lessons from the Case of Shell and the Ogoni," *ABI/INFORM Kluwer Academic Publishers Group Journal of Business Ethics* (September, 2002) Section: 39, no. 3, pp. 297–318.
5. Shell Nigeria, "Environment of the Niger Delta: Devastation?" http://www.shellnigeria.com.
6. John Vidal, " 'Let People See Our Plight' – Ogoni," *Africa Today*, September/October 1995, pp. 18–19.

Chapter One

1. "Good morning." "Hi." Jim Hudgens and Richard Trillo, *West Africa: The Rough Guide* (London: Rough Guides Ltd., 1999), 928.
2. Ken Saro-Wiwa, *The Ogoni Nation Today and Tomorrow* (Port Harcourt, Nigeria: Saros International Publishers, 1993), 8.
3. http://purpleplanetmedia.com/bhp/pages/jaja.shtml.

4. To support this, the Ogonis point to the *Polyglotta Africana*. This major early work on African languages by the nineteenth-century missionary-teacher Sigismund W. Koelle contains a list of some 300 words and phrases in 156 different African dialects. Koelle compiled this information after interviewing thousands of freed slaves living in the British West African protectorate of Sierra Leone. Koelle's book contains nothing that can be traced to an Ogoni language root. Had there been European contact with the Ogoni, Koelle would have included Ogoni language references in the *Polyglotta*. Joseph McLaren, "A Political Assessment: Genocide in Nigeria: The Ogoni Tragedy," in *Ken Saro-Wiwa: Writer and Political Activist*, Ed. Craig W. McLuckie and Aubrey McPhail (Boulder, Colo.: Lynne Reinner Publishers, 2000), 19.

5. Ken Saro-Wiwa, *Genocide in Nigeria: The Ogoni Tragedy* (Port Harcourt, Nigeria: Saros International Publishers, 1992), 11.

6. Saro-Wiwa, *Genocide in Nigeria*, 12–13.

7. Ken Wiwa, *In the Shadow of a Saint* (Toronto: Knopf Canada, 2000), 74.

8. The Ogonis have traditionally used the rotation of farm crops to calculate the passage of time and the ages of their children. Each year, a farmer would typically plant only one of his several fields, leaving the others fallow. If one were to ask a mother how old her son was, she would say something like "This one was born when we entered that farm." If her family had four fields and they had worked the particular field she was indicating three times since she gave birth, her child would be about twelve years old.

9. Saro-Wiwa, *Genocide in Nigeria*, 16.

10. Technically, Ken's birth name was Kenule Beeson Nwiwa. According to Ken Wiwa, Jr., in his book *In the Shadow of a Saint* (Toronto: Knopf Canada, 2000), as a young man Ken convinced his family to drop the N from Nwiwa to avoid them being mistaken for Igbos, whose last names commonly begin with Nw. On entering university in the 1960s, he began using the last name "Tsaro-Wiwa" but in 1982 simplified the spelling to "Saro-Wiwa" after tiring of the constant misspellings and mispronunciations. According to Owens, Ken had three nicknames. His mother called Ken "Inadum"; his father called him "Gbole"; other family intimates called him "Dede."

11. Eckhard Breitinger, "Ken Saro-Wiwa: Writer and Cultural Manager," in *Ogoni's Agonies*, ed. Abdul Rasheed Na'Allah (Trenton, N.J.: Africa World Press, 1998), 245.

12. Jackson Gaius-Obaseki, "Investment and Trade Opportunities: A Sectoral Approach – Chemical and Petrochemical" (September 2000), transcript of speech found at http://www.nigerianembassy.nl/Obaseki.htm.

13. O. Okechukwu Ibeanu, "Insurgent Civil Society and Democracy in Nigeria: Ogoni Encounters With the State, 1990–1998," (http://www.ids.ac.uk/ids/civsoc/final/nigeria/Nga8.doc).

14. Saro-Wiwa, *Genocide in Nigeria*, 26.

15. http://www.nigeriangalleria.com/portrait/bios/zik.htm.

16. "Nigeria's Political History," Nigeria@Hyperia, http://www.hyperia.com/politics.htm.

17. "Nigeria's Political History."

18. Saro-Wiwa, *Genocide in Nigeria*, 20–27.

19. Ken Saro-Wiwa, *On a Darkling Plain* (Port Harcourt, Nigeria: Saros International, 1989), 118.

20. Ken Saro-Wiwa, "Complete Statement by Ken Saro-Wiwa to Ogoni Civil Disturbances Tribunal," www.ratical.org/corporations/KSWstmt.pdf.

21. Laura Neame, "Chronology of Ken Saro-Wiwa's Life," in *Ken Saro-Wiwa*, McLuckie and McPhail, 234.

22. Saro-Wiwa, *On a Darkling Plain*, 55.

23. Saro-Wiwa, *On a Darkling Plain*, 55.

24. Saro-Wiwa, *On a Darkling Plain*, 114.

25. Saro-Wiwa, *On a Darkling Plain*, 112.

26. Saro-Wiwa, *On a Darkling Plain*, 111.

27. Ken Saro-Wiwa, *Songs in a Time of War* (Port Harcourt, Nigeria: Saros International, 1985), 13.

28. Saro-Wiwa, *On a Darkling Plain*, 113–115.

29. Ken Wiwa, *In the Shadow of a Saint* (Toronto: Knopf Canada, 2000), 39.

30. Saro-Wiwa, *On a Darkling Plain*, 187.

31. Neame, "Chronology of Ken Saro-Wiwa's Life," 234.

32. Saro-Wiwa, *On a Darkling Plain*, 146.

33. Saro-Wiwa, *On a Darkling Plain*, 149–150.

34. http://www.biafraland.com/Deception.rtf and http://www.biafraland.com/Part%201.htm.

35. Saro-Wiwa, *On a Darkling Plain*, 199.

36. Saro-Wiwa, *On a Darkling Plain*, 173.

37. Saro-Wiwa, *On a Darkling Plain*, 174.

38. Saro-Wiwa, *On a Darkling Plain*, 174.

39. Saro-Wiwa, *On a Darkling Plain*, 188.

40. Saro-Wiwa, *The Ogoni Nation Today and Tomorrow*.

41. Ken Saro-Wiwa, *A Month and a Day: A Detention Diary* (London: Penguin Books, 1995), 51.

42. Saro-Wiwa, *The Ogoni Nation Today and Tomorrow*, 21.

43. Neame, "Chronology of Ken Saro-Wiwa's Life," 234.

44. Saro-Wiwa, *A Month and a Day*, 50.

45. Saro-Wiwa, *On a Darkling Plain*, 199.

Chapter Two

1. Petition from Chief T.N. Adda-Kobani, et al., in Ken Saro-Wiwa, *Genocide in Nigeria: The Ogoni Tragedy* (Port Harcourt, Nigeria: Saros International, 1992), 45–46.

2. Letter to the Military Governor of Rivers State by J. Spinks, June 1970, in Saro-Wiwa, *Genocide in Nigeria*, 50–51.

3. Ken Saro-Wiwa, *On a Darkling Plain* (Port Harcourt, Nigeria: Saros International, 1989), 241.

4. Petition from Chief T.N. Adda-Kobani, et al., 47.

5. Karl Maier, *This House Has Fallen* (New York: PublicAffairs, 2000), 83.

6. Ken Saro-Wiwa, *Similia: Essays on Anomic Nigeria* (Port Harcourt, Nigeria: Saros International, 1991), 47.

7. Andrew Rowell and Stephen Kretzmann, *All for Shell: The Ogoni Struggle – A Project Underground Report*, http://www.moles.org/ProjectUnderground/motherlode/shell/timeline.html (Yale Law School, 1996).

8. Bronwen Manby, *The Price of Oil: Corporate Responsibility and Human Rights Violations in Nigeria's Oil Producing Communities* (New York: Human Rights Watch, 1999), 70.

9. Manby, *The Price of Oil*, 69.

10. Manby, *The Price of Oil*, 69–70.

11. World Bank, "Defining an Environmental Development Strategy for the Niger Delta, 1995," http://www.earthlife.org.za/factsheets/fs-nigeria.htm.

12. Ken Saro-Wiwa, statement to the CDST, 1995, http://www.earthlife.org.za/factsheets/fs-nigeria.htm.

13. Professor Claude Ake, "Tell," January 29, 1996, http://www.earthlife.org.za/factsheets/fs-nigeria.htm.

14. Godson Ukpevo, "Agony of the Ogoni," *Newswatch*, January 25, 1993, p. 10.

15. Saro-Wiwa, *Genocide in Nigeria*, 17.

16. Ken Saro-Wiwa, *A Month and a Day: A Detention Diary* (London: Penguin Books, 1995) 24.

17. Saro-Wiwa, *A Month and a Day*, 25–26.

18. Saro-Wiwa, *A Month and a Day*, 51.

19. Saro-Wiwa, *A Month and a Day*, 52.

20. Olesegun Obasanjo biography. Federal Republic of Nigeria Web site, http://www.nopa.net/president_obasanjo/.

21. Ken Wiwa, *In the Shadow of a Saint* (Toronto: Knopf Canada, 2000), 41.
22. Eghosa E. Osaghae, *Crippled Giant: Nigeria Since Independence* (Bloomington, Ind.: Indiana University Press, 1998), 77–78.
23. Osaghae, *Crippled Giant*, 81.
24. Osaghae, *Crippled Giant*, 81.
25. Osaghae, *Crippled Giant*, 83.
26. Osaghae, *Crippled Giant*, 88–89.
27. Maier, *This House Has Fallen*, 57.
28. Osaghae, *Crippled Giant*, 89.
29. Wiwa, *In the Shadow of a Saint*, 56.
30. Osaghae, *Crippled Giant*, 95.
31. Wiwa, *In the Shadow of a Saint*, 74.
32. William Boyd, "Death of a Writer," *The New Yorker*, November 27, 1995, pp. 51–55.
33. *Ogoni Review*, June 1993, Saros Publishing.
34. Eckhard Breitinger, "Ken Saro-Wiwa: Writer and Cultural Manager," in *Ogoni's Agonies*, ed. Abdul Rasheed Na'Allah (Trenton, N.J.: Africa World Press 1998), 247.
35. Osaghae, *Crippled Giant*, 149.
36. Maier, *This House Has Fallen*, 61.
37. Maier, *This House Has Fallen*, 43.

Chapter Three

1. Eckhard Breitinger, "Ken Saro-Wiwa: Writer and Cultural Manager," in *Ogoni's Agonies*, ed. Abdul Rasheed Na'Allah (Trenton, N.J.: Africa World Press, 1998), 246.
2. Ken Saro-Wiwa, *A Month and a Day: A Detention Diary* (London: Penguin Books, 1995), 65.
3. Saro-Wiwa, *A Month and a Day*, 65.
4. "The Ogoni Bill of Rights," in Saro-Wiwa, *A Month and a Day*, 67–70.
5. Saro-Wiwa, *A Month and a Day*, 78.
6. Saro-Wiwa, *A Month and a Day*, 102.
7. http://archive.greenpeace.org/~comms/ken/refer.html.
8. Ken Saro-Wiwa, *Second Letter to Ogoni Youth* (Port Harcourt, Nigeria: Saros International, 1993), 16–17.
9. Robert Evans, "Southern Minorities Plan Strategies for Third Republic," *Guardian* (Lagos), March 26, 1992, p. 1. Quoted in Clifford Bob, "Global Linkages and the Emergence of Indigenous Collective Action in Nigeria's Niger River Delta," unpublished research paper, 14.
10. Bob, "Global Linkages and the Emergence of Indigenous Collective Action," 14.

11. Human Rights Watch/Africa, *The Ogoni Crisis: A Case Study of Military Repression in Southeastern Nigeria* (New York: Human Rights Watch, 1995), 33.

12. Ike Okonta and Oronto Douglas, *Where Vultures Feast: Shell, Human Rights and Oil in the Niger Delta* (San Francisco: Sierra Club Books, 2001), 142.

13. Eghosa E. Osaghae, *Crippled Giant: Nigeria since Independence* (Bloomington, Ind.: Indiana University Press, 1998), 245.

14. Godson Ukpevo, "Agony of the Ogoni," *Newswatch*, January 25, 1993, p. 15.

15. Andrew Rowell and Stephen Kretzmann, "All for Shell: The Ogoni Struggle – A Project Underground Report," http://www.moles.org/ProjectUnderground/motherlode/shell/timeline.html (Yale Law School, 1996).

16. Saro-Wiwa, *A Month and a Day*, 102.

17. Saro-Wiwa, *A Month and a Day*, 102.

18. http://www.ciesin.org/docs/010-000a/Year_Worlds_Indig.html.

19. Saro-Wiwa, *A Month and a Day*, 104–105.

20. Saro-Wiwa, *A Month and a Day*, 104–105.

21. Saro-Wiwa, *A Month and a Day*, 106.

22. Saro-Wiwa, *A Month and a Day*, 107.

23. Saro-Wiwa, *A Month and a Day*, 115.

24. Saro-Wiwa, *A Month and a Day*, 109.

Chapter Four

1. Ken Saro-Wiwa, *Second Letter to Ogoni Youth* (Port Harcourt, Nigeria: Saros International, 1993).

2. Ken Saro-Wiwa, *A Month and a Day: A Detention Diary* (London: Penguin Books, 1995), 122.

3. Saro-Wiwa, *A Month and a Day*, 122.

4. Ledum Mitee read the entire Chapter 5 from the book of Lamentations, beginning, "Remember, O Lord, what has befallen us; behold, and see our disgrace!" per Saro-Wiwa, *A Month and a Day*, 120.

5. Godson Ukpevo, "Agony of the Ogoni," *Newswatch*, January 25, 1993, p. 10.

6. Cris McGreal, "Ken Saro-Wiwa: Not Entirely Innocent?" *Weekly Mail and Guardian* (Johannesburg, South Africa), April 4, 1996, http://www.sn.apc.org/wmail/issues/960404/NEWS25.html. Note: McGreal mistakenly says Kobani made this speech on Ogoni Day, January 4, 1993. Saro-Wiwa in *A Month and a Day*, p. 121, says it was made at Birabi's grave after the church service on January 3, 1993. Note also that most of McGreal's article is based on government propaganda and unreliable sources.

7. Saro-Wiwa, *A Month and a Day*, 125.

8. Saro-Wiwa, *A Month and a Day*, 127.

9. Slogans from Godson Ukpevo, "Agony of the Ogoni," *Newswatch*, January 25, 1993, p. 9 and *Delta Force*. A Catma Films production for Channel Four Television Corp., 1995.

10. Saro-Wiwa, *A Month and a Day*, 130–132.

11. *Delta Force*.

12. Minutes of meeting of the Expanded MOSOP Executive Committee held at the permanent domicile of Chief Edward Kobani, in Bodo on January 5, 1993.

13. Saro-Wiwa, *A Month and a Day*, 137–138.

14. Saro-Wiwa, *A Month and a Day*, 143–144.

15. *Delta Force*.

16. Andrew Rowell and Stephen Kretzmann, "All for Shell: The Ogoni Struggle – A Project Underground Report," http://www.moles.org/ProjectUnderground/motherlode/shell/timeline.html (Yale Law School, 1996).

17. N.A. Achebe, Memo "Meeting at Central Offices on Community Relations and Environment (16/17th February in London, 18th February in The Hague)." To: MD, DMD, GMC, GMH, GME, GMW. From: GMB. Dated "22nd February, 1993."

18. *Delta Force*.

19. Draft minutes, February 15/16, 1993, in London, February 18, 1993, in The Hague, reprinted in Rowell and Kretzmann, "All for Shell."

20. Saro-Wiwa, *A Month and a Day*, 145.

21. Saro-Wiwa, *A Month and a Day*, 146.

22. The use of the ONOSUF as an accurate census of the Ogoni people was somewhat unreliable. Some people donated more than one naira, some people donated one naira multiple times under different names, and some donated for Ogonis yet to be born. At the Ken-Khana ONOSUF rally, Owens, for example, donated five naira personally and ten thousand naira in honour of his clinic and staff.

23. Saro-Wiwa, *A Month and a Day*, 150.

24. Saro-Wiwa, *A Month and a Day*, 152.

25. Laura Neame, "Chronology of Ken Saro-Wiwa's Life," in *Ken Saro-Wiwa: Writer and Political Activist*, ed. Craig W. McLuckie and Aubrey McPhail (Boulder, Colo.: Lynne Rienner Publishers, 2000), 235. Note: Neame says Tedum died in 1992, but Ken Wiwa, Jr., explicitly states in *Shadow of a Saint* that Tedum died on March 15, 1993.

26. Ken Wiwa, *In the Shadow of a Saint* (Toronto: Knopf Canada, 2000), 94.

27. Saro-Wiwa, *A Month and a Day*, 153.

28. Saro-Wiwa, *A Month and a Day*, 173.

29. Richard Boele, *Ogoni – Report of the UNPO Mission to Investigate the Situation of the Ogoni of Nigeria, 17–26 February, 1995* (UNPO, May 1, 1995), 49. Reprinted in Rowell and Kretzmann, "All for Shell."

30. Boele, *Ogoni – Report of the UNPO Mission to Investigate the Situation of the Ogoni of Nigeria*, 23, in Rowell and Kretzmann, "All for Shell."

31. Saro-Wiwa, *A Month and a Day*, 156.

32. Muyiwa Adekeye, "The Shooting Field: Soldiers Shoot Eleven Ogoni Marchers Protesting Ecological Ravage," *The News* (Nigeria), May 17, 1993, p. 18.

33. Boele, *Ogoni – Report of the UNPO Mission to Investigate the Situation of the Ogoni of Nigeria*, 49, in Rowell and Kretzmann, "All for Shell."

34. Human Rights Watch/Africa, *The Ogoni Crisis: A Case-Study of Military Repression in Southeastern Nigeria* 7, no. 5 (July 1995): 10.

35. Letter from J.K. Tillery, Divisional Manager, Willbros West Africa, Inc., to J.R. Udofia, SPCD General Manager (East), May 3, 1993. This account was contained in a document, "Review of events leading to the withdrawal of workforce from the Bomu area," which was attached to the letter. Reprinted in Human Rights Watch/Africa, *The Ogoni Crisis*, p. 11.

36. Saro-Wiwa, *A Month and a Day*, 157.

37. Boele, *Ogoni – Report of the UNPO Mission to Investigate the Situation of the Ogoni of Nigeria*, 49.

38. American University, Washington DC, Trade and Environment Database, Case number 149. "TED Case Studies: Ogoni and Oil." http://www.american.edu/TED/OGONI.HTM.

39. *Delta Force*.

40. Amnesty International, "Possible Extrajudicial Execution/Legal Concern," in Rowell and Kretzmann, "All for Shell."

41. J. Udofia, "Disruption of Work on the 36" Rumuekpe-Bomu Trunkline," in Rowell and Kretzmann, "All for Shell."

42. Rowell and Kretzmann, "All for Shell."

43. Saro-Wiwa, *A Month and a Day*, 161.

44. Photocopy of original document titled "The Activities of the Movement for Survival of the Ogoni People (MOSOP)," presented to Governor Rufus Ada George, signed by I.S. Kogbara, Chief S.N. Orage, Chief Hon. Kemte Giadom, Chief J.K. Kiponi, HRH W.Z.P. Nziidee, HRH J.P. Bagia, HRH M.S.H. Eguru, HRH M.T. Igbara, HRH G.N.K. Giniwa, Dr. N.A. Ndegwe, Elder Lekue Lah Loolo (not dated, but in reference to events of "4th May 1993").

45. Saro-Wiwa, *A Month and a Day*, 142.

46. Saro-Wiwa, *A Month and a Day*, 171.

47. Ike Okonta Oronto Douglas, *Where Vultures Feast: Shell, Human Rights and Oil in the Niger Delta* (San Francisco: Sierra Club Books, 2001), 121.

48. Boele, *Ogoni – Report of the UNPO Mission to Investigate the Situation of the Ogoni of Nigeria*, 49, in Rowell and Kretzmann, "All for Shell."

49. Saro-Wiwa, *A Month and a Day*, 174.

50. Saro-Wiwa, *A Month and a Day*, 174.

51. Okonta and Douglas, *Where Vultures Feast*, 121.

52. Saro-Wiwa, *A Month and a Day*, 175.

53. Boele, *Ogoni – Report of the UNPO Mission to Investigate the Situation of the Ogoni of Nigeria*, 49, in Rowell and Kretzmann, "All for Shell."

54. Saro-Wiwa, *A Month and a Day*, 175–176.

Chapter Five

1. Richard Boele, *Ogoni – Report of the UNPO Mission to Investigate the Situation of the Ogoni of Nigeria, 17–26 February, 1995* (UNPO, May 1, 1995), 50, in Andrew Rowell and Stephen Kretzmann, "All for Shell: The Ogoni Struggle – A Project Underground Report," http://www.moles.org/ProjectUnderground/motherlode/shell/timeline.html (Yale Law School, 1996).

2. Ken Saro-Wiwa, *A Month and a Day: A Detention Diary* (London: Penguin Books, 1995), 178.

3. Saro-Wiwa, *A Month and a Day*, 181.

4. Saro-Wiwa, *A Month and a Day*, 181.

5. Distance from Port Harcourt to Lagos from Saro-Wiwa, *A Month and a Day*, 6.

6. Handwritten letter by Bridget Birabi addressed to "Dede" dated 13/6/93. Return address is 27 Obagi Street, G.R.A. Phase 1, Port Harcourt.

7. Saro-Wiwa, *A Month and a Day*, 181.

8. Saro-Wiwa, *A Month and a Day*, 7.

9. Saro-Wiwa, *A Month and a Day*, 13.

10. Saro-Wiwa, *A Month and a Day*, 18.

11. Saro-Wiwa, *A Month and a Day*, 29.

12. Michael Birnbaum, QC. *Nigeria: Fundamental Rights Denied: Report of the Trial of Ken Saro-Wiwa and Others.* Article 19 in association with the Bar Human Rights Committee of England and Wales and the Law Society of England and Wales, June 1995, p. 15. http://www.article19.org/docimages/654.htm.

13. Rowell and Kretzmann, "All for Shell."

14. Boele, *Ogoni – Report of the UNPO Mission to Investigate the Situation of the Ogoni of Nigeria*, 50, in Rowell and Kretzmann, "All for Shell."

15. Joshua Hammer, "Nigeria Crude: A Hanged Man and an Oil-Fouled Landscape," *Harper's*, June 1996, 58–68.

16. Boele, *Ogoni – Report of the UNPO Mission to Investigate the Situation of the Ogoni of Nigeria*, 50.

17. Saro-Wiwa, *A Month and a Day*, 47.

18. Handwritten memorandum on National Assembly letterhead from "Senator B. Birabi" to Chief T.D. Gbanane: *Subject: Sarowiwa's Arrest.* Dated "2/7/93."

19. Saro-Wiwa, *A Month and a Day*, 205.

20. Boele, *Ogoni – Report of the UNPO Mission to Investigate the Situation of the Ogoni of Nigeria*, 50, in Rowell and Kretzmann, "All for Shell."

21. Saro-Wiwa, *A Month and a Day*, 209.

22. The approval for Ken's transfer was on July 8, but Ken wasn't moved until the next day (*A Month and a Day*, 211).

23. Boele, *Ogoni – Report of the UNPO Mission to Investigate the Situation of the Ogoni of Nigeria*, 50, in Rowell and Kretzmann, "All for Shell."

24. Saro-Wiwa, *A Month and a Day*, 218–220.

25. Saro-Wiwa, *A Month and a Day*, 208.

26. Saro-Wiwa, *A Month and a Day*, 233–234.

27. Saro-Wiwa, *A Month and a Day*, 235.

28. Saro-Wiwa, *A Month and a Day*, 236.

Chapter Six

1. Ike Okonta and Oronto Douglas, *Where Vultures Feast: Shell, Human Rights and Oil in the Niger Delta* (San Francisco: Sierra Club Books, 2001), 124.

2. Moffat Ekoriko and Sam Olukoya, "Brothers at War: In Rivers and Akwa Ibom States, Various Communities Engage in Ethnic Clashes with Extensive Loss of Lives and Property." *Newswatch Magazine*, August 16, 1993, 17.

3. Ekoriko and Olukoya, "Brothers at War," 17.

4. Karl Maier, *This House Has Fallen* (New York: PublicAffairs, 2000), 101.

5. Human Rights Watch/Africa, *The Ogoni Crisis: A Case Study of Military Repression in Southeastern Nigeria* (New York: Human Rights Watch, 1995).

6. Maier, *This House Has Fallen*, 100.

7. American University, Washington DC, Trade and Environment Database, Case number 149, "TED Case Studies: Ogoni and Oil," http://www.american.edu/TED/OGONI.HTM.

8. *Delta Force* (Catma Films production for Channel Four Television Corp., 1995).

9. Human Rights Watch/Africa, *Nigeria: The Ogoni Crisis*, 12.

10. Adotei Akwei, "And Justice for All? The Two Faces of Nigeria," in *Ogoni's Agonies*, ed. Abdul Rasheed Na'Allah (Trenton, N.J.: Africa World Press, 1998), 35.

11. Andrew Rowell and Stephen Kretzmann, "All for Shell: The Ogoni Struggle – A Project Underground Report,"

http://www.moles.org/ProjectUnderground/motherlode/shell/timeline.html
(Yale Law School, 1996).

12. Maier, *This House Has Fallen*, 86.

13. Ken Wiwa, *In the Shadow of a Saint* (Toronto: Knopf Canada, 2000), 56.

14. Maier, *This House Has Fallen*, 102.

15. It is interesting to note that a similar phenomenon occurred among the Plains
    Indians of the United States in the late 1800s. Young warriors desperate for
    the restoration of a life free of subjugation by white men, epidemic disease,
    and hunger joined a messianic movement called the Ghost Dance. The fol-
    lowers of this new religion believed they could render themselves impervious
    to the white man's bullets by painting their shirts with powerful medicine
    symbols such as eagles, buffalo, and morning-stars. In 1890, during the battle
    of Wounded Knee, two hundred Sioux warriors, women, and children were
    massacred while wearing these "Ghost Shirts." "Ghost Dance Movement,"
    http://www.geocities.com/BourbonStreet/Bayou/6029/Wolf/gdance.html.

16. Photocopy of original Ogoni/Andoni Peace Accord, in *The Ogoni Question in
    Nigeria: The Reality of the Situation* (Marine Communications Ltd., 1995),
    Appendix 12, 126.

17. Photocopy of original Ogoni/Andoni Peace Accord, in *The Ogoni Question in
    Nigeria*, Appendix 12, 126–127.

18. Photocopy of original notice "Andoni/Ogoni Peace Resolution" from the
    Daily Champion. *The Ogoni Question in Nigeria*, Appendix 15, 135.

19. Ken Saro-Wiwa, Personal *curriculum vitae* dated "14.04.94" stating
    "RELIGION: Occasional Christian (Anglican)."

20. Sir Awhelebe Uhor and Nkereuwem Akpan, "Forty-day Oil Spill Devastates
    Community," *Sunray*, July 29, 1993.

21. Shell Nigeria Web site.
    http://www.shellnigeria.com/info/info_display.asp?Id=136#korok.

22. Shell Nigeria Web site.

23. Environmental Rights Action, "ERA Monitor Report No. 8: Six Year Old
    Spillage in Botem-Tai" (1993) reported on Sierra Club Web site,
    http://www.sierraclub.org/human-rights/nigeria/background/spill.asp.

24. Environmental Rights Action, "ERA Monitor Report No. 8: Six Year Old
    Spillage in Botem-Tai."

25. Nigeria Independent Media Centre,
    http://www.nigeria.indymedia.org/front.php3?article_id=38&group=webcast.

26. http://www.ratical.org/corporations/OgoniFactS.html.

27. Shell Nigeria Web site.

28. Environmental Rights Action, "ERA Monitor Report No. 8: Six Year Old
    Spillage in Botem-Tai."

29. Oronto Douglas, "Niger Delta: The Battle with Corporate Rule," http://www.ijawlink.com/Forum/Articles_Reports_Interviews/Battle_with_ corporate_rule/battle_with_corporate_rule.html.
30. Okonta and Douglas, *Where Vultures Feast*, 136.
31. Okonta and Douglas, *Where Vultures Feast*, 127.
32. Okonta and Douglas, *Where Vultures Feast*, 136.
33. Okonta and Douglas, *Where Vultures Feast*, 127–128.
34. "The Flames of Shell: Oil, Nigeria, & the Ogoni," http://www.thirdworldtraveler.com/Boycotts/Flames_Shell.html.
35. Environmental Rights Action report to the Ecumenical Community for Corporate Responsibility reported on Sierra Club Web site, http://www.sierraclub.org/human-rights/nigeria/background/spill.asp.
36. Maier, *This House Has Fallen*, 72–73.
37. Maier, *This House Has Fallen*, 17.
38. Rowell and Kretzmann, "All for Shell."
39. Stevie Waidor-Pregbagha, "Brothers at War," *TELL Magazine* (Nigeria), January 10, 1994, p. 25.
40. *Delta Force.*

Chapter Seven

1. Gabriel Arisa, "Saro-Wiwa Blames Govt for Disturbances," *Daily Sunray* (Nigeria), January 18, 1994, p. 1.
2. *Delta Force* (Catma Films production for Channel Four Television Corp., 1995). (Beginning of part 3.)
3. *Delta Force.*
4. Cyril Bakwuye, "FG Wades into Ogoni Problems," *Daily Sunray* (Nigeria), January 20, 1994, p. 1.
5. Winters Negbenebor, "Statement to the Ogoni Civil Disturbances Tribunal," Nigeria Independent Media Centre, http://www.nigeria.indymedia.org/ (November 15, 2001, modified on August 31, 2002). Basically, the article is a transcript of Ken Saro-Wiwa's speech to the tribunal.
6. Geraldine Brooks, "Slick Alliance: Shell's Nigerian Fields Produce Few Benefits for Region's Villagers: Despite Huge Oil Revenues, Firm and Government Neglect the Impoverished: How Troops Handle Protests," *The Wall Street Journal* (May 6–7, 1994), p. 1.
7. Negbenebor, "Statement to the Ogoni Civil Disturbances Tribunal."
8. Tayo Lukula, "MOSOP Leaders Released on Bail," *The Guardian*, April 22, 1994.
9. Negbenebor, "Statement to the Ogoni Civil Disturbances Tribunal."
10. Sister Majella McCarron, handwritten letter to Ken Saro-Wiwa dated "10-5-94" quoting a section of a letter she received from the coordinator of

the Catholic Relief Team of the Catholic Diocese of Port Harcourt. The letter excerpt from the unnamed nun is dated "24-4-94."

11. http://multinationalmonitor.org/hyper/issues/1995/01/mm0195_06.html.

12. http://www.sierraclub.org/human-rights/nigeria/background/memo.asp.

13. Photocopy of original "restricted" memo on Government House stationery – From: The Chairman Rivers State Internal Security (RSIS), To: His Excellency the Military Administrator Rivers State, Subject: RSIS Operations: Law and Order in Ogoni, Etc. May 12, 1994.

14. http://nativenet.uthscsa.edu/archive/nl/9502/0169.html.

15. "Statement by the Chiefs and Leaders of Gokana: The Giokoo Accord," *Sunray*, May 14, 1994, in *The Ogoni Question in Nigeria: The Reality of the Situation* (Publisher/Date unknown), Appendix 7d, 113.

Chapter Eight

1. Chris McGreal, "Ken Saro-Wiwa: Not Entirely Innocent?" *Weekly Mail and Guardian*, April 4, 1996.

2. CNN Interactive. "Newsmaker Profiles: Sani Abacha," http://www.cnn.com/resources/newsmakers/world/africa/abacha.html.

3. Edetaen Ojo, "Dress Rehearsal for the Constitutional Conference," *The Guardian* (Nigeria), March 10, 1994.

4. *The Ogoni Question in Nigeria: The Reality of the Situation* (Publisher/Date unknown), 30.

5. Ken Wiwa, *In the Shadow of a Saint* (Toronto: Knopf Canada, 2000), 115.

6. Ike Okonta and Oronto Douglas, *Where Vultures Feast: Shell, Human Rights and Oil in the Niger Delta* (San Francisco: Sierra Club Books, 2001), 129–130.

7. For obvious reasons, the doctor has requested anonymity.

8. Wiwa, *In the Shadow of a Saint*, 113–114.

9. *Delta Force* (Catma Films production for Channel Four Television Corp., 1995).

10. Simon Ebare, "11 Ogoni Leaders Declared Wanted," *Daily Times* (Lagos, Nigeria), June 3, 1994, p. A1.

11. Wiwa, *In the Shadow of a Saint* (Toronto: Knopf Canada, 2000), 134.

12. http://www.africeur.de/afroview.htm.

13. Human Rights Watch Africa. *The Ogoni Crisis: A Case Study of Military Repression in Southeastern Nigeria* (Human Rights Watch, 1995), 13.

14. *Delta Force*.

15. http://www.nigeria.indymedia.org/front.php3?article_id=35.

16. *The Ogoni Review: Newsletter of the Movement for the Survival of Ogoni People (Mosop)* 1, no. 6 (October 1993), http://www.mosopcanada.org/text/review1.6.html.

Chapter Nine

1. Jim Hudgens and Richard Trillo, *West Africa: The Rough Guide* (London: Rough Guides Ltd., 1999), 954.
2. Hudgens and Trillo, *West Africa: The Rough Guide*, 997.
3. Right Livelihood Award Official Web site, http://www.rightlivelihood.se/about.html.
4. Ken Wiwa, *In the Shadow of a Saint* (Toronto: Knopf Canada, 2000), 130.
5. Desmond Lera Orage, "The Ogoni Question and the Role of the International Community in Nigeria," in *Ogoni's Agonies*, ed. Abdul Rasheed Na'Allah (Trenton, N.J.: Africa World Press 1998), 41–43.
6. Luke Fisher, "Double Take: Flora MacDonald," *Maclean's*, June 15, 1998, 12.

Chapter Ten

1. BBC Monitoring, Radio Nigeria-Lagos, October 24, 1996.
2. Handwritten letter from Ken Saro-Wiwa to Owens Wiwa dated "13-5" (i.e., May 13, 1995).
3. Ike Okonta and Oronto Douglas, *Where Vultures Feast* (San Francisco: Sierra Club Books, 2001), 166–167.
4. Baby Milk Action Web site. "PR For Beginners," www.babymilkaction.org/pdfs/spinpdfs/appendices/PR_for_beginners.pdf and Okonta and Douglas, *Where Vultures Feast*, 166.
5. *Delta Force* (Catma Films production for Channel Four Television Corp., 1995).
6. Melissa Crow et al., *The Ogoni Crisis: A Case-Study of Military Repression in Southeastern Nigeria* (Trenton, N.J.: Human Rights Watch/Africa, Vol. 7, No. 5a, July 1995), 31.
7. Charles Hoff, "Nigeria Executes 9 Activists; World Outraged," CNN World News (November 10, 1995), http://www-cgi.cnn.com/WORLD/9511/nigeria/.
8. Amnesty International, "Appeal for Action: Olusegun Obasanjo, Shehu Musa yar'Adua," http://www.amnesty.org/ailib/intcam/nigeria/obasanjo.htm.
9. Crow, *The Ogoni Crisis*, 26.
10. Ima Niboro, "Ogoni Verdicts: Abacha's Final Revenge: How Saro-Wiwa Snubbed the General," *Tell* Magazine (No. 46, November 13, 1995), p. 14.

Chapter Eleven

1. Shell Nigeria Web site, http://www.shellnigeria.com/info/info_display.asp?Id=150.
2. Shell Web site, http://www.shell.com/html/investor-en/shellreport01/reports2001/rde/htm/ginBoa.html.
3. Handwritten letter from Ken Saro-Wiwa to Owens Wiwa, dated "53" (i.e, March 5, 1995).

4. Andrew Rowell and Stephen Kretzmann, "All for Shell: The Ogoni Struggle – A Project Underground Report," http://www.moles.org/ProjectUnderground/motherlode/shell/timeline.html (Yale Law School, 1996).

5. Rory O'Connor and Danny Schechter, "The Power of Corporations" episode of *Globalization and Human Rights*, Globalvision Inc. Transcript available at http://www.pbs.org/globalization/.

6. *Delta Force* (Catma Films production for Channel Four Television Corp., 1995).

7. Greenpeace Web site, "Greenpeace *Brent Spar* Protest in the North Sea" (http://archive.greenpeace.org/~comms/brent/brent.html.

8. BBC Online Network, "WORLD: Europe *Brent Spar* Gets Chop," November 25, 1998, http://news.bbc.co.uk/1/hi/world/europe/221508.stm.

9. *Time*, "Six Who Do Care," April 24, 1995, http://www.goldmanprize.org/press/press.html.

10. Rowell and Kretzmann, "All for Shell."

11. Ima Niboro, "Ogoni Verdicts: Abacha's Final Revenge: How Saro-Wiwa Snubbed the General," *Tell* Magazine (No. 46, November 13, 1995), p. 14.

12. Front-page headline, "OGONI TRIAL SHOCKER: I Was Bribed to Lie Against Saro-Wiwa – Says Prosecution Witness," *The Masses*, February 21, 1995. Close-up of paper seen in *Delta Force*.

13. Niboro, "Ogoni Verdicts," 14.

14. Niboro, "Ogoni Verdicts," 14–15.

15. *Delta Force*.

16. "United States Policy Toward Nigeria." Transcripts of the Hearing before the Subcommittee on Africa of the Committee on International Relations, U.S. House of Representatives. 105th Congress, 1st Session (September 18, 1997), 45–971 CC.

17. Amnesty International, "Appeal for Action: Olusegun Obasanjo, Shehu Musa yar'Adua," http://www.amnesty.org/ailib/intcam/nigeria/obasanjo.htm.

18. Eghosa E. Osaghae, *Crippled Giant: Nigeria Since Independence* (Bloomington: Indiana University Press, 1998), 303.

19. Niboro, "Ogoni Verdicts," 15–16.

20. Agence France-Presse, "Writer and Campaigner for Ethnic Rights Condemned," *Irish Times*, November 1, 1995, City edition, p. 9.

21. Michael Birnbaum, Q.C., *Fundamental Rights Denied: Report of the Trial of Ken Saro-Wiwa and Others* (London: Article 19, June 1999).

22. Agence France-Presse, "Writer and Campaigner for Ethnic Rights Condemned," p. 9.

23. Andy Rowell, "Oil, Shell & Nigeria," *Ecologist* 25 no. 6 (November/December 1995), http://members.lycos.nl/kaalslag/arm/arm_uk_nigeria.htm.

24. Ken Wiwa, *In the Shadow of a Saint* (Toronto: Knopf Canada, 2000),
    144–155.

Chapter Twelve

1. Frank Kane, "Body Shop Helps Saro-Wiwa's Brother Flee Nigeria to Britain,"
   The Sunday Times, November 26, 1995, Edition 1, p. 1.
2. Eniwoke Ibagare, "On the Buses in Lagos," "BBC News Online," February 28,
   2001, http://news.bbc.co.uk/1/low/world/africa/1186572.stm.
3. Jim Hudgens and Richard Trillo, *West Africa: The Rough Guide* (London:
   Rough Guides Ltd., 1999), 830–831.
4. Hudgens and Trillo, *West Africa: The Rough Guide*, 927.
5. David Else et al., *West Africa*, 4th ed. (Melbourne: Lonely Planet Publications,
   1999), 871.
6. Joshua Hammer, "Nigeria Crude: A Hanged Man and an Oil-Fouled
   Landscape," *Harper's*, June 1996, 58–71.
7. Hammer, "Nigeria Crude."
8. "Commonwealth Votes to Suspend Nigeria," *Washington Post*,
   November 12, 1995, p. A34, http://www.washingtonpost.com/wp-
   srv/inatl/longterm/nigeria/stories/common111295.htm.
9. Karl Maier, *This House Has Fallen* (New York: PublicAffairs, 2000), 109.
10. Michelle Kort, "The Fierce Kind: In Business and in Life, Anita Roddick
    Proves It's Not Hard to Be Soft," *Ms. Magazine Online*, September 2003,
    http://www.msmagazine.com/sept03/kort.asp.
11. Ken Saro-Wiwa, "For Anita Roddick," *Ogoni Review* 1, no. 17 (October 1998),
    AFRIDA.org, http://www.afrida.org/index.cfm.

Chapter Thirteen

1. Frank Kane, "Body Shop Helps Saro-Wiwa's Brother Flee Nigeria to Britain,"
   Sunday Times, November 26, 1995, pp. 1 and 3.
2. *Delta Force*. A Catma Films production for Channel Four Television Corp., 1995.
3. *New York Times* paid advertisement quoted in Joshua Hammer, "Nigeria
   Crude: A Hanged Man and an Oil-Fouled Landscape," *Harper's*, June 1996.
4. Frank Kane, Steven Haynes, and Christina Lamb, "Shell Axes 'Corrupt'
   Nigeria Staff," *The Sunday Times*, December 17, 1995, Edition 1W, p. 1.
5. Kane, Haynes, and Lamb, "Shell Axes Corrupt' Nigeria Staff," 1.
6. Kane, Haynes, and Lamb, "Shell Axes 'Corrupt' Nigeria Staff," 1.
7. Ike Okonta and Oronto Douglas, *Where Vultures Feast: Shell, Human Rights
   and Oil in the Niger Delta* (San Francisco: Sierra Club Books, 2001) 269.
8. Bob Herbert, "Unholy Alliance in Nigeria: A Doctor Fails to Save His Brother
   from Execution," *The New York Times*, January 26, 1996, Section A, p. 27.

9. Ken Wiwa, *In the Shadow of a Saint* (Toronto: Knopf Canada, 2000), 151.
10. John Stremlau, "Sharpening International Sanctions," http://wwics.si.edu/subsites/ccpdc/pubs/sum/3.htm.
11. Joyce Carol Oates, "On Doris Lessing," *Southern Review* 9 no. 4 (October 1973), http://www.usfca.edu/fac-staff/southerr/lessing.html.
12. *Benét's Reader's Encyclopedia*, 3rd Ed. (New York: Harper & Row, 1987), 563.

Chapter Fourteen

1. http://www.dcmsoft.com/zuba/zuba/maryam.htm.
2. Tom Masland and Jeffrey Bartholet, "The Lost Billions: The Inside Story of the Hunt – from Lagos to New York to Geneva – for an African Dictator's Stolen Loot," *Newsweek*, March 13, 2000, U.S. Edition, Section: International, p. 38.
3. http://www.transnationale.org/anglais/sources/tiersmonde/dirigeants_abacha.htm.
4. http://www.dcmsoft.com/zuba/zuba/maryam.htm.
5. "Abiola's Family Doubts Official Version of Death," CNN.com, July 8, 1998, http://www.cnn.com/WORLD/africa/9807/08/nigeria.abiola.01/.
6. "Nigeria Riots after Death of Abiola," *Electronic Mail & Guardian* (Johannesburg, South Africa), July 8, 1998.
7. MediaNet Briefing: Bulletin on Nigeria, "Ken Saro-Wiwa: the Grim Story of His Death Revealed," http://www.panosinst.org/Nigeria/Nig3.shtml.
8. Barrisuatam N. Deeyeh, *Authentic Report of the Home coming from Four Year Exile of Dr. Owens Wiwa, Co-Ordinator of MOSOP Canada*, MOSOP report (July 14, 1999), 13.
9. Deeyeh, *Authentic Report of the Home coming from Four Year Exile of Dr. Owens Wiwa, Co-Ordinator of MOSOP Canada*, 2.
10. Deeyeh, *Authentic Report of the Home coming from Four Year Exile of Dr. Owens Wiwa, Co-Ordinator of MOSOP Canada*, 3.
11. Deeyeh, *Authentic Report of the Home coming from Four Year Exile of Dr. Owens Wiwa, Co-Ordinator of MOSOP Canada*, 3.
12. Deeyeh, *Authentic Report of the Home coming from Four Year Exile of Dr. Owens Wiwa, Co-Ordinator of MOSOP Canada*, 4.
13. Deeyeh, *Authentic Report of the Home coming from Four Year Exile of Dr. Owens Wiwa, Co-Ordinator of MOSOP Canada*, 13.
14. Deeyeh, *Authentic Report of the Home coming from Four Year Exile of Dr. Owens Wiwa, Co-Ordinator of MOSOP Canada*, 13.
15. Deeyeh, *Authentic Report of the Home coming from Four Year Exile of Dr. Owens Wiwa, Co-Ordinator of MOSOP Canada*, 5.
16. Deeyeh, *Authentic Report of the Home coming from Four Year Exile of Dr. Owens Wiwa, Co-Ordinator of MOSOP Canada*, 6.

17. Deeyeh, *Authentic Report of the Home coming from Four Year Exile of Dr. Owens Wiwa, Co-Ordinator of MOSOP Canada*, 11.
18. "Pathologists Believe Nigeria's Abiola Died Natural Death," CNN.com, July 11, 1998, http://www.cnn.com/WORLD/africa/9807/11/nigeria.autopsy.02/.
19. Owens Wiwa, "Chronology of Events for the Exhumation of Ken Saro-Wiwa and Eight Ogoni Activists." Unpublished, undated document.
20. Wiwa, "Chronology of Events for the Exhumation of Ken Saro-Wiwa and Eight Ogoni Activists."
21. Catma Films, "The Journey" (First broadcast in U.K. on Channel Four News, April 10, 2000). Transcript, p. 1.
22. Catma Films, "The Journey," p. 1.
23. Catma Films, "The Journey," p. 2.
24. Catma Films, "The Journey," p. 4.
25. Catma Films, "The Journey," p. 8.
26. Catma Films, "The Journey," p. 9.
27. Catma Films, "The Journey," p. 9.

Chapter Fifteen

1. Ken Wiwa, *In the Shadow of a Saint* (Toronto: Knopf Canada, 2000), 172.
2. Wiwa, *In the Shadow of a Saint*, 173.
3. Wiwa, *In the Shadow of a Saint*, 174.

INDEX

377